APPLICATION OF MUTATION BREEDING METHODS IN THE IMPROVEMENT OF VEGETATIVELY PROPAGATED CROPS

AN INTERPRETIVE LITERATURE REVIEW

DEVELOPMENTS IN CROP SCIENCE (2)

APPLICATION OF MUTATION BREEDING METHODS IN THE IMPROVEMENT OF VEGETATIVELY PROPAGATED CROPS

AN INTERPRETIVE LITERATURE REVIEW

C. BROERTJES

Institute for Atomic Sciences in Agriculture (ITAL), Association Euratom-ITAL, Wageningen (The Netherlands)

A.M. VAN HARTEN

Institute of Plant Breeding, Agricultural University, Wageningen (The Netherlands)

ELSEVIER SCIENTIFIC PUBLISHING COMPANY

Amsterdam — Oxford — New York 1978

ELSEVIER SCIENTIFIC PUBLISHING COMPANY
335 Jan van Galenstraat
P.O. Box 211, Amsterdam, The Netherlands

Distributors for the United States and Canada:

ELSEVIER NORTH-HOLLAND INC.
52, Vanderbilt Avenue
New York, N.Y. 10017

Library of Congress Cataloging in Publication Data

Broertjes, C.
Application of mutation breeding methods in the improvement of vegetatively propagated crops.

(Developments in crop science ; 2)
Bibliography:
Includes index.
1. Plant mutation breeding. 2. Plant propagation. 3. Plant cell culture. I. Harten, A. M. van, joint author. II. Title. III. Series.
SB123.B697 631.5'3 78-475
ISBN 0-444-56959-6

ISBN 0-444-41617-X (series)

ISBN 0-444-56959-6 (vol. 2)

Printed in The Netherlands

CONTENTS

PREFACE

The significance of artificially induced mutations in cultivated plants has long been a point of controversy among (commercial) plant breeders. Today, mutation breeding, notably that of vegetatively propagated plants, is one of the accepted methods available to the breeder to improve his crop. This is particularly demonstrated by the relatively large and steadily increasing number of mutants which have been put on the market so far (Table 1).

It seemed appropriate to us to make up the balance and examine what has been achieved so far, while trying to come to conclusions about the value of the method, its possibilities, limitations and shortcomings and also to determine what difficulties are in the way of further application in various crops.

Earlier attempts to evaluate the literature on mutation induction in (certain groups of) vegetatively propagated crops have been published in general articles, e.g. Broertjes (1969a, 1977a), Donini (1976c), Lapins (1973), Nishida (1973), Nybom (1961, 1970), Nybom and Koch (1965), Ohba (1971a) and Privalov (1968b), but these articles all covered, generally on purpose, only part of the literature. This is not at all surprising in view of the large number of publications in this field in many different languages over a prolonged period. Moreover, they are concerned with a great diversity of plants: root and tuber crops, temperate and tropical fruit crops, ornamentals and various other crops.

The main aim of the authors of this book was to bring together all available and accessible references and to discuss them crop by crop. The discussions of individual crops were to be preceded by a few specific chapters, necessary to make the interaction between mutagenic treatment and plant material and its consequences for plant breeding understood.

It is obvious that a number of important questions must be answered before one should start a mutation breeding programme with a given crop. Such questions are, for example:

(1) What problems are to be solved?

(2) What is the best strategy? Can it be done (easier, faster, cheaper) by mutation induction alone or in combination with other methods?

(3) On what kind and number of genes is the character to be mutated based? Is the cultivar heterozygous for the character(s) in question? Consequently, what cultivar(s) appear(s) to be the best starting material?

(4) What is the best kind of plant material to treat? Is any adventitious bud technique, in vivo or in vitro, available?

(5) What mutagen, radiation (type, dose, dose rate), or chemical (type, concentration, duration of treatment, etc.) should be used? What is available, obtainable?

TABLE 1

Commercial mutants of various vegetatively propagated crops (December 1st, 1977)

Plant group	Number of commercial mutants (crop)	Total number	Total number of commercial mutants in the period			
			Before 1950	1950—1960	1960—1970	1970—1977
Root and tuber crops	1 (potato)	1				1
Ornamentals						
Tuber and bulb crops	23 (*Dahlia*); 2 (*Lilium*); 2 (tulip)	27	1	1	13	12
Pot plants	21 (*Begonia*); 1 (*Guzmania*); 8 (*Streptocarpus*); 8 (*Achimenes*); 10 (*Azalea*)	48			7	41
Cut flowers	14 (*Alstroemeria*); 2 (carnation); 35 * (*Chrysanthemum*); 3 (roses)	54			12	42
Garden plants	1 (*Portulaca*); 1(*Bougainvillea*) **	2				2
Woody plants	1 (*Abelia*)	1				1
Fruit crops						
Temperate tree fruits	4 (apple); 1 (apricot); 3 (cherry); 1 (peach)	9			3	6
Small fruits	1 (black currant)	1			1	
Tropical fruits	1 (grapefruit)	1				1
Other crops	2 (peppermint)	2				2
Totals		146	1	1	36	108

* Since mutation breeding has been practised already for a long time by commercial *Chrysanthemum* breeders, the number of mutants that were put on the market must be many more, probably even several hundreds.

** Just before this book was printed, a radiation-induced *Bougainvillea* mutant was reported by V. Abraham and B.M. Desai (Curr. Sci., 1977, vol. 46(10): 351—352).

(6) How should the material be handled (managed) after treatment?

(7) How will mutants be selected?

Some of the answers can be found in the literature. (For a detailed discussion of the afore-mentioned questions see Broertjes (1977a).) However, the crop specialist has to answer the more specific questions in consultation with colleagues before being able to cherish some hope that the application of artificial mutation induction will contribute to solving the breeding problems of the crop in question.

Literature

Broertjes, C., 1969a. Mutation breeding of vegetatively propagated crops. In: G.C. Chisci and G. Haussmann (Editors), Proc. 5th Eucarpia Congr., Milan, 1968. Genet. Agrar.: 139—165.

Broertjes, C., 1977a. Induced-mutant techniques in breeding asexually propagated plants. In: Manual of Mutation Breeding. IAEA, Vienna, 2nd edn, pp. 159—166.

Donini, B., 1976c. Breeding methods and applied mutagenesis in fruit plants. In: Assoc. Euratom-ITAL (Editor), Proc. Workshop Eur. Comm. — Israel, Use of Ionizing Radiation. Wageningen, pp. 445—478.

Lapins, K.O., 1973. Induced mutations in fruit trees. In: Induced Mutations in Vegetatively Propagated Plants. IAEA, Vienna, pp. 1—19.

Nishida, T., 1973. Induction of somatic mutations in deciduous fruit trees by gamma irradiation. Gamma Field Symp., 12: 1—15.

Nybom, N., 1961. The use of induced mutations for the improvement of vegetatively propagated plants. In: Proc. Symp., Mutations and Plant Breeding, Cornell Univ., Ithaca, N.Y. NAS—NRC, Publ. 891, pp. 252—294.

Nybom, N., 1970. Mutation breeding of vegetatively propagated plants. In: Manual on Mutation Breeding. Tech. Rep. Ser., No. 119. IAEA, Vienna, pp. 141—147.

Nybom, N. and Koch, A., 1965. Induced mutations and breeding methods in vegetatively propagated plants. In: The Use of Induced Mutations in Plant Breeding. Radiat. Bot., 5 (suppl.): 661—678.

Ohba, K., 1971a. Studies on the radiation breeding of forest trees. Bull. Inst. Radiat. Breed., 2: 1—102.

Privalov, G.F., 1968b. Investigations of experimental mutagenesis in arboreous plants. Genetika, 4 (6): 144—158 (in Russian with English summary).

INTRODUCTION

In plant breeding several methods may be used to increase the genetic variability within a crop followed by selection procedures which attempt to identify useful genotypes (those showing higher yield, disease resistance, improved flower colour). One of the main methods is cross-breeding, with which the plant breeder tries to combine beneficial characters from different sources into one genotype.

Mutation breeding, on the other hand, makes use of the possibility of altering genes by exposing seeds or other plant parts to chemical or physical mutagens. These artificially induced genetic events are, like spontaneous ones, various in nature and can be divided into gene mutations (involving only one gene) and chromosome aberrations (involving a few or many genes). Generally, the plant breeder is more interested in the (recessive) gene mutations than in the chromosome aberrations which often cause reduced fertility, a serious drawback in seed-propagated species.

Mutations can be induced by chemicals, such as EMS (ethyl methanesulphonate), or various types of ionizing radiation (X-rays, neutrons, ultraviolet light, etc.). Although the interaction of these mutagens with biological matter varies greatly, the final result is, unfortunately, rather similar in the sense that neither mutation frequency nor mutation spectrum is basically different using any of the various mutagenic treatments. However, the (slight) differences in the ratio between gene mutations and chromosome aberrations, as well as the probably distinct gene mutation spectra, force the mutation breeder to consider carefully which mutagen(s) to apply and in what way (see Chapter 1).

For various reasons, vegetatively propagated crops are a very suitable group of plants for the application of mutation breeding methods. The generally high degree of heterozygosity which causes a complex inheritance of genetic factors as well as the frequent polyploidy, both serious handicaps in conventional breeding, are advantageous in mutation breeding as large variations can often be observed in the irradiated plants. Mutations are the only source of variability in sterile plants or in obligate apomicts.

The main advantage of mutation induction in vegetatively propagated crops is the ability to change one or a few characters of an otherwise outstanding cultivar without altering the remaining and often unique part of the genotype. It is therefore considered the obvious means to perfect the leading products of conventional plant breeding, which are often the result of many years of painstaking work. A recent example of radiation-induced mutants of great commercial value are those of *Chrysanthemum morifolium* cv. Horim (Plate 1).

The interest in the application of mutation induction in vegetatively propagated crops is, in addition, a consequence of the large economic importance of many species of this group. Plants like sugarcane, potato, sweet potato, several fruits, including grapes and cassava, are among the approximately 15 world crops with the highest fresh-weight production (Anonymous, 1976a) (Table 2). In The Netherlands, the production value of vegetatively propagated plants is over 50% of that of the total agricultural and horticultural production value (dairy and poultry products excluded) (Anonymous, 1976b) (Table 3).

The main bottle-necks in mutation breeding of vegetatively propagated plants, where vegetative plant parts have to be irradiated, are chimera formation and diplontic selection, both being complications caused by the multicellular nature of the bud-apex and the fact that a mutation is a one-cell event. The result is a relatively low mutation frequency and probably a limited mutation spectrum, while selection procedures cannot be applied before the stable periclinal chimera stage has been reached (see Chapter 2).

These difficulties can be largely restricted or avoided by the use of an in vivo or in vitro adventitious bud technique, with which large numbers of solid, non-chimeral mutants can be produced if detached leaves or explants, respectively, are irradiated before regeneration of the adventitious shoots. Early detection, early selection and vegetative propagation are then without problems, provided good selection procedures are available (see Chapter 3).

TABLE 2

World production in 1975 of various crops (fresh weight in 10^6 metric tons) (Anonymous, 1976a)

Sugarcane *	652
Sugar	23.3
Roots and tubers *	562
Potatoes *	291
Sweet potato *	136
Cassava *	105
Total cereals	1359
Wheat	355
Rice	344
Maize	322
Barley	155
Fruits (grapes, citrus, bananas, apples) *	255
Grapes, total *	60
Grapes, for wine *	31
Pulses	46
Fibres	18
Cotton	17

* Vegetatively propagated.

TABLE 3

Production value (10^6 Dfl.) of horticultural and agricultural products in The Netherlands, 1975 (Anonymous, 1976b)

Potatoes	1061
Various fruit crops	379
Ornamentals	
Tuber and bulb crops	324
Cut flowers and pot plants	1394
Woody plants	198
Total horticultural and agricultural products (dairy and poultry excluded)	6578
Total vegetatively propagated plants	3356 (51%)
Total ornamental crops	1916 (29%)

It is, therefore, not surprising that a great number of mutation breeding projects started with vegetatively propagated crops, not only economically important food crops (see Chapters 4 and 7), but also ornamentals, which in some countries are as economically important, even though they cannot be consumed (see Chapter 5).

So far, relatively few projects have resulted in mutants which are being commercially exploited, and those which are belong for the most part to the group of ornamentals (see Table 1 and the tables in the appropriate chapters). This occasionally has been used as an argument against the use of mutation breeding. Opponents forget that most mutation breeding projects are too young to have resulted in a more convincing number of commercial mutants. Moreover, the scale on which conventional cross breeding has to be operated to obtain improved commercial seedlings is often not realized.

The breeder of any given vegetatively propagated crop has to decide, using present-day knowledge, which problems are most likely to be solved by cross breeding and which by mutation induction, taking into account the possibilities as well as the limitations of each method (Broertjes, 1977a).

The literature lists in this book contain most of the important books and articles which have been published in all the languages accessible to the authors. Most of them have been read by them, but some could not be obtained. Unfortunately, articles in Chinese, Japanese, Russian or similar languages, which were published without an English summary, are in general not included in the literature reviews. The authors would be very grateful if any book, article or summary, published in an "accessible" language, and which is not included in the lists of references, were brought to their attention.

Although strictly speaking not quite correct, most plant species have been indicated by their generic names only, as this is common practice in, for instance, the world of ornamental plant growers and is also observed in many publications concerning mutation breeding.

Literature

Anonymous, 1976a. World production of various crops (1975). In: FAO Production Yearbook, Vol. 29(2). FAO, Rome, Table 11.

Anonymous, 1976b. De waarde van de land- en tuinbouwproduktie. (The value of the agricultural and horticultural production.) Cent. Bur. Stat., Maandstat. Landbouw, 24(7): Table 33.

Broertjes, C., 1977a. Induced-mutant techniques in breeding asexually propagated plants. In: Manual on Mutation Breeding. IAEA, Vienna, 2nd edn, pp. 159—166.

Plate 1. Radiation-induced flower colour mutants of *Chrysanthemum morifolium* cv. Horim: Milonka ®, white; Mikrop ®, pink; Milava ®, pale yellow, and Middelry ®, dark yellow. (The original colour of cv. Horim was very pale pink to almost white.)
(Slides from Fa. Chryveco, 's-Gravenzande, The Netherlands, who sponsored this colour plate.)

Chapter 1

VARIOUS ASPECTS OF MUTAGENIC TREATMENT

For the induction of mutational events in plant material the mutation breeder can choose between two groups of mutagenic agents, namely, chemical and physical mutagens. The latter have been used for many decades, whereas the use of chemicals is relatively recent, starting in approximately 1940.

Especially in the beginning, scientists were very hopeful that, by the use of various physical and, more particularly, chemical mutagens, it would be possible to induce specific mutations at will. The expectation that mutagenic chemicals have a gene-specific effect, however, seems unrealistic in view of our present knowledge about the structure of DNA, in which only four bases form the "genetic alphabet".

The best way to reflect present opinions on this matter, is by quoting one of the leading mutation specialists, C. Auerbach, who recently said in her introduction to the 13th International Congress of Genetics (Auerbach, 1974): ". . . one of the exciting aspects of modern mutation research is the breaking down of the divisions between the induction of mutations by ionizing and non-ionizing radiations, and by various groups of chemical mutagens . . ." and later she came to the conclusion that ". . . if there is any chance at all of attaining the kind of mutagen specificity in which plant breeders are interested, I can only see two possible ways to it. One, which has already been tried with some success by several workers, is the use of mutagens that act specifically on replicating or transcribing genes. The other is a search for treatments or treatment conditions which, like caffeine in *Ophiostoma* or diepoxybutane in *Neurospora*, create internal sieves that select for or against certain types of mutation".

Other scientists, however, have a somewhat different opinion. C.H. Naumann et al. (1976) claim that chemicals like EMS (ethyl methanesulphonate) are ". . . known to mutate active genes with a higher frequency than inactive genes in *E. coli*. In addition repetitive or heterochromatic DNA regions are thought to be hot spots for effects of certain chemicals". They and others claim that different mutation spectra are obtained with different chemicals, and between chemicals, as compared with X-rays, for instance. The debate is also influenced by the fact that very few chromosome aberrations are produced by alkylating agents and that chemicals are often more efficient in inducing mutations at specific loci.

Chemical mutagens. — Detailed information about chemical mutagens can be found, for instance, in the Manual on Mutation Breeding (IAEA, 1977);

details concerning different groups of compounds (e.g. alkylating agents, base analogues), their mode of action, methods of treatment, modifying factors, and techniques to be used before and after treatment are included. The compounds considered exclude chemicals like colchicine (an effective means to induce polyploidy), radioisotopes for internal irradiation (used, for instance, by Ehrenberg and Granhall, 1952) and those to induce specific effects, such as plastom mutants (Hentrich and Beger, 1974; Pohlheim and Pohlheim, 1976).

For inducing mutations in vegetatively propagated plants, chemical mutagens are not usually considered, mainly because the number of cases in which they have been applied successfully (and in which they were better than radiation) has been small. The lack of success is probably a consequence of poor uptake and penetration of the chemical compound (Bowen, 1965; Moës, 1966; Nybom, 1961). Moreover, bulky material, like bulbs, scions for grafting and plants, is difficult to treat with chemicals in a reproducible way. It is somewhat surprising, therefore, that in some countries, for example, the U.S.S.R., mutation breeders concentrate on, or even use exclusively, chemicals for the induction of mutations, as demonstrated by the work of Amirov (1974), Dryagina (1974), Dryagina and Limberger (1974) (Fig. 1), Kolesnikova and Maksimova (1974), Kolontaev (1974b,c,d), Kopan (1974) and Rapoport (1973). They claim, as do other investigators, that "chemical mutagens have a higher efficiency and output of mutations (in trees and shrubs) if the duration of treatment and the concentrations are well adjusted" (IAEA,

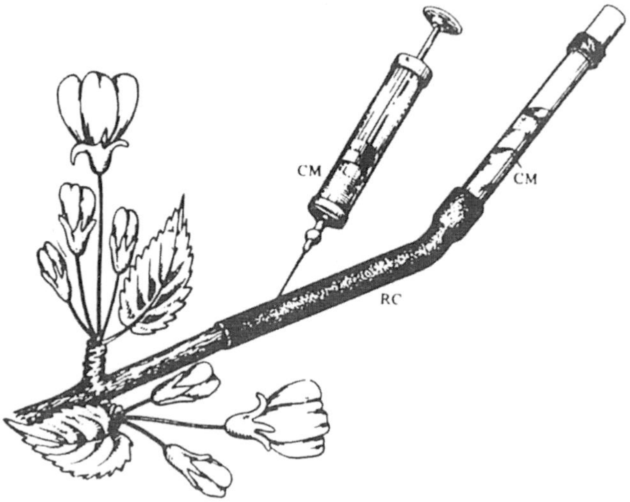

Fig. 1. Diagram showing a new method for treatment of perennial trees with chemical mutagens. (RC, rubber chamber; CM, chemical mutagen.) (Dryagina and Limberger, 1974, Fig. 1.)

1973, p. 91). Whether this statement is justified by the facts is hard to judge because of the poor accessibility of most Russian literature in this field of research.

Other authors report the successful application of chemicals to various crops: Singh and Iyer (1974) successfully applied chemical mutagens to pineapple, Kaul and Kak (1973, 1975) to peppermint, and Mee et al. (1969) to sugarcane. In sugarcane, chemicals were applied to in vitro cultures, a technique which makes much more sense than the treatment of (woody) plant parts, such as cuttings (Heinz, 1973; Jagathesan, personal communication, 1976). Whether the in vitro technique itself or the chemical treatment was responsible for the mutations is hard to say, since without the application of mutagenic chemicals appreciably high mutation frequencies are often found in vitro, especially if prolonged and repeated callus sub-culturing is involved. Negative results, such as those of Burton (1976), have been published as well. He treated stolon sections of Bermuda grass with EMS without result, in contrast with his successful use of γ-irradiation, to mention only one (recent) example.

In general, most mutation breeders prefer ionizing radiation (easily applicable; clean; good penetration and reproducibility; high mutation frequency). We feel that, with present knowledge, one should consider chemicals only if the necessary tools for radiation are not available or if there is clear-cut evidence that more mutations per locus or a better mutation spectrum can be obtained using chemicals (i.e. better in the sense that fewer drastic chromosome aberrations are induced in favour of gene mutations). On the other hand, the increasing possibilities of in vitro techniques (which may give better penetration of chemicals) may tip the balance towards the use of chemical agents.

Physical mutagens. — These are widely used to induce mutations in all kinds of plant parts. In theory, one can choose between a variety of ionizing radiation types; in practice, only an X-ray machine and/or a γ-source are generally available. Sometimes, thermal or fast neutrons can be used.

For practical mutation breeding, acute irradiation with fairly hard X-rays (Fig. 2) or γ-rays (Fig. 3) is recommended using a dose rate of 100—1 000 rad/min (and short irradiation times). If neutrons are available, fast neutrons are preferable to thermal neutrons, although both are useful. (The reproducibility of results with fast neutrons should be better, since the outcome is independent of the chemical composition of the material.) For more details about radiation types and sources, radiobiology, dosimetry etc. the reader is again advised to consult the Manual on Mutation Breeding (IAEA, 1970) and the 2nd revised edition (IAEA, 1977).

The dose to apply depends on the radiosensitivity of the species in general and that of the plant part and the stage of development in particular. Plant parts which still have to produce new (adventitious) roots and shoots (un-

Fig. 2. X-ray beam pointing down to pot plants on a turntable. To the right of the plants on the table is an ionization chamber connected to the dosimeter by which the dose to be given is controlled. (Photograph, ITAL, Wageningen.)

rooted cuttings, freshly detached leaves etc.) are more sensitive than plant parts with existing root and shoot meristems. By shielding the basal part of a non-rooted bud-stick or cutting, better results are obtained and sometimes a (much) higher dose can be applied, for example, to apple (Lacey, 1977), black currant (Bauer, 1957), carnation (L.D. Sparnaay and C. Broertjes, unpublished data) or potato (A.M. van Harten, unpublished data). Radiosensitivity varies between plant species and, according to Sparrow et al. (1963a, 1968), depends mainly on the nuclear volume (the greater the DNA content, the more sensitive), the number of chromosomes (plants with fewer chromosomes, given a certain nuclear volume, are more sensitive than plants with more, smaller chromosomes) and the ploidy level (the higher it is, the less

Fig. 3. (A) Tomato plants at different distances from a ^{137}Cs source in the γ-green house of the Institute for Atomic Sciences in Agriculture, Wageningen, The Netherlands.(Photograph, ITAL, Wageningen.) (B) Flowering plants of sugar beet and grasses at different distances from a ^{137}Cs source in the γ-greenhouse of the Institute for Atomic Sciences in Agriculture, Wageningen, The Netherlands. (Photograph, ITAL, Wageningen.)

radiosensitive). At the same time, genetic factors, climatic and other environmental conditions before and after treatment of the plant part, as well as the stage of development of root or shoot seem to be of importance.

Since so many factors are involved in radiosensitivity (including dosimetry), it is impossible simply to apply a dose that is mentioned in the literature. Therefore, the best practice is the empirical one: the starting material should first be exposed to a range of doses, centred round the optimum one, either that found in the literature or a calculated optimum (Sparrow et al., 1967). This should be done very carefully, preferably under controlled conditions in order to ensure reproducibility of the treatment. Generally a second, and often a third, range of doses may be necessary to determine the optimum dose of the cultivar in question. An advantage of this approach is that in the meantime one gains experience with the special properties of irradiated material, such as growth retardation and the occurrence of abnormalities (bifurcation, speckled or deformed leaves etc.).

Another question is what should be considered to be an *optimum dose*. Mutation frequency increases with increasing dose (linearly with X- and γ-rays, more exponentially with neutrons), but survival and capacity to regenerate decrease with increasing dose. One must therefore choose some point between a low dose (100% survival; low mutation frequency) and a high dose (low survival; high(er) mutation frequency). At high doses, too many mutational events per cell may be induced, with increased risk that a favourable mutation is accompanied by one or more unfavourable genetic changes. In vegetatively propagated crops it is impossible to separate favourable from unfavourable mutations by cross-breeding mutants among each other or by back-crossing with the original material. Therefore, the determination of the optimum dose is much more decisive than in seed propagated crops, where separation can easily be achieved by crossing, selfing etc.

What dose level should or can be applied depends on the crop, the method of propagation available, the numbers which can be handled and the selection method. For example, in peppermint, when very large numbers of plant parts were irradiated, relatively high doses were preferred (see Chapter 8). In general, it seems better to start with a moderate dose, that permits good growth and propagation of the material. (For a given plant, please consult the section on the plant in question and the literature references discussed.)

The mutagenic treatment of the material is only the starting point of a breeding programme. The first problem after treatment is to provide the mutated cell(s) with optimum chances of contributing to the formation of the largest possible part of a shoot or plant. If one kind or another of adventitious bud technique, in vivo or in vitro, is applicable, many of the mutants will be solid and early selection is then possible. If, however, multicellular apices have been irradiated, measures should be taken to promote an increase in sector size and to obtain complete periclinal chimeras as soon as possible. Selection and further propagation can then begin. When clearly visible mutations are involved (e.g. flower colour), selection can be started earlier. (For more details, please consult Chapters 2 and 3).

Literature

Amirov, Z.S., 1974. Chemical mutagenesis in interspecific hybrids of potato. Eksp. Mutagenez. Rast., 2: 115—117 (in Russian); Pl. Breed. Abstr., 47: No. 1354.

Auerbach, C., 1974. Mutagenesis. Introduction by the chairman. In: E.A. Carlson (Editor), Proc. 13th Int. Congr. Genet. Genetics, 78: 77—79.

Bauer, R., 1957. The induction of vegetative mutations in *Ribes nigrum*. Hereditas, 43: 323—337.

Bowen, H.J.M., 1965. Mutations in horticultural chrysanthemums. In: The Use of Induced Mutations in Plant Breeding. IAEA/FAO Tech. Meet., Rome, 1964. Radiat. Bot., 5 (Suppl.): 695—700.

Burton, G.W., 1976. Using gamma irradiation to improve sterile turf and forage bermudagrasses. In: Improvement of Vegetatively Propagated Plants and Tree Crops through Induced Mutations. Wageningen, 1976. IAEA, Vienna, pp. 25—32.

Dryagina, I.V., 1974. The use of physical and chemical mutagens in breeding horticultural plants. Biofiz. Fiziol. Biokhim. Issled. Plodovod. Yagod. Kult., 1974: 146—154 (in Russian); Plant Breed. Abstr., 46: No. 1731.

Dryagina, I.V. and Limberger, G.E., 1974. A new method for treating perennial fruit trees with chemical mutagens. Moscow Univ. Biol. Sci. Bull., 29(6): 50—53.

Ehrenberg, L. and Granhall, I., 1952. Effects of beta-radiating isotopes in fruit trees. Hereditas, 38: 385—419.

Heinz, D.J., 1973. Sugar-cane improvement through induced mutations using vegetative propagules and cell culture techniques. In: Induced Mutations in Vegetatively Propagated Plants. IAEA, Vienna, pp. 53—61.

Hentrich, W. and Berger, B., 1974. Untersuchungen über die mutagene Effizienz von N-nitroso-N-methylharnstoff bei *Saintpaulia ionantha* H. Wendl. Arch. Züchtungsforsch., 4(1): 29—43.

IAEA, 1970. Manual on Mutation Breeding. Technical Reports Series No. 119, IAEA, Vienna, 237 pp.

IAEA, 1973. Isotopes and Radiation in Agricultural Research in the Soviet Union. Report of Study Tours, 1971 and 1972. IAEA, Vienna, 108 pp.

IAEA, 1977. Manual on Mutation Breeding. Technical Reports Series No. 119, IAEA, Vienna, 2nd edn, 288 pp.

Kaul, B.L. and Kak, S.N., 1973. Improvement of *Mentha arvensis* L. through induced mutations. In: Advancing Frontiers in Cytogenetics. Hindustan Publishing Corp. (India), Delhi, pp. 189—195.

Kaul, B.L. and Kak, S.N., 1975. Use of radiations and radiomimetic chemicals in the breeding of vegetatively propagated plants. I. *Mentha arvensis*. Plant Breed. Abstr., 45: No. 1002.

Kolesnikova, L.G. and Maksimova, A.D., 1974. Effect of chemical mutagens on biochemical characters in potato. Usp. Khim. Mutageneza Sel, 1974: 259—264 (in Russian); Plant Breed. Abstr., 47: No. 1353.

Kolontaev, V.M., 1974b. Experimental mutagenesis in apple II. Characteristics of development and cytogenetic modifications in *Malus domestica* L. induced by dimethyl sulphate. Sov. Genet., 10(3): 291—299.

Kolontaev, V.M., 1974c. Experimental mutagenesis in apple III. Morphological and genetic effects in *Malus domestica*, induced by N-nitroso-N-methylurea. Sov. Genet., 10(4): 427—433.

Kolontaev, V.M., 1974d. Experimental mutagenesis in apple IV. Phenotypic and genotypic changes in *Malus domestica*, induced by low doses of X-rays and nitrosomethylurea. Sov. Genet., 10(5): 578—583.

Kopan, V.P., 1974. The problem of radiation and chemical mutagenesis of apples. Biofiz. Fiziol-Biokhim. Issled Plodovod. Yagod. Kult., 1974: 165—167 (in Russian).

Lacey, C.N.D., 1977. The mutation spectrum of Cox's Orange Pippin resulting from gamma irradiation. In: Eucarpia Meeting on Tree Fruit Breeding. Inst. Hortic. Plant Breed., Wageningen, The Netherlands (1976), pp. 46—52.

Mee, G.W.P., Nickel, L.G. and Heinz, D.J., 1969. Chemical mutagens — their effects on cells in suspension culture. Hawaii. Sugar Plant. Assoc. Exp. Stn Annu. Rep., 7—8.

Moës, A., 1966. Mutations induites chez le glaïeul (*Gladiolus*). Bull. Rech. Agron. Gembloux, 1: 76—95.

Naumann, C.H., Sparrow, A.H. and Schairer, L.A., 1976. Comparative effects of ionizing radiation and two gaseous chemical mutagens on somatic mutation induction in one mutable and two non-mutable clones of *Tradescantia*. Mutat. Res., 38(1): 53—70.

Nybom, N., 1961. The use of induced mutations for the improvement of vegetatively propagated plants. In: Mutations and Plant Breeding, Cornell Univ., Ithaca, N.Y., NAS—NRC, Publ. 891: 252—294.

Pohlheim, F. and Pohlheim, E., 1976. Herstellung von Plastommutanten bei *Saintpaulia ionantha* H. Wendl. Biochem. Physiol. Pflanz., 169: 377—383.

Rapoport, I.A., 1973. Chemical mutagenesis. In: Isotopes and Radiation in Agricultural Research in the Soviet Union. Report of Study Tours, 1971 and 1972. IAEA, Vienna, pp. 24—25.

Singh, R. and Iyer, C.P.A., 1974. Chemical mutagenesis in pineapple (*Ananas comosus*). Proc. Int. Hortic. Congr., XIX(1): 108.

Sparrow, A.H., Schairer, L.A. and Sparrow, R.C., 1963a. Relationship between nuclear volumes, chromosome numbers and relative radiosensitivities. Science, 141(3576): 163—166.

Sparrow, A.H., Underbrink, A.G. and Sparrow, R.C., 1967. Chromosomes and cellular radiosensitivity. I. The relationship of D_0 to chromosome volume and complexity in seventy-nine different organisms. Radiat. Res., 32: 915—945.

Sparrow, A.H., Rogers, A.F. and Schwemmer, S.S., 1968. Radiosensitivity studies with woody plants — I. Acute gamma irradiation survival data for 28 species and predictions for 190 species. Radiat. Bot., 8: 149—186.

Chapter 2

SHOOT APICES: ORGANIZATION AND POST-IRRADIATION BEHAVIOUR

STRUCTURE AND FUNCTIONING OF SHOOT APICES

In this section some previous and present opinions about the organization of apices are summarized. For a fairly recent and more detailed review, the reader is referred for example to Gifford and Corson (1971).

When discussing the structure of shoot apices axillary buds and adventitious buds are included unless otherwise stated. Different words are often used to indicate the apical area of a shoot. Dermen (1960), for example, mentions growing point, shoot tip, shoot apex, apical meristem and apical dome. Most often the term shoot apex is used in a rather general sense to indicate that part of the stem lying distal to the youngest leaf primordium, together with a few lower leaf primordia. The shoot apex is the area from which new growth originates.

A long and animated discussion still continues concerning "initial cells" in an apex. Strictly speaking, an initial cell is a cell which divides into two daughter cells, one of which remains in the fixed (apical) position, the other being added to the meristematic tissue (Esau, 1965). The question of whether such initial cells exist is of direct importance to mutation work, because induction of a mutation in such an initial cell means that a permanent lineage of mutated daughter cells will be produced by that cell. In the case where occupation of the apical position is not permanent, i.e. if a mutated initial cell shifts to the flanks, another non-mutated initial cell will take its position and the production of a lineage of mutated cells will be terminated unless permanent initial cells exist at the flanks (see below).

In 1852, Hofmeister (1852) reported the existence of a single initial cell in Phanerogams. After that, many investigators tried to provide microscopical proof of this, but they failed to do so as most cells in the apical region look very similar. However, Hofmeister's concept has received considerable support over the years. In 1868 Hanstein (1868) presented his Histogen Theory. According to his theory, some kind of stratification exists in apices of angiopermous plants. A central core of irregularly arranged cells was reported to be surrounded by a number of regular, mantle-like layers. Each layer and the corpus are said to be derived from initial cells which are found vertically superimposed on each other, at the ultimate tip of the apical dome. At present, the general idea of stratified shoot apices with a limited number of initial cells per layer is generally accepted for angiosperms, but most authors no longer accept the strongly predestined role of the different growth layers or histogens attributed to them by Hanstein. Newman (1965), on the other hand, pointed out that Hanstein's original concept was not as rigid as his later critics believed it to be.

A new and relatively flexible theory concerning the organization of shoot apices was put forward by Buder (1928) and Schmidt (1924). In this "Tunica—Corpus Theory" several tunica layers are distinguished, the layers being characterized by anticlinal cell divisions. The plane of division causes a certain independence of each layer. Below the tunica layers is found a core, the corpus, which consists of cells dividing in all directions. The Tunica—Corpus Theory does not imply that the tunica layers remain discrete throughout their lives as periclinal divisions in tunica layers are sometimes observed.

At present most authors still consider the Tunica—Corpus Theory to be valid. Support for the general correctness of the opinions of Buder and Schmidt is usually derived from histological work and from studies on chimeras, notably periclinal ploidy chimeras (or cyto-chimeras). Over the years a number of modifications have been suggested and some authors, e.g. Von Guttenberg (1960), completely reject the theory because it does not relate apical activity to the origin of the tissues.

Satina et al. (1940) and Satina and Blakeslee (1941) proposed the now generally accepted expressions L-I, L-II and L-III to indicate the three independent germ layers of the apex (Fig. 4). Dermen (1951) refers to primary "histogenic layers" or "histogens" which he believes to be present in perhaps all angiosperms.

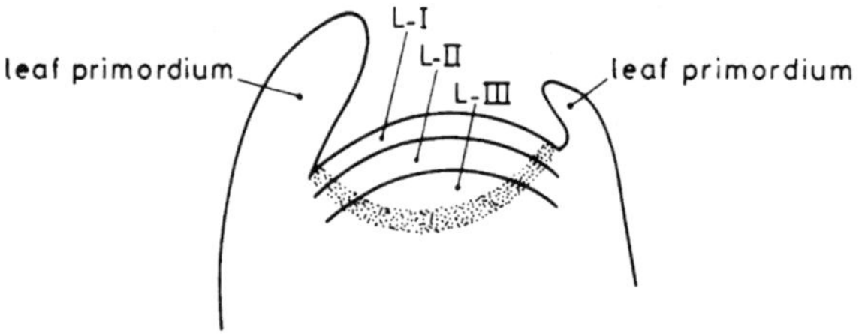

Fig. 4. Shoot apex showing the three histogenic layers and cyto-histological zonation (dotted area). (Institute of Plant Breeding, Agricultural University, Wageningen.)

Another concept of zonation was proposed by French scientists. According to Plantefol (1947), foliar helices terminate in meristematic tissue some distance from the apical tip. The author claimed that these cells, which are called the "anneau initial", are the real initial cells. No true initials were said to be present at the summit of the axis during the vegetative phase of development in dicotyledonous plants. Buvat (1952) calls this area "méristéme d'attente" or "waiting meristem" (Fig. 5) because, according to French

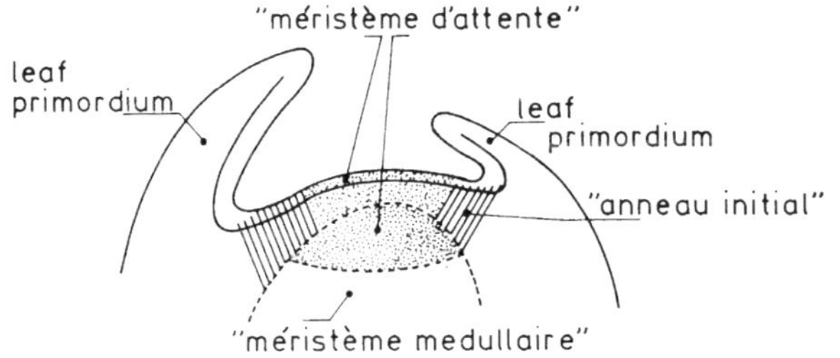

Fig. 5. Shoot apex organization according to Buvat (N.B. both the "méristème d'attente" as well as the "anneau initial" consist of several layers). (Institute of Plant Breeding, Agricultural University, Wageningen.)

opinion, no division occurs there during the vegetative phase. Later on, however, Buvat had to admit that vegetative summital cells sometimes show divisions.

The French ideas were not entirely new. For example, Foster (1941) and Majumdar (1942) had already demonstrated that shoot apices are not as homogeneous as they were previously believed to be. Both authors referred to some kind of cyto-histological zonation (Fig. 4). Even earlier, Gregoire (1938) reported that different stages of development may lead to differences in the zonal pattern. The French ideas have been criticized on several occasions, e.g. by Gifford (1954), but since Buvat (1952) has admitted that the original concept was too rigid, it seems that the different views can be reasonably compatible.

It is now generally accepted that cells at the summit of vegetative apices also divide, but at a much lower rate than those along the flanks. Much controversy is still attached to such things as the existence, role, position and number of initial cells. It would take us too long to discuss all relevant literature here, but some additional remarks must be made. Most authors now agree that "real" initial cells do not exist (Dulieu, 1970). The apical meristem is said to be made up of a dynamic population of cells and to have self-organizing properties (Soma, 1973). It cannot be denied, however, that some ultimate source of primary growth must exist. The most reliable method to investigate the position of this source is to use marked cells which carry mutations that affect the phenotype of the individual. In this respect, Stewart and Dermen (1970c) studied the dimension of mericlinal stripes in periclinal chimeras and thus determined the number of initial cells in shoot apices of different plant species. Their conclusion for *Ligustrum* was that two or three initial cells per layer exist. An initial cell was found to divide only once in approximately 12 days, after which the most distal daughter cell should be considered as the new initial cell. According to Stewart and Dermen (1970c) the low division rate mentioned suffices to make initial cells the ultimate source of growth.

BEHAVIOUR OF AXILLARY AND ADVENTITIOUS BUDS

Axillary buds are formed on the stem in close association with the leaves, normally at some distance from the apical summit, i.e. not in the axils of the youngest leaves.

A controversial point is whether axillary buds under normal conditions reproduce the layered structure of the main apex. Howard et al. (1963), for example, suggested that axillary potato buds are often derived from L-I and L-II only. Klopfer (1965), on the other hand, demonstrated an identical axillary reproduction of the three-layered structure in a number of cases in potatoes. Schmidt (1924) mentioned that the relative contribution of the different layers does not need to be the same for bud and leaf initiation.

Axillary buds are arrested at a certain point in time by the influence of the apical dominance of the main apex of the same shoot. This dormancy may be terminated by seasonal effects or by artificial means.

Adventitious buds are initiated outside the primary meristematic regions or meristematic remnants of plants. They can occur on different parts of plants, often from wound callus or without a previous callus phase from more or less mature undamaged tissue. They can be of exogenous as well as of endogenous origin. The number of cells from which the primordia of adventitious buds arise varies from one to many. This is, of course, of importance for mutation breeding (see Introduction and Chapter 3).

In practice, it is often difficult to make a proper distinction between adventitious and axillary buds. The use of periclinal chimeras as a starting point may help in this respect.

THE POSITION OF A MUTATED CELL WITHIN A PLANT; CHIMERISM

Mutations (spontaneous as well as artificially induced ones) take place in all kinds of tissue, but a single mutational event is normally restricted to one cell only. The position of this mutated cell is important for its further fate. Sometimes a distinction is made between (intra-)apical and extra-apical mutations (Bergann, 1967b), but at present this distinction has no practical meaning as extra-apical mutations in many plant species can also be of value to the plant breeder, thanks to the adventitious bud technique (Broertjes et al., 1968).

A plant consisting of two or more genetically different somatic tissues is called a chimera, after Baur (1909), who expanded the original definition used by Winkler (1907) for graft hybrids, to include all organisms which are genetically dissimilar in their somatic tissue. Chimerism automatically occurs if more than one cell is present at the moment of a mutagenic event.

If a non-lethal mutation is induced in a dividing cell of the apical meristem, a lineage of mutated cells is produced, the dimensions of which depend on the position of the mutated cell, the total number of dividing apical cells and the fitness of the mutated cell. A distinction is normally made between sectorial chimeras, mericlinal chimeras and periclinal chimeras (Fig. 6). The word "sector" is often used erroneously in mutation work and should be restricted to "real" sectors, i.e. to chimeras in which a united group of cells, representing all histogenic layers and their derivatives, are mutated. If only a part of one histogenic layer and its derivatives are mutated, the expression mericlinal chimera should be used, according to Jørgensen and Crane (1927). After one or a few cycles of vegetative propagation, periclinal chimerism is the common situation in mutated (vegetatively propagated) plants (Dermen, 1947b). Bergann (1954) further distinguished and named different types of periclinal chimeras.

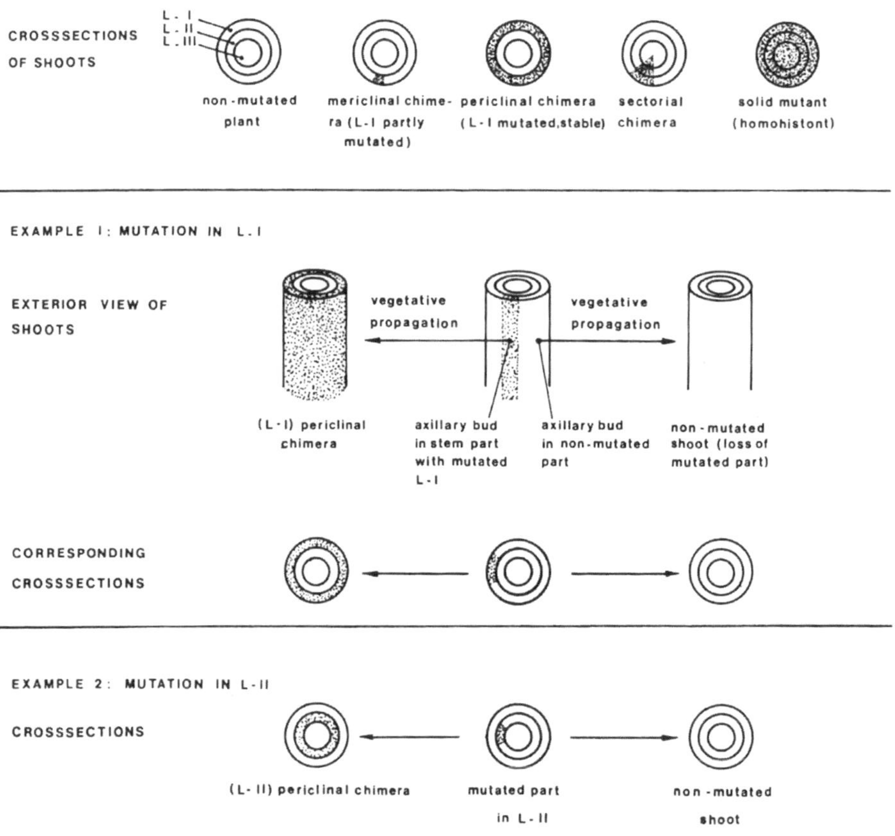

Fig. 6. Chimerism in shoots as observed at some distance from the apical region. In examples 1 and 2 it is shown how, starting from a mericlinal chimera (L-I or L-II partly mutated), after vegetative propagation either a stable periclinal climera or a completely non-mutated shoot is obtained. In example 2 the mutation may remain unobserved, depending on the genetic character involved. (Institute of Plant Breeding, Agricultural University, Wageningen.)

In all three types of chimera, the mutated cells remain united and occupy a certain area of the plant. If the mutated cells are dispersed throughout a plant, chimerism will be mosaic in nature.

REARRANGEMENTS OF CELL LAYERS

It has already been mentioned that shoot apex layers are not fully stable. The instability, mostly caused by occasional periclinal divisions in L-I or L-II, may lead to a rearrangement of the layers. If the shoot apex is already genetically heterogeneous (e.g. periclinal chimeric for flower colour), such rearrangements may be observed at the phaenotype of the plant. This kind of histological effect was often taken erroneously for the occurrence of a (real) mutagenic effect.

Different kinds of tissue rearrangements have been documented. Bergann and Bergann (1959, 1962) distinguish between perforations (i.e. tissue from L-II and/or L-III penetrating the outer layers), reduplications (the reverse) and transgression (lateral replacement within one layer). A combination of these three effects may lead to so-called translocations, i.e. to complete layer inversion within the apex. The choice of words by the authors is not a very lucky one as most words have older established meanings in genetics. Another drawback of this work is that it does not provide figures concerning the frequency of the different effects. Other words, replacement instead of duplication and displacement instead of perforation, were proposed by Dermen (1960).

Rates of cell exchange have been given by Stewart and Burk (1970). They estimate the frequency of periclinal divisions in L-I in nature to be 1/3 100 divisions. Their figures are based on calculations of dark green spots on the surface of white tobacco leaves with the original constitution L-I green, L-II + L-III white. Dark green patches occurred when green L-I tissue replaced white L-II cells.

SHOOT APICES AFTER IRRADIATION

Normally, effects of irradiation are divided into primary effects (often, though not completely correctly, indicated as physiological damage) and heritable changes or mutations (Sax and Swanson, 1941). Both groups include a number of very different phenomena. The group of primary effects includes, for example, retardation or inhibition of cell division, cell death, induction of mitotic activity, effects on growth rate or growth habit and changes in reproductive capacity (Sparrow, 1961), and increased frequency of tissue rearrangements (Bergann, 1967a). (N.B. This last effect can be used to transfer a mutation, which is already present in one layer, to another layer within the apex. The method may be useful in breeding work. For example, one could transfer a valuable mutation for flower colour which is only present in L-III (and therefore remains unobserved) to L-I, where the mutation will be expressed.)

Not all cells within a shoot apex display the same physiological activity or are at the same stage of the mitotic cycle. Therefore, cells within an apex may react to radiation in different ways.

The radiosensitivity of cells, plant parts, various stages of development or plant species has been an important subject of investigation. For a number of plants it has been demonstrated that radiosensitivity may be, at least partly, genetically determined (see, e.g. Smith, 1942; Yamagata et al., 1969; Blixt, 1970). On the other hand, Sparrow and Christensen (1953) and Sparrow et al. (1961, 1965) explained differences in radiosensitivity between different plant species mainly on the basis of differences in nuclear size and average interphase chromosome volume (ICV). Recent Russian work (Semerdzhyan and Nor-Arevyan, 1971) suggests that radioresistance is directly correlated with sulphydryl content in the meristematic regions of *Vicia faba* and *Triticum*.

Concerning differences in radiosensitivity at different stages of development, it has been found that younger stages of development are generally more sensitive and that the actively dividing cells are more sensitive than non-dividing or dormant cells. Meristematic zones are more sensitive than more differentiated parts. There are indications, however, that not all meristematic cells are radiosensitive (Gunckel, 1957) and that actively dividing cells are not necessarily the most radiosensitive ones (Lapins and Hough, 1970). Such findings are of importance in relation to irradiation of growing points.

Whole plants or organized plant parts are not an ideal material for radiosensitivity studies. However, not much is known of the behaviour of cell suspensions or callus after irradiation. Bajaj (1971) stated that callus tissue has a higher radioresistance than seedlings.

PATTERNS OF RADIATION-INDUCED MORPHOLOGICAL/HISTOLOGICAL DAMAGE AND RECOVERY

In principle, a mutation can be induced in any of the (three) histogenic layers of the apical meristem (Renner, 1936). It is not known whether the frequency of mutations (mutation rate per cell) is the same in each layer. Recovery of a shoot apex after irradiation happens via regeneration either from within the apex or from outside this area. In the latter case axillary buds or adventitious buds may take over. The height of the dose applied may be important in this respect, because a difference in dose may alter the place and type of regeneration.

The stage of development and activity of existing axillary buds and tissues from which adventitious buds may arise may affect the final mutation frequency of the irradiated material. Moreover, the relative participation of the different histogenic layers may vary. As mentioned above, axillary buds do not always copy the layered structure of the main bud exactly and adventitious buds may develop from corpus tissue (L-III) or epidermal tissue (L-I) only. If periclinal chimeras are irradiated this may have practical consequences for the stability of the phenotypic characters which are involved.

Within shoot apices different kinds of regeneration types are recognized. Pratt (1963) distinguishes between (a) direct healing of the apical meristem leading, for example, to compression of the non-functioning damaged zone between L-I and the inner layers, (b) formation of substitute meristems (in cells laterally to the damaged zone) and (c) development of axillary meristems, separated from the top meristem by at least one leaf.

A point which has not yet been sufficiently studied is whether any regularity can be found with respect to the observed differences in radiosensitivity within shoot apices. Miksche et al. (1962) reported an outer to inner pattern of decreasing radiosensitivity in *Taxus media* (a gymnospermous plant). Several authors have tried to explain their results in terms of this pattern, but have failed to do so.

It has been suggested in a number of publications that not all layers have the same radiosensitivity. A simple explanation for this may be that there is differential mitotic activity, because of different organogenic properties of

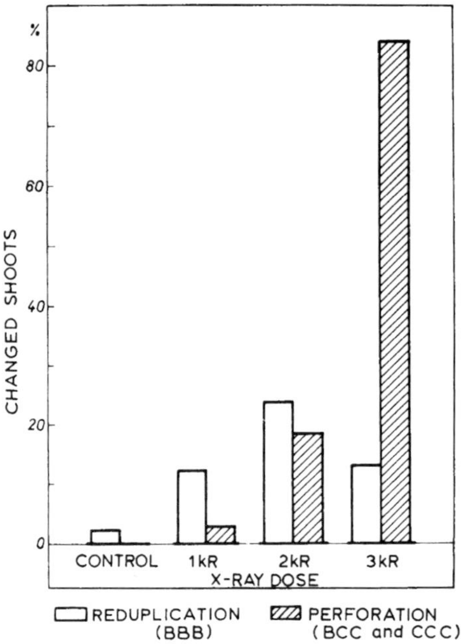

Fig. 7. Dose response of different radiation-induced histogenetic effects (reduplication and performation) in *Pelargonium zonale* (original constitution: BBC). (Pötsch, 1969, Fig. 4.)

Plate 2. A mutation is a one-cell event (1). Therefore, a mutated cell has to grow out into a group of cells (2) and a layer of cells (3) before it can be observed as a mericlinal chimera (4), in this case of *Dahlia*.

(Slides from Brookhaven National Laboratory, Upton, Long Island, N.Y.)

By cutting back, complete, stable periclinal chimeras (sports) can be obtained, as demonstrated in the *Chrysanthemum morifolium* cv. Hortensien Rose, shown here with a yellow "sector" (5), a partly mutated plant (6) and a complete "sport" (7).

(Slides, ITAL, Wageningen.)

the distinct layers, in which the overall stage of development (e.g. flowering stage or vegetative stage) also plays a role.

It has been reported that the frequencies of different categories of histogenic reaction are dose dependent, as was demonstrated, for example, by the difference in relative frequencies of reduplications and perforations in *Euphorbia pulcherrima* and *Abutilon* (Pötsch, 1966a,b; 1967) and in *Pelargonium zonale* (Pötsch, 1969) (Fig. 7).

THE FATE OF A MUTATED APICAL CELL

Dommergues (1962) pointed out that the fate of a mutated cell during organogenesis depends on its position within the apex. Mutations in cells which are predetermined to give rise to, or to become part of, leaves will normally be lost. Mutations, which by definition are one-cell events, are especially desirable if they are induced in apical initial cells, as they may produce mutated "sectors" (Plate 2). Stable initial cells produce long sectors. The occurrence of long sectors increases the chance that an axillary bud will be formed which is partly or completely situated in the mutated tissue.

It has already been pointed out that the apical layer in which a mutation is induced largely determines whether or not the mutation can be easily detected and utilized. Many apparently solid mutants in vegetatively propagated plants are in fact periclinal chimeras, developed after one or more cycles of vegetative propagation from plants carrying mutated "sectors" (mericlinal chimeras) (Plate 2). From a breeder's point of view, it would be very useful in those cases if plants could be grown from a number of cells, or even better from one cell only, derived from the mutated layer. This is frequently attained if the adventitious bud technique is applied.

DIPLONTIC SELECTION

In the previous sections it has been assumed that a mutated cell within (apparently) non-mutated tissue has the same chance of survival as normal cells, and has an equal rate of division etc. This is not necessarily so. A kind of competition between mutated and non-mutated cells is reported to exist, which often leads to (sooner or later) elimination of the mutated cells. This phenomenon is called diplontic selection, intra-individual selection, intra-somatic or somatic elimination. The first studies on this competition were published by Kaplan (1951) and Gaul (1959, 1961).

Opinions differ as to the degree of selective disadvantage of specific mutational events in cells. It seems logical to assume that cells with heavy physiological or chromosomal damage may have less chance to pass through the mitotic sieve than non-damaged cells. It is not necessarily so, however, that cells, carrying, for example, a single-gene mutation for leaf colour or a minor chromosomal deletion, should have a selective disadvantage for this reason.

Other cells within the same irradiated area, which may not be easily recognizable as such because of the absence of an easily observable marker, may have been damaged much more severely and eliminated. Nevertheless, the disappearance after a period of time of, for example, a mutated stripe of a lighter colour during the course of the further development of a plant has often been considered as an example of diplontic selection.

Balkema (1971), who made a detailed study of this subject, came to the conclusion that many of the examples quoted in relation to the occurrence of diplontic selection can be explained equally well, or even better, in other ways. According to her, most of the observed loss of chimerism is inherent to the random occurrence of sectors in the apex. She further concludes that mutated cells will be at a disadvantage only in those cases where there are many potential organs of which only a few can develop, for example, the development of side shoots and axillary or adventitious buds.

This opinion was opposed by Broertjes (1972c) in the case of adventitious buds on detached leaves of *Saintpaulia*. He found relatively high percentages of slowly growing dwarf mutants, even where low X-ray doses had been applied and sufficient undamaged cells probably were present to suppress the outgrowth of the slowly growing mutant cells.

Loss of chimerism related to the normal development of a plant has been called "diplontic drift" by Balkema (1971).

Literature shoot apices

Bajaj, Y.P.S., 1971. Direct and indirect effects of gamma irradiation on plant tissue cultures. In: Int. Symp. Use of Isotopes and Radiation in Agricultural and Animal Husbandry Research, New Dehli, pp. 285—301.

Balkema, G.H., 1971. Chimerism and Diplontic Selection. Thesis, Landbouwhogeschool, Wageningen. A.A. Balkema, Rotterdam, Cape Town, 173 pp.

Baur, E., 1909. Das Wesen und die Erblichkeitsverhältnisse der "Varietates albomarginatae hort." von *Pelargonium zonale*. Z. Indukt. Abstamm. Vererbungsl., 1: 330—351.

Bergann, F., 1954. Praktische Konsequenzen der Chimärenbildung für die Pflanzenzüchtung. Wiss. Z. Karl Marx Univ. Leipzig, Math. Naturwiss. Reihe, 4: 281—291.

Bergann, F., 1967a. The relative instability of chimerical clones — the basis of further breeding. In: H. Stubbe (Editor), Induced Mutations and their Utilization. Abh. Dtsch. Akad. Wiss. Berlin Kl. Medizin, 2: 287—300.

Bergann, F., 1967b. Mutations-chimären: Rohmaterial züchterischer Weiterbehandlung. Umschau, 24: 791—797.

Bergann, F. and Bergann, L., 1959. Uber experimentell ausgelöste vegetative Spaltungen und Umlagerungen an chimärischen Klonen zugleich als Beispiel erfolgreicher Staudenauslese. I. *Pelargonium zonale* Ait. 'Madame Salleron'. Züchter, 29: 361—374.

Bergann, F. and Bergann, L., 1962. Über Umschichtungen (Translokationen) an den Sprossscheiteln periklinaler Chimären. Züchter, 32: 110—119.

Blixt, S., 1970. Studies of induced mutations in peas. XXV. Genetically conditioned differences in radiation sensitivity. 4. Agric. Hort. Genet., 28: 55—116.

Broertjes, C., 1972c. Use in Plant Breeding of Acute, Chronic or Fractionated Doses of X-rays or Fast Neutrons as Illustrated with Leaves of *Saintpaulia*. Agric. Res. Rep. 776. Pudoc, Wageningen, 74 pp.

Broertjes, C., Haccius, B. and Weidlich, S., 1968. Adventitious bud formation on isolated leaves and its significance for mutation breeding. Euphytica, 17: 321—344.
Buder, J., 1928. Der Bau des phanerogamen Sprossvegetationspunktes und seine Bedeutung für die Chimärentheorie. Ber. Dtsch. Bot. Ges., 46: 20—21.
Buvat, R., 1952. Structure, évolution et fonctionnement du méristème apical de quelques dicotylédones. Ann. Sci. Nat. Bot. Biol. Veg. (11 Ser.), 13: 202—203.
Dermen, H., 1947b. Histogenesis of some bud sports and variegations. Proc. Am. Soc. Hortic. Sci., 50: 51—73.
Dermen, H., 1951. Ontogeny of tissues in stem and leaf of cytochimeral apples. Am. J. Bot., 38: 753—760.
Dermen, H., 1960. Nature of plant sports. Am. Hortic. Mag., 39: 123—173.
Dommergues, P., 1962. La destinée de la cellule mutée: Conséquences dans le cas des plantes a multiplication végétative et dans le cas des plantes a reproduction sexuée. Eucarpia Congr. Assoc. Eur. Amélior. Plant. (Paris), 1962: 115—139.
Dulieu, H., 1970. Les mutations somatiques induites et l'ontogénie de la pousse feuillée. Ann. Amélior. Plant., 20: 27—44.
Esau, K., 1965. Plant Anatomy. Wiley and Sons, New York, N.Y., 2nd edn, 767 pp.
Foster, A.S., 1941. Comparative studies on the structure of the shoot apex in seed plants. Bull. Torrey Bot. Club, 68: 339—350.
Gaul, H., 1959. Über die Chimärenbildung in Gerstenpflanzen nach Röntgenbestrahlung von Samen. Flora (Jena) Abt. B, 47: 209—241.
Gaul, H., 1961. Studies on diplontic selection after X-irradiation of barley seeds. In: Proc. Symp. Effects of Ionizing Radiation on Seeds, Karlsruhe, 1960. IAEA, Vienna, 117—136.
Gifford, Jr, E.M., 1954. The shoot apex in angiosperms. Bot. Rev., 20: 477—529.
Gifford, Jr, E.M. and Corson, Jr, G.E., 1971. The shoot apex in seed plants. Bot. Rev., 37: 143—229.
Gregoire, V., 1938. La morphogénèse et l'autonomie morphologique de l'appareil floral. Cellule, 47: 287—452.
Gunckel, J.E., 1957. The effects of ionizing radiation on plants. Morphological effects. In: Symp. Effects of Ionizing Radiation on Plants. Q. Rev. Biol., 32(1): 46—56.
Hanstein, J., 1868. Die Scheitelzellgruppe im Vegetationspunkt der Phanerogamen. In: Niederrhein. Ges. Nat. Heilkd. Festschr. zum 50-jährigen Jubiläum Univ. Bonn, pp. 109—143.
Hofmeister, W., 1852. Zur Entwicklungsgeschichte der *Zostera*. Bot. Ztg., 10: 121—131; 137—149; 157—158.
Howard, H.W., Wainwright, J. and Fuller, J.M., 1963. The number of independent layers at the stem apex in potatoes. Genetica, 34: 113—120.
Jφrgensen, C.A. and Crane, M.B., 1927. Formation and morphology of *Solanum* chimeras. J. Genet., 18: 247—273.
Kaplan, R.W., 1951. Chromosomen und Faktormutationsraten in Gerstenkörnern bei verschiedenartigen Quellungsbehandlungen oder Kälte während oder nach der Röntgenbestrahlung sowie bei Dosisfraktionierung. Z. Indukt. Abstamm. Vererbungsl., 83: 347—382.
Klopfer, K., 1965. Über den Nachweis von drei selbständigen Schichten im Sprosscheitel der Kartoffel. Z. Pflanzenzücht., 53: 67—87.
Lapins, K.O. and Hough, L.F., 1970. Effects of gamma-rays on apple and peach leaf buds at different stages of development. II. Injury to apical and axillary meristems and regeneration of shoot apices. Radiat. Bot., 10: 59—68.
Majumdar, G.P., 1942. The organization of the shoot in *Heracleum*. Ann. Bot. (London), 6: 49—81.
Miksche, J.P., Sparrow, A.H. and Rogers, A.F., 1962. The effects of chronic gamma irra-

diation on the apical meristem and bud formation of *Taxus media*. Radiat. Bot., 2: 125—129.

Newman, I.V., 1965. Patterns in the meristems of vascular plants. III. Pursuing the patterns where no cell is a permanent cell. J. Linn. Soc. London, Bot., 59: 185—214.

Plantefol, L., 1947. Hélice foliaire, point végétatif et stéle chez les dicotylédones. La notion d'anneau initial. Rev. Gen. Bot., 54: 49—80.

Pötsch, J., 1966a. Über die Auslösung extramutativer Strahlungseffekte an Klonsorten von *Euphorbia pulcherrima* Willd. Züchter, 36: 12—25.

Pötsch, J., 1966b. Das Verhalten von *Abutilon hybridum* hort. 'Andenken an Bonn' nach einmaliger und fraktionierter Röntgenbestrahlung. Z. Pflanzenzücht., 55: 183—200.

Pötsch, J., 1967. On the dissociation of chimerical shoot-variants by the use of X-rays. In: H. Stubbe (Editor), Induced Mutations and Their Utilization. Abh. Dtsch. Akad. Wiss. Berlin Kl. Medizin, 2: 301—304.

Pötsch, J., 1969. Die Abhängigkeit röntgeninduzierter Histogeneseanomalien von der Höhe der Bestrahlungsdosis bei *Pelargonium zonale* Ait "Madame Salleron". Wiss. Z. Pädagog. Hochsch. Potsdam, 13(1): 129—137.

Pratt, C., 1963. Radiation damage and recovery in diploid and cytochimerical varieties of apples. Radiat. Bot., 3: 193—206.

Renner, O., 1936. Zur Entwicklungsgeschichte randpanaschierter und reingrüner Blätter von *Sambucus, Veronica, Pelargonium, Spiraea, Chlorophytum*. Flora (Jena), 30: 454—466.

Satina, S. and Blakeslee, A.F., 1941. Periclinal chimeras in *Datura stramonium* in relation to development of leaf and flowers. Am. J. Bot., 28: 862—871.

Satina, S., Blakeslee, A.F. and Avery, A.G., 1940. Demonstration of the three germlayers in the shoot apex of *Datura* by means of induced polyploidy in periclinal chimeras. Am. J. Bot., 27: 895—905.

Sax, K. and Swanson, C.P., 1941. Differential sensitivity of cells to X-rays. Am. J. Bot., 28: 52—59.

Schmidt, A., 1924. Histologische Studien an phanerogamen Vegetationspunkten. Bot. Arch., 7/8: 345—404.

Semerdzhyan, S.P. and Nor-Arevyan, N.G., 1971. On the role of sulphydryls in the determination of natural radiosensitivity of *Vicia faba* and *Triticum* sprouts. Radiobiologiya, 11: 278—281 (in Russian).

Smith, L., 1942. Hereditary susceptibility to X-ray injury in *Triticum monococcum*. Am. J. Bot., 29: 189—191.

Soma, K., 1973. Experimental studies on the morphogenesis in the vegetative shoot apex. Induced mutation and chimera in woody plants. Gamma Field Symp., 12: 83—94.

Sparrow, A.H., 1961. Types of ionizing radiation and their cytogenetic effects. In: Mutation and Plant Breeding. NAS—NRC, Publ. 891: 55—119.

Sparrow, A.H. and Christensen, E., 1953. Tolerance of certain higher plants to chronic exposure to gamma-radiation from cobalt-60. Science, 118: 697—698.

Sparrow, A.H., Cuany, R.L., Miksche, J.P. and Schairer, L.A., 1961. Response of plants to acute and chronic radiation exposures. Radiat. Bot., 1: 10—34.

Sparrow, A.H., Sparrow, R.C., Thompson, K.H. and Shairer, L.A., 1965. The use of nuclear and chromosomal variables in determining and predicting radiosensitivities. In: The Use of Induced Mutations in Plant Breeding. FAO/IAEA Meeting, Rome (1964). Radiat. Bot., 5 (Suppl.): 101—132.

Stewart, R.N. and Burk, L.G., 1970. Independence of tissues derived from apical layers in ontogeny of the tobacco leaf and ovary. Am. J. Bot., 57: 1010—1016.

Stewart, R.N. and Dermen, H., 1970c. Determination of number and mitotic activity of shoot apical initial cells by analysis of mericlinal chimeras. Am. J. Bot., 57: 816—826.

Von Guttenberg, H., 1960. Grundzüge der Histogenese höherer Pflanzen. I. Die Angiospermen. In: Handbuch der Pflanzenanatomie III, 3. Borntraeger, Berlin-Nikolassee.

Winkler, H., 1907. Über Propfbastarde und pflanzliche Chimären. Ber. Dtsch. Bot. Ges., 25: 568.

Yamagata, H., Kowyama, Y. and Syakudo, K., 1969. Radiosensitivity and polyploidy in some non-tuber bearing *Solanum* species. Radiat. Bot., 9: 509—521.

Chapter 3

ADVENTITIOUS BUD TECHNIQUES AND OTHER IN VIVO OR IN VITRO METHODS OF ASEXUAL PROPAGATION

INTRODUCTION

As has already been mentioned, irradiation of bulbs, tubers, rhizomes, cuttings, grafts, other plant parts or whole plants, all having buds with multicellular apices composed of a number of fairly autonomous cell layers, automatically leads to the formation of chimeras. This is the main obstacle in mutation breeding, especially in species in which there is no ontogenetically young stage of buds to be irradiated. Radiating such buds produces small-sectored mericlinal chimeras which have relatively low chances for the recovery of the mutations induced. Other causes of reduced efficiency when irradiating multicellular buds are found in the structure of the apex and diplontic selection (see Chapter 2).

This reduced efficiency can be improved by offering a mutated cell optimum chances to form the largest possible portion of a plant (part). Obviously, the most ideal situation is for a plant (part) to originate from a single cell, in vivo or in vitro. Such methods reduce or avoid chimera formation, increase the mutation frequency and probably widen the mutation spectrum as well. Moreover, the recognition of favourable mutations as well as their recovery becomes easier. It is therefore of the utmost importance to look for such methods and to develop them.

IN VIVO TECHNIQUES

Adventitious buds originating from one cell

At present, an important method for the commercial plant breeder is the adventitious bud technique, which makes use of the phenomenon that (the apices of) adventitious buds, e.g. on detached leaves, ultimately originate from only one (epidermal) cell (Broertjes et al., 1968; Haccius and Hausner, 1974). This has been demonstrated in *Saintpaulia* (Naylor and Johnson, 1937; Sparrow et al., 1960; Broertjes, 1968), *Achimenes* (Broertjes, 1972b), *Kalanchoë* (Broertjes and Leffring, 1972), *Streptocarpus* (Broertjes, 1969b) and *Nicotiana* (De Nettancourt et al., 1971). In all these cases it was shown, by mutation induction and cytological—histological observations, that L-I cells, which had become dedifferentiated, developed into bud primordia after the formation of a meristem.

Such adventitious buds were formed exclusively by cells at the lowest part of the petiole near the cut end, as has also been found to be the case in other plants, such as *Dendrophthoe* (Nag and Johri, 1970). In all our experiments with *Saintpaulia* and other species, the basal 5 mm of the petiole was cut off after completion of irradiation and prior to planting. In this way, epidermal cells situated higher on the petiole, which had no previous stimulus to divide, were activated to do so. Consequently, all such epidermal cells underwent the radiation treatment in a resting (non-dividing) stage of development, either G_1 or G_2 (Broertjes, 1972c).

An important phenomenon is the fact that the overwhelming majority of the adventitious mutants thus produced and observed in *Saintpaulia, Achimenes, Streptocarpus, Kalanchoë, Begonia* and other species, were solid and non-chimeral. Therefore, their apices must ultimately have originated from a single (epidermal) cell (Plate 3).

Adventitious bud formation on leaves in general. Many plants can be propagated by adventitious plantlet formation on detached leaves (Fig. 8). Broertjes et al. (1968) listed over 350 of such species, belonging to various plant families, which are on record in the literature; for example *Begonia*, various Gesneriaceae and *Peperomia*, both being propagated on a commercial scale using leaves or parts of leaves. Many, however, are not listed, including a number of species in economically very important plant families, such as Gramineae (cereals) and Solanaceae (potato, tomato). Many such plants have been tried without success. Many more have never been tried, however, and the breeder should always experiment with the cultivar(s) in which he is

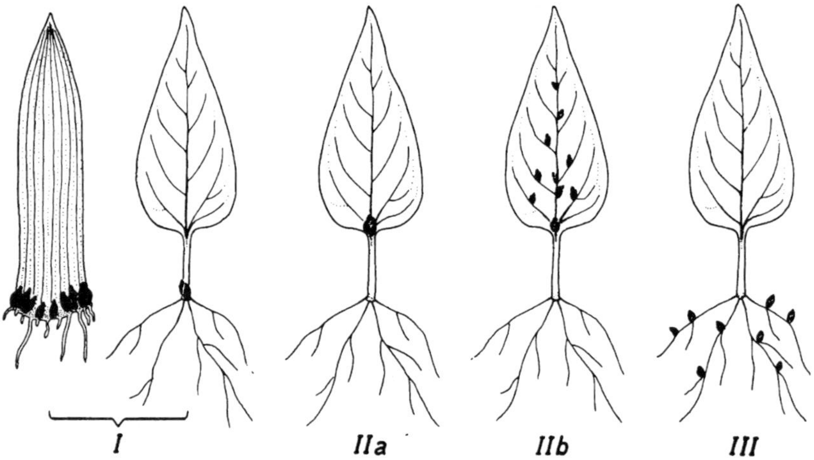

Fig. 8. Location types of adventitious bud formation on detached leaves. (Broertjes et al., 1968, Fig. 1.)

interested, since it has been shown that the capacity to produce adventitious plantlets varies greatly among different cultivars of one species. In *Kalanchoë*, for instance, the cultivar Annette produces fewer plantlets than the cultivar Josine (Table 4) (Broertjes and Leffring, 1972). Similar differences were found using various *Chrysanthemum* cultivars (S. Roest, unpublished data). One should therefore not conclude that a given species cannot produce adventitious shoots if only one cultivar has been tested.

Factors affecting adventitious bud formation on leaves. When testing the ability of plants to form adventitious buds, variations in such things as rooting substrate (mixture of, e.g. sand, peat, vermiculite), climatic conditions (temperature, relative humidity, light) and the use of plant hormones to promote root and shoot formation, should be included in the experiment. One should not overlook the history and quality of the stock plant or leaf factors, such as the physiological age of the leaves; this latter factor proved decisive in *Kalanchoë*. Broertjes and Leffring (1972) compared physiologically old leaves, fully-grown leaves and young leaves of *Kalanchoë*. Although all leaves rooted readily, the young leaves produced adventitious plantlets at the base of the petiole sooner than the older leaves, and in much greater numbers. In *Chrysanthemum* more buds were formed on callus either on the upper parts of roots or at the base of the petiole in light than in darkness (S. Roest, unpublished data). In *Streptocarpus*, however, old leaves produced more but smaller plantlets in comparison with young leaves (Brown, 1971b). The petiole also seems to play a part: leaves of the *Kalanchoë* cultivar Annette rarely produced plantlets when the petiole was cut off, whereas cv. Josine produced some plantlets on leaves at all three developmental stages, although many more grew from those with a petiole (see Table 4). (Leaves without a petiole rooted and developed plantlets almost exclusively near the midrib.)

The size of the leaf as well as of the leaf part is an important factor for the

TABLE 4

Number of adventitious plantlets produced on detached leaves of two *Kalanchoë* cultivars *

	Cv. Annette		Cv. Josine	
	With petiole	Without petiole	With petiole	Without petiole
Young leaves	27	—	73	24
Fully grown leaves	20	3	40	5
Old leaves	—	3	47	13

* In all cases 20 leaves were used. All leaves were irradiated with 2 krad X-rays. The number of plantlets was determined 7 months after irradiation of detached leaves.

number of adventitious buds formed in monocotyledons. In *Ornithogalum*, leaves of approximately 20 cm in length were compared with leaf tops, leaf bases and mid-leaf pieces, all approximately 10 cm long. In Table 5 it is shown that whole leaves produced twice as many adventitious bulblets as any of the leaf pieces. Basal parts produced the lowest number of bulblets, but those produced were larger (Broertjes, 1972a).

The number of adventitious plantlets is also influenced by the way the leaves are used: in *Streptocarpus*, more plantlets are obtained when two halves of a leaf are planted after removing the midrib, than when a whole leaf is planted without petiole (Broertjes, 1969b).

The use of plant hormones is another possible way of influencing the outcome. Rooting is often promoted by the application of auxins, but little is known about the factors which control adventitious bud formation. Some cytokinins and other growth regulators play a role, as was demonstrated in *Begonia* (Heide, 1965). However, for plants, not spontaneously forming adventitious plantlets, no practical application of these hormones is so far in sight (Bhojwani and Johri, 1970; Menhenett, 1970; Nag and Johri, 1970; Plummer and Leopold, 1957).

Adventitious bud formation on (artificially made) bulb scales. In *Lilium* hybrids, such as the Midcentury hybrid cultivars Enchantment and Tabasco, solid mutants can be obtained by irradiating bulb scales immediately after scaling the bulbs or, even better, by scaling after irradiation of the bulbs. Since clonal propagation of selected mutants, especially during the early years, is slow, it also takes approximately 5 years before enough plants are available and enough experience of the mutants has been gained to justify commercial exploitation of a mutant.

Adventitious bulblets are obtained in some bulb crops from scooping the bulbs, a traditional way of vegetative propagation of, for instance, hyacinth. This method may have a wider application, e.g. in *Muscari*, since the detrimental effects of fungi can now be prevented very efficiently by the use of new fungicides, such as benomyl (Benlate) (Broertjes and Alkema, 1970).

TABLE 5

Number of adventitious bulblets produced on detached leaves and leaf parts of *Ornithogalum thyrsoïdes* Jacq., 4 months after detachment of leaves

	Number of leaves	Length (cm)	Number of bulblets per leaf or leaf part
Whole length	10	20	16.2 ± 1.4
Basal part	20	10	6.7 ± 0.4 (larger bulblets)
Mid-leaf piece	20	10	8.4 ± 0.6
Top	20	10	8.8 ± 0.5

Artificially made bulb scales, obtained by cutting bulbs into pieces, also offer possibilities in propagation, for example, application of the one-scale or two-scale method (Luyten, 1935; Traub, 1935; Alkema, 1971a,b; Sytsema, 1977) to species like *Amaryllus* (*Hippeastrum*), hyacinth, *Iris*, *Nerine*, tulip and daffodil. Whether adventitious bulblets in such cases originate from one cell to produce non-chimeral mutants after irradiation is not (yet) known, but it seems highly probable. Preliminary results with *Amaryllus* show that apparently solid mutants are produced on scale pieces of irradiated bulbs (C. Broertjes, unpublished data).

Significance for mutation breeding. A convincing example of the value of the above-mentioned adventitious bud technique for mutation breeding in vegetatively propagated ornamental plants has been demonstrated in *Streptocarpus*, using the cultivar Constant Nymph. Irradiated or colchicine-treated half-leaves produced a strikingly high percentage of mutants or polyploids, respectively: over 30% of the adventitious plantlets produced were solid mutants after irradiation with the optimum X-ray dose of 3 krad; over 30% plantlets after treatment of half-leaves for 6 h with 0.1% colchicine were non-cytochimeral tetraploids. In total, 857 mutants were obtained, five of which were sold commercially within 3 years of the start of the experiment (Broertjes, 1969b).

Similar results were obtained with *Achimenes*. Though fewer mutants were produced than in *Streptocarpus*, irradiation of leaves of the cultivar Paul Arnold with 3 krad X-rays produced a number of solid mutants. Within slightly more than 3 years of irradiation, several mutants were being commercially exploited, namely 'Springtime', and 'Early Arnold', both of which flowering approximately 3 weeks earlier than 'Paul Arnold' and other mutants (see Chapter 5, section on Gesneriaceae) (Broertjes, 1972b).

Brown (1971a) reported 30 solid and two chimeral mutants among 204 plants using similar cultivars and the same X-ray dose as that mentioned here.

Broertjes and Leffring (1972) obtained exclusively solid mutants after the irradiation of *Kalanchoë* leaves with X-rays. Adventitious plantlet production was much slower than in Gesneriaceae, for instance. Moreover, *Kalanchoë* plants require short-day treatment to flower which makes progress with the programme relatively slow. Consequently, it takes somewhat longer (4—5 years) before the first mutants can be put on the market.

Hundreds of solid mutants of *Ipomoea batatas* (sweet potato) have been produced by irradiation of detached leaves. Several of them are promising with respect to various characters, such as dry matter and protein content, skin and flesh colour (B-carotene) of the tuberous roots, type of plant, and yield (Love, 1968; Wilson, 1973) (see section on sweet potato, Chapter 4).

In *Begonia* too, hundreds of solid mutants have been obtained by the irradiation of detached leaves from a few, triploid hybrids. Several of the mutants were commercialized within 4 years (Doorenbos and Karper, 1975;

Mikkelsen et al., 1975) (see section on *Begonia*, Chapter 5).

Peperomia also propagates readily by irradiation of detached leaves (Bigot and Brulfert, 1968). If *Peperomia obtusifolia variegata* is thus propagated, either completely green or (almost) completely chlorophyll-less (yellow) plantlets are formed. This is an indication that the apices of adventitious plantlets develop from one original cell. The outcome of a small-scale experiment with *P. resediflora* was in line with these results, but the number of (solid) mutants was far from being sufficient to give real proof (C. Broertjes, unpublished data).

The detached leaves of several Liliaceae, for example, *Endymion*, *Hyacinthus*, *Muscari*, *Ornithogalum* and *Scilla*, readily produce adventitious bulblets. However, even following an optimum X-ray dose, the mutation frequency in these species is (very) low as a consequence of their limited heterozygosity. The mutants obtained were non-chimeral, which is in agreement with cytological and histological observations carried out so far and revealing generally identical pictures to those observed on several Gesneriaceae (C. Broertjes and H.Y. Alkema, 1970, unpublished data).

In *Lilium*, hundreds of solid mutants have been obtained after irradiation of bulb scales immediately after scaling, or after irradiation of bulbs just before scaling. Several promising mutants or tetraploids (obtained after colchicine treatment of bulb scales) are being propagated for further testing. One is released, namely cv. Mies Bouwman, an orange-flowering mutant of cv. Tabasco (C. Broertjes and A.J. Bischoff Tulleken, unpublished data).

The adventitious bud technique was applied in an experiment with *Nicotiana alata*. Solid mutants were obtained, again after irradiation of detached leaves. By using this method, mutant individuals of a self-incompatible clone were obtained, which were characterized by the same genetic background as the stock plant and were consequently free from inbreeding effects (De Nettancourt et al., 1971).

Adventitious buds originating from more than one cell

In several cases it is not obvious, or even not to be expected, that the apices of adventitious buds originate from one cell. It is known that in many cases buds on stems and roots develop endogenously from more than one cell, probably from a restricted number. In *Chrysanthemum*, adventitious buds developing on stems of de-budded young periclinal plants, were either still periclinal or genetically uniform (Stewart and Dermen, 1970a,b). The authors concluded that the shoots had originated from two or more cells which may or may not have derived from different apical layers. In *Poinsettia* (*Euphorbia pulcherrima*) much larger sectors and even complete (periclinal) mutants were obtained by adventitious bud formation after removal of the terminal bud as well as all lateral buds on the irradiated plants (Love, 1972).

In *Gerbera* a method, in which mature plants are pruned of all leaves and large parts of the stems and roots, has been used. After irradiation, these "plants" were placed in a warm and humid greenhouse. Within 3—4 months, a great number of dormant buds had developed into shoots; the number ranged from 10 to 25, dependent, amongst other things, on the cultivar. The small shoots were cut off, rooted and planted (Breutmann, 1976).

Large sectored and apparently solid mutants of potato can be obtained in various ways. Ferwerda (1965) irradiated single-eye pieces and among the mutants obtained, a relatively high percentage (30%) seemed to be solid. Dormant eyes or tubers irradiated directly after harvest, de-eyed tuber pieces and very small, prematurely harvested tubers (3—5 mm) respectively showed gradually declining chimerism (Ferwerda, 1965).

Chimerism can also be limited by growing adventitious sprouts on the roots of irradiated tuber slices which, as has been mentioned by Howard (1961), develop from a restricted number of cells (Miedema, 1967, 1973a,b). However, the best way known at present to obtain solid (non-chimeral) mutants is to de-eye tuber parts shortly (1—2 h) before irradiation, to apply relatively high doses and dose rates, and to use adventitious sprouts which develop 3 months after irradiation; any shoot developing earlier is discarded (Van Harten et al., 1972).

Dormant stolons of peppermint, *Mentha* × *piperita* were irradiated, cut into 4—7-cm pieces and planted. Later, two categories of mutants were observed: complete mutants and chimeras; it is believed that the former derived from single cells (as adventitious buds), while the latter developed from preformed, uninjured multicellular meristems (Murray, 1969). Using this method, several *Verticillium*-resistant mutants were produced.

The use of callus, in general, may be another valuable method for reducing chimerism. *Chrysanthemum* leaves can be rooted easily; they produce callus at the base of the petiole or on the upper part of roots on which one or a few adventitious shoots are occasionally formed (S. Roest, unpublished data). This is essentially different from the direct production of adventitious shoot apices, originating from one (epidermal) cell at the base of a petiole, as described earlier for several species. In contrast, the majority of the mutants obtained in *Chrysanthemum* were chimeras, although several (approximately 20%) apparently solid mutants were observed as well (Broertjes et al., 1976). In a later experiment, using a slightly different procedure, a much higher percentage of solid-looking mutants was observed (C. Broertjes, unpublished data). The low and variable average number of plantlets per leaf and the long time interval between the irradiation of the leaves and the appearance of the shoots, irrespective of the number of solid mutants obtained by this method, are the reasons for this method not being preferred to the relatively easy and fast method of producing mutants by using rooted cuttings in vivo or pedicel explants in vitro as the starting material (Broertjes et al., 1976).

In other species, however, where no quick year-round propagation method is available, the use of callus might offer possibilities. *Populus*, for example, can easily be propagated by splitting a branch into two parts or by making a long cut through it. If the pieces are kept under humid and warm conditions, callus develops and adventitious plantlets are eventually formed (see section on *Populus*, Chapter 6). The speed of regeneration as well as the number of shoots depend on the species and sometimes the clone (H.M. Heybroek and S. Shapiro, unpublished data). To our knowledge, this method has not yet been applied either to *Populus* or to other species.

Buds in ontogenetically young stages of development

A situation similar to adventitious bud formation on stems, roots or callus is found when existing bud meristems are irradiated in an ontogenetically (very) young stage of development. In this case, chimerism can also be restricted or even partly avoided.

The optimum time for irradiating *Dahlia* tubers is immediately after harvest, when no visible buds are present, whereas after winter storage the tubers carry several relatively large sprouts. The irradiation of freshly harvested tubers results in chimeras as well as apparently solid mutants — roughly 50% of each. Numerous apparently solid mutants have been commercialized and grown on a relatively large scale. None showed any sign of chimerism, indirectly proving that they must have been solid (Broertjes and Ballego, 1967) (see section on *Dahlia*, Chapter 5).

This method may also be of value for other tuber crops as well as for bulb crops. In tulip, for instance, irradiation is carried out either immediately after lifting, with the intention of affecting the developing clisters in the bulbs, or in the autumn, to affect future clisters in the existing ones (see section on tulip, Chapter 5).

In *Iris*, the best irradiation time is immediately after lifting, when the growing point is at the youngest possible stage of development (Hekstra and Broertjes, 1968).

Various other methods

Although generally not appropriate for vegetatively propagated plants, some other plant parts, for example, decapitated seedlings on which buds originate from one or a few (?) cells of the hypocotyl, are among the potentially promising material that will or might result in a relatively high percentage of solid mutants, provided mutagenic treatment is given just before or immediately after decapitation as demonstrated, for example, in *Vaccinium* (Bain, 1940) and *Linum* (Link and Eggers, 1946). This also applies to the use of detached cotyledons in Petri dishes; plants used in this way include *Borago*, *Cucumus*, *Luffa*, *Phaseolus* and *Pisum* (Kowalewska, 1927).

Alstroemeria is a special case. When actively growing young rhizomes of various of its cultivars were irradiated, almost exclusively apparently solid and most probably non-chimeral mutants were produced. At first sight, this phenomenon, unexpected but advantageous, could not be explained, but, on second thoughts, may be related to the sympodial growth of this species. Several mutants were commercialized within a relatively short period (Broertjes and Verboom, 1974) (see section on *Alstroemeria*, Chapter 5). Whether this would also occur in other species which reproduce by rhizomes is not known, but it appears to be worth trying.

The production of solid polyploids

What has been stated about mutated cells also applies to polyploid cells, which result, for instance, from colchide treatment of a multicellular apex. The generally slow growing, polyploid cells often lose the competition with the faster growing, original (diploid) cells. If a polyploid cell survives, the end product is often a mixoploid or a cyto-chimera although, in contrast with what happens in mutation induction, genome doubling is not necessarily restricted to just one cell. Colchicine treatment in combination with the adventitious bud technique, in which freshly detached leaves were used, resulted in a very high percentage of solid, non-cyto-chimeral polyploids (Broertjes, 1974).

It is obvious that an adventitious bud technique in which the buds preferably develop directly from one or a restricted number of de-differentiated (epidermal) cells is very useful in producing mutants which are either solid or have large mutated sectors, respectively. As was stated earlier, it enables early and relatively easy selection of valuable mutant genotypes which can then be vegetatively propagated fairly easily for further testing.

If no adventitious bud technique is available, irradiation of buds which are in an ontogenetically very early stage of development is preferable to treatment of well-developed buds.

Literature in vivo adventitious bud techniques

Alkema, H.Y., 1971a. Nieuwe vermeerderingsmethoden bij bolgewassen. Weekbl. Bloembollencult., 81(46): 1211.

Alkema, H.Y., 1971b. Veredeling van bolgewassen door bestraling. Weekbl. Bloembollencult., 81(48): 1262—1263.

Bain, H.F., 1940. Origin of adventitious shoots in decapitated cranberry seedlings. Bot. Gaz. (Chicago), 101: 872—880.

Bhojwani, S.S. and Johri, B.M., 1970. Cytokinin-induced shoot bud differentiation in mature endosperm of *Scurrula pulverulenta*. Z. Pflanzenphysiol., 63: 269—275.

Bigot, C. and Brulfert, J., 1968. Dynamics of the natural appearance of neoformations of roots and buds from isolated leaves of *Peperomia griseo-argentea* (Yunchen). C.R. Acad. Sci. Sér. D, 267: 1586—1588.

Breutmann, B., 1976. Vegetative Vermehrung von *Gerbera jamesonii*. Gartenbauliche Versuchsber. Rheinland, 15: 216—219.

Broertjes, C., 1968. Dose-rate effects in *Saintpaulia*. In: Mutations in Plant Breeding, Vol. 2. IAEA, Vienna, pp. 63—71.
Broertjes, C., 1969b. Mutation breeding of *Streptocarpus*. Euphytica, 18: 333—339.
Broertjes, C., 1972a. Improvement of vegetatively propagated plants by ionizing radiation. In: Induced Mutations and Plant Improvement. IAEA, Vienna, pp. 293—299.
Broertjes, C., 1972b. Mutation breeding of *Achimenes*. Euphytica, 21: 48—63.
Broertjes, C., 1972c. Use in Plant Breeding of Acute, Chronic or Fractionated Doses of X-rays or Fast Neutrons as Illustrated with Leaves of *Saintpaulia*. Thesis, Centr. Agric. Publ. Doc., Wageningen, Agric. Res. Rep. 776, 74 pp.
Broertjes, C., 1974. The production of polyploids using the adventitious bud technique. In: FAO/IAEA/EUCARPIA Conf. Mutation and Polyploidy, Bari, 1972. IAEA, Vienna, pp. 29—35.
Broertjes, C. and Alkema, H.Y., 1970. Mutation breeding of flowerbulbs. In: First Int. Symp. Flower Bulbs, Vol. 2. Noordwijk/Lisse, pp. 407—411.
Broertjes, C. and Ballego, J.M., 1967. Mutation breeding of *Dahlia variabilis*. Euphytica, 16: 171—176.
Broertjes, C. and Leffring, L., 1972. Mutation breeding of *Kalanchoë*. Euphytica, 21: 415—423.
Broertjes, C. and Verboom, H., 1974. Mutation breeding of *Alstroemeria*. Euphytica, 23: 39—44.
Broertjes, C., Haccius, B. and Weidlich, S., 1968. Adventitious bud formation on isolated leaves and its significance for mutation breeding. Euphytica, 17: 321—344.
Broertjes, C., Roest, S. and Bokelmann, G.S., 1976. Mutation breeding in *Chrysanthemum morifolium* Ram. using in vivo and in vitro adventitious bud techniques. Euphytica, 25(1): 11—19.
Brown, A.G., 1971a. X-ray treatment of F_1 hybrid plants from *Streptocarpus johannis* × *S. hybridus*. John Innes Inst. Annu. Rep., 62: 49.
Brown, A.G., 1971b. The effect of supplementary light on *Streptocarpus* 'Constant Nymph'. John Innes Inst. Annu. Rep., 62: 50—51.
De Nettancourt, D., Dijkhuis, P., Van Gastel, A.J.G. and Broertjes, C., 1971. The combined use of leaf irradiation and of the adventitious bud technique for inducing and detecting polyploidy, marker mutations and self-compatibility in clonal populations of *Nicotiana alata* Link and Otto. Euphytica, 20: 508—521.
Doorenbos, J. and Karper, K., 1975. X-ray induced mutations in *Begonia* × *hiemalis*. Euphytica, 24: 13—19.
Ferwerda, F.P., 1965. Mutagenic effects of X-rays and ethyl methane sulphonate on potato sprouts. In: The Use of Induced Mutations in Plant Breeding. Rep. Meet. FAO/IAEA, Rome, 1964. Pergamon Press, Oxford, pp. 687—690.
Haccius, B. and Hausner, G., 1974. Adventivknospen und nicht-zygotische Embryonen — Grundlagen und Anwendung. Vortr. 22. Vortragstag. Ges. Arzneipflanzenforsch., Tübingen, pp. 1—10.
Heide, O.M., 1965. Interaction of temperature, auxins and kinins in the regeneration ability of begonia leaf-cuttings. Physiol. Plant., 18: 891—920.
Hekstra, G. and Broertjes, C., 1968. Mutation breeding in bulbous *Iris*. Euphytica, 17: 345—351.
Howard, H.W., 1961. An octoploid potato from eye excision experiments. J. Hered., 52: 191—192.
Kowalewska, Z., 1927. Über Sprossregenerate an isolierten Keimblättern von Bohnen und Erbsen. Bull. Acad. Pol. Sci. Lett., pp. 713—718.
Link, G.K.K. and Eggers, V., 1946. Mode, site and time of initiation of hypocotyledonary bud primordia in *Linum usitatissimum* L. Bot. Gaz. (Chicago), 107: 441—454.
Love, J.E., 1968. Mutation breeding in the sweet potato by means of fast-neutron irradiation. IAEA-SM-121/34. IAEA, Vienna, pp. 331—336.

Love, J.E., 1972. Somatic mutation induction in poinsettia and sweet potato. In: J. Constantin (Editor), Mutation Breeding Workshop, Knoxville, Tenn. University of Tennessee, Knoxville, Tenn.
Luyten, L., 1935. Vegetative propagation of *Hippeastrum*. Yearb. Am. *Amaryllus* Soc., 2: 115—122.
Menhenett, R., 1970. Effects of adenine on the formation of roots and buds in leaf squares. New Phytol., 69: 537—547.
Miedema, P., 1967. Induction of adventitious buds on roots of the potato. Euphytica, 16: 163—166.
Miedema, P., 1973a. A Physiological Study of Adventitious Bud Formation in Potato. Thesis; Cent. Agric. Publ. Doc., Wageningen, Agric. Res. Rep. 787, 66 pp.
Miedema, P., 1973b. The use of adventitious buds to prevent chimerism in mutation breeding of potato. Euphytica, 22: 209—219.
Mikkelsen, J.C., Ryan, J. and Constantin, M.J., 1975. Mutation breeding of Rieger's *elatior* begonias. Am. Hortic., 54(3): 18—21.
Murray, M.J., 1969. Successful use of irradiation breeding to obtain *Verticillium*-resistant strains of peppermint, *Mentha piperita* L. In: Induced Mutations in Plants. IAEA, Vienna, pp. 345—370.
Nag, K.K. and Johri, B.M., 1970. Effects of cytokinins and injury on the formation of shoots by leaves of *Dendrophthoe falcata*. Planta, 90: 360—364.
Naylor, E. and Johnson, B., 1937. A histological study of vegetative reproduction in *Saintpaulia ionantha*. Am. J. Bot., 24: 673—678.
Plummer, T.H. and Leopold, A.C., 1957. Chemical treatment for bud formation in *Saintpaulia*. Proc. Am. Soc. Hortic. Sci., 70: 442—444.
Sparrow, A.H., Sparrow, R.C. and Schairer, L.A., 1960. The use of X-rays to induce somatic mutations in *Saintpaulia*. Afr. Violet Mag., 13: 32—37.
Stewart, R.N. and Dermen, J., 1970a. The origin of adventitious buds in *Chrysanthemum*. Am. J. Bot., 57: 734—735.
Stewart, R.N. and Dermen, H., 1970b. Somatic genetic analysis of the apical layers of chimeral sports in *Chrysanthemum* by experimental production of adventitious shoots. Am. J. Bot., 57: 1061—1071.
Sytsema, W., 1977. Schubvermeerdering van *Nerine*. Vakbl. Bloemisterij, 32(6/7): 21.
Traub, H.P., 1935. Propagation of *Amaryllis* by stem cuttage. Yearb. Am. *Amaryllis* Soc., 2: 123—126.
Van Harten, A.M., Bouter, H. and Van Ommeren, A., 1972. Preventing chimerism in potato (*Solanum tuberosum* L.) Euphytica, 21: 11—21.
Wilson, L.A., 1973. Stimulation of adventitious bud production in detached sweet potato leaves by high levels of nitrogen supply. Euphytica, 22: 324—326.

IN VITRO TECHNIQUES

During the last decade there has been a rapidly increasing stream of publications concerning in vitro propagation techniques. Within the framework of this book, it would be impossible to discuss all articles describing the in vitro aseptic cultivation of embryos, seeds, plants, organs, tissues, callus, cells and protoplasts on an artificial medium as Binding (1974), Carlson and Polacco (1975), D'Amato (1972), Devreux (1973), Durzan and Campbell (1974a), Melchers (1975), Murashige (1974), Pierik (1975a,b) and Sagawa (1976) have done. These publications indicate that in vitro techniques are becoming more and more important, not only for rapid and virus-free multiplication but also for use in mutation breeding to prevent or restrict chimera formation (Broertjes et al., 1976; Nitsch, 1976; Roest, 1976b; Spiegel-Roy and Kochba, 1976). Moreover, they can widen the application of chemical mutagens, and increase the production of isogenic lines from mutated sectors, single cells or even protoplasts (Kameya, 1975), while in vitro culture of meristems from infected tissue can produce virus-free plants (Asjes, 1974; Asjes et al., 1974). In addition, autodiploids and autotetraploids can be obtained by colchicine treatment of haploids and di(ha)ploids, respectively. Other advanced techniques are somatic (inter- and intra-specific) hybridization by protoplast fusion, the isolation of biochemical mutants in cell cultures and hetero-specific transformation (Carlson, 1973; Holl, 1975; Scowcroft, 1975; Smith, 1974; Smith et al., 1976).

One of the disadvantages of in vitro techniques, particularly those methods using callus and cell suspensions, is that they in themselves may be mutagenic, in the sense that either mutated cells get a larger chance to survive (the organization is less strict then that in an apex) or mutations may be induced by the auxins (2,4-D) and cytokinins added to the medium. This might seem advantageous at first sight but it will in general, be disadvantageous since in vitro propagation often serves a dual purpose in mutation breeding: the production of solid mutants, and subsequently, fast propagation of valuable genotypes which should maintain genetic stability (integrity).

Mono-cell cultures

For our purpose, to avoid or restrict the consequences of chimera formation, the most ideal method would be the irradiation of mono-cell cultures (or irradiation of the plant tissue from which they derive) with subsequent formation and growth of meristems and regeneration of plantlets. So far, however, this method can only be applied in very few plant species and, consequently, it is not (yet) available for the commercial plant breeder (Heinz and Mee, 1971; Vasil and Hildebrandt, 1965, 1966).

Callus cultures

Another possibility is a method in which plants are produced in vitro from undifferentiated callus. Compared with a cell in a well-organized apex, a mutated cell in a mass of callus seems to have a larger chance of surviving competition with other cells. The majority of the mutants produced by this method will be solid, especially if shoots are regenerated from a repeated sub-culture of irradiated callus, since the number of cells from which adventitious shoots originate in vitro seems to be (very) restricted. D'Amato (1972) even reports that, in cauliflower and *Pelargonium*, the plantlets which developed on irradiated callus grown from petals (or callus from irradiated petals?), had probably originated from one "epidermal" (= external) cell, as

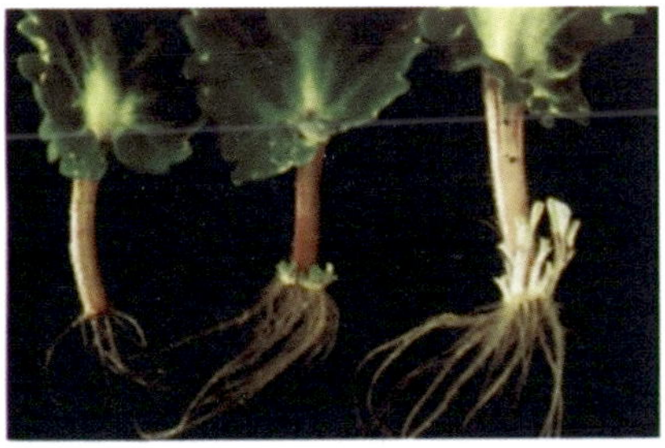

1

2

3

4

Plate 3. The apex of an adventitious plantlet developing on a detached leaf of *Saintpaulia ionantha* cv. Utrecht (1), originates from a single epidermal cell. Irradiated leaves consequently may produce solid, non-chimeral mutants as shown in photographs (2) (left is control), (3) (centre is control) and (4) which shows 2 *albino* mutants, attached to a third adventitious plantlet.
(Slides, ITAL, Wageningen.)

do the adventitious plantlets produced on detached leaves in vivo. This conclusion was drawn because the majority of the mutants were solid (non-chimeral). However, it may not be correct: a high ratio of non-chimeric to chimeric plants can also be obtained if (a) the number of initial cells of an adventitious bud is very small and (b) the ratio between the total area of mutated and non-mutated callus tissue to that of the border between the two, is large. Even so, in some crops the method might be the obvious means of producing a relatively high proportion of solid mutants.

Many other species, and the number is increasing all the time, can be propagated from callus in vitro, for example, alfalfa (Saunders and Bingham, 1972), *Allium sativum* (Havranek and Novak, 1973), *Anthurium scherzerianum* (Pierik and Steegmans, 1976a), *Asparagus officinalis* (Steward and Mapes, 1971b; Wilmar and Hellendoorn, 1968), *Chrysanthemum morifolium* (Ben-Jaacov and Langhans, 1970; Hill, 1968), *Cichorium endivia* (Vasil and Hildebrandt, 1966), *Coptis* spp. (Syono and Furuya, 1972), *Cymbidium* spp. (Steward and Mapes, 1971a), *Datura innoxia* (Engvild, 1973), *Daucus carota* (Steward et al., 1958), *Dendrophthoe falcata* (Nag and Johri, 1970), *Freesia* spp. (Davies and Griffiths, 1971), *Gladiolus* spp. (Simonsen and Hildebrandt, 1971), *Nicotiana* spp. (Feng and Linck, 1970; Ross and Thorpe, 1973; Vasil and Hildebrandt, 1965), *Petroselium hortense* (Vasil and Hildebrandt, 1966), *Petunia* spp. (Binding, 1971), sugarcane (Heinz and Mee, 1969), to mention only a few.

Explant cultures

From the mutation breeder's point of view, there are other, probably more promising in vitro methods than the regeneration of shoots from undifferentiated callus tissue, such as the regeneration of adventitious shoots on plant organs or plant parts, similar to that in some examples presented under in vivo techniques. We especially refer here to plant organs or explants on which adventitious shoots are expected to originate from one cell. They may originate from the following:

(small) leaves of

Begonia	(Arora et al., 1970);
Cattleya	(Champagnat et al., 1970);
Echeveria	(Raju and Hines, 1973);
Heloniopsis	(Kato, 1975; Kato and Kawahara, 1972);
Petunia	(Rao et al., 1973), and
Torenia	(Bajaj, 1972).

leaf disc explants (pieces, discs, squares, tips) of:

Begonia	(Wirth, 1960);
Cichorium	(Legrand, 1972);
Heloniopsis	(Kato and Kawahara, 1972);

orchids	(Churchill et al., 1973);
Peperomia	(Harris and Hart, 1964);
Streptocarpus	(Appelgren and Heide, 1972)
epidermis strips of:	
Nautilocalyx	(Tran Thanh Van and Drira, 1970; Venverloo, 1974)
Ranunculus	(Konar and Nataraja, 1965)

Explants of petioles, peduncles, pedicels, bulb scales and capitula have also demonstrated the ability to produce adventitious shoots, in, for example, *Alstroemeria* (Ziv et al., 1973), *Daucus* and *Brassica* (Margara, 1969), *Chrysanthemum cinerariaefolium* (Bokelmann, 1972; Roest, 1976a; Roest and Bokelmann, 1973), *Chrysanthemum morifolium* (Roest and Bokelmann, 1975), *Freesia* (Davies and Allen, 1973; Pierik, 1976a; Pierik and Steegmans, 1975b), *Hyacinth* (Pierik and Woets, 1971; Pierik and Ruibing, 1973; Pierik and Post, 1975), *Gerbera* (from dormant buds?) (Pierik et al., 1973), potato (Roest and Bokelmann, 1976) and sugar beet (Margara, 1970).

Adventitious shoots developing on root or stem segments (*Gladiolus*: Ziv et al., 1970; *Petunia*: Rao et al., 1973; *Plumbago*: Nitsch and Nitsch, 1967; *Verbascum*: Caruso, 1971) or on tuber explants (*Cyclamen persicum*: Stichel, 1959; Morel, 1975) probably originate endogeneously from more than one cell, with two exceptions, namely *Striga senegalensis* (an angiospermous obligate root parasite), in which some of the adventitious buds originate from one hypodermal or epidermal cell (Okonkwo, 1970), and *Heloniopsis orientalis* (Liliaceae), in which epidermal cells contribute to bud formation as well (Kato, 1975). Consequently, exclusively chimeral mutants are to be expected after mutagenic treatment of segments or explants of these. However, although the method may not be ideal, in many cases, this material will certainly be useful, as was demonstrated for comparable plant material in vivo (see section on adventitious buds originating from more than one cell), since large-sectored mericlinal chimeras, apparently solid periclinal chimeras or even solid (non-chimeral) mutants may be produced.

What has been said about decapitated seedlings and detached cotyledons (see section on various other in vivo methods), is also appropriate for endosperms, pollen, zygotes and embryos. The heterozygosity of vegetatively propagated plants is generally (very) large, which causes the above mentioned material to be unsuitable for use in mutation breeding. In true-breeding botanical species, it may perhaps be used in conjunction with self-fertilization to avoid loss of mutations in L-I and L-III and unfavourable segregation in M_2 because of chimerism in the L-II of the treated material.

Why adventitious plantlets originate exogenously from one or a few epidermal cells in one case while in another they develop endogenously from several parenchymal cells is obscure. But one thing seems to be very clear, namely, that the mutation breeder should investigate the crop of interest with regard to its possibilities for vegetative propagation both in vivo and in vitro. He should preferably look for methods in which (a) as few cells as possible take part in the formation of the apex of an adventitious shoot (higher mutation frequency, wider mutation spectrum, easier selection and easier clonal propagation) and (b) a direct regeneration takes place (fast propagation, maintenance of genetic integrity).

This does not necessarily mean that solid mutants are always preferable. It is a fact that most mutational events in vegetatively propagated plants result from chromosome aberrations, rather than from single gene mutations. In other words, more than one character is in general mutated. In solid mutants all mutated characters are expressed including the unfavourable ones, whereas in a periclinal chimera, where for instance the L-I is mutated, only some of the mutated characters are expressed (Broertjes et al., 1976) and "true-to-type" vegetative propagation, using in vivo or in vitro adventitious bud techniques, will in most cases be impossible.

Literature in vitro adventitious bud techniques

Appelgren, M. and Heide, O.M., 1972. Regeneration in *Streptocarpus* leaf discs and its regulation by temperature and growth substances. Physiol. Plant., 27: 417—424.

Arora, Y.K., Nakao, S. and Nakajima, T., 1970. Perpetuation of *Begonia rex* by aseptic culture with micro-leaf cuttings under various conditions of auxins and cytokinin. Jpn. J. Breed., 20(5): 275—281.

Asjes, C.J., 1974. The prospect of eliminating virus diseases in lilies in The Netherlands. In: Lilies 1974 and Other Liliaceae. R. Hort. Soc., London, pp. 2—12.

Asjes, C.J., Bunt, M.H. and Van Slogteren, D.H.M., 1974. Production of hyacinth mosaic virus-free hyacinths and lily symptomless virus-free lilies by meristem tip culture. Acta Hortic., 36: 223—228.

Bajaj, Y.P.S., 1972. Effects of some growth regulators on bud formation by excised leaves of *Torenia fournieri*. Z. Pflanzenphysiol., 66: 284—287.

Ben-Jaacov, J. and Langhans, R.W., 1970. A tissue culture technique for rapid multiplication of *Chrysanthemum morifolium*. Proc. Int. Hortic. Congr., 18(1): 211—255.

Binding, H., 1971. Organogenesis in callus cultures of *Petunia hybrida*. Z. Pflanzenphysiol., 65: 359—364.

Binding, H., 1974. Mutation in haploid cell cultures. In: Int. Symp. Haploids in Higher Plants; Guelph, Ont. pp. 323—337.

Bokelmann, G.S., 1972. Vegetatieve Vermeerdering van *Chrysanthemum cinerariaefolium* in vitro. Report, ITAL, Wageningen, 42 pp.

Broertjes, C., Roest, S. and Bokelmann, G.S., 1976. Mutation breeding of *Chrysanthemum morifolium* Ram. using in vivo and in vitro adventitious bud techniques. Euphytica, 25: 11—19.

Carlson, P.S., 1973. Somatic cell genetics as a tool for plant breeding. In: Induced Mutations in Vegetatively Propagated Plants. IAEA, Vienna, pp. 167—179.

Carlson, P.S. and Polacco, J.C., 1975. Plant cell cultures: genetic aspects of crop improvement. Science, 188(4188): 622—625.

Caruso, J.L., 1971. Bud formation in excised stem segments of *Verbascum thapsus*. Am. J. Bot., 58: 429—431.

Champagnat, M., Morel, G. and Mounetou, B., 1970. Vegetative multiplication of *Cattleya* in vitro from aseptic culture of young leaves. Ann. Sci. Nat. Bot. Biol. Veg., 11: 97—113.

Churchill, M.E., Ball, E.A. and Arditti, J., 1973. Tissue culture of orchids. I. Method for leaf tips. New Phytol., 72: 161—166.
D'Amato, F., 1972. Significato teoretico e pratico delle colture in vitro di tessuti e cellule vegetali. Acad. Naz. Lincei, Quad., 173: 1—55.
Davies, D.R. and Allen, Y., 1973. In vitro propagation of *Freesia*. Annu. Rep. John Innes Hortic. Inst., 64: 67—68.
Davies, D.R. and Griffiths, P., 1971. In vitro propagation of *Freesia*. Annu. Rep. John Innes Hortic. Inst., 62: 45.
Devreux, M., 1973. In vitro culture and mutation breeding. In: Induced Mutations in Vegetatively Propagated Plants. IAEA, Vienna, pp. 41—52.
Durzan, D.J. and Campbell, R.A., 1974a. Prospects for the mass production of improved stock of forest trees by cell and tissue culture. Can. J. For. Res., 4(2): 151—174.
Engvild, K.C., 1973. Shoot differentiation in callus cultures of *Datura innoxia*. Physiol. Plant., 28: 155—159.
Feng, K.A. and Linck, A.J., 1970. Effects of N-1-naphthylphtalamic acid and the growth and bud formation of tobacco callus grown in vitro. Plant Cell Physiol., 11: 589—598.
Harris, G.P. and Hart, M.H., 1964. Regeneration from leaf squares of *Peperomia sandersii*: a relationship between rooting and budding. Ann. Bot. (London), 28: 509—526.
Havranek, P. and Novak, F.J., 1973. The bud formation in the callus cultures of *Allium sativum* L. Z. Pflanzenphysiol., 68: 308—318.
Heinz, D.J. and Mee, G.W.P., 1969. Plant differentiation from callus tissue of *Saccharum* species. Crop Sci., 9: 346—348.
Heinz, D.J. and Mee, G.W.P., 1971. Morphologic, cytogenetic and enzymatic variation in *Saccharum* species hybrid clones derived from callus tissue. Am. J. Bot., 58(3): 257—262.
Hill, G.P., 1968. Shoot formation in tissue cultures of *Chrysanthemum* 'Bronze Pride'. Physiol. Plant., 21: 386—389.
Holl, F.B., 1975. Innovative approaches to genetics in agriculture. Can. J. Genet. Cytol., 17: 517—524.
Kameya, T., 1975. Culture of protoplasts from chimeral plant tissue of nature. Jpn. J. Genet., 50(5): 417—420.
Kato, Y., 1975. Adventitious bud formation in etiolated stem segments and leaf callus of *Heloniopsis orientalis* (Liliaceae). Z. Pflanzenphysiol., 75: 211—216.
Kato, Y. and Kawahara, D., 1972. Bud formation in leaves, leaf fragments and midrib pieces of *Heloniopsis orientalis* (Liliaceae). Planta, 107: 111—120.
Konar, R.N. and Nataraja, K., 1965. Experimental studies in *Ranunculus sceleratus* L. Development of embryos from the stem epidermis. Phytomorphology, 15: 132—137.
Legrand, B., 1972. Influence des conditions initiales d'éclairement sur le bourgeonnement et la polarité de tissus de feuilles d'endive cultivés in vitro. C.R. Acad. Sci. Sér. D, 275: 31—32.
Margara, J., 1969. On the capacity of tissue for budding and floral initiation in vitro. C. R. Acad. Sci. Sér. D, 268: 803—805.
Margara, J., 1970. Neo-formation de bourgeons in vitro chez la Betterave sucrière, *Beta vulgaris* L. C. R. Acad. Sci. Sér. D, 270: 698—701.
Melchers, G., 1975. Genetik und Pflanzenzüchtung mit mikrobiologischen Methoden. Planta Med., (Suppl.): 5—34.
Morel, G., 1975. La multiplication végétative du cyclamen à partir de petiole foliaire permettra-elle une nouvelle application de la culture 'in vitro' à l'horticulture. Pépinieristes, Hortic. Maraichers, 158: 25—28.
Murashige, T., 1974. Plant propagation through tissue cultures. Annu. Rev. Plant Physiol., 25: 135—166.
Nag, K.K. and Johri, B.M., 1970. Effects of cytokinins and injury on the formation of

shoot buds by leaves of *Dendrophtoe falcata*. Planta, 90: 360—364.
Nitsch, C., 1976. The use of in vitro microspore culture for the modification of plants. In: Proc. Workshop Eur. Comm. — Israel, The Use of Ionizing Radiation, Wageningen, 22—24 March. Assoc. Euratom—ITAL, Wageningen, pp. 411—416.
Nitsch, C. and Nitsch, J.P., 1967. The induction of flowering in vitro in stem segments of *Plumbago indica* L. I. The production of vegetative buds. Planta, 72: 355—370.
Okonkwo, S.N.C., 1970. Studies on *Striga senegalensis* V. Origin and development of buds from roots of seedlings reared in vitro. Phytomorphology, 20(2): 145—150.
Pierik, R.L.M., 1975a. Plantenteelt in Kweekbuizen. Thieme, Zutphen, 164 pp.
Pierik, R.L.M., 1975b. Vegetative propagation of horticultural crops in vitro with special attention to shrubs and trees. Acta Hortic., 54: 71—82.
Pierik, R.L.M., 1976a. Nieuwe methoden voor vegetatieve vermeerdering van *Freesia* in kweekbuizen. Vakbl. Bloemisterij, 31(38): 61.
Pierik, R.L.M. and Post, A.J.M., 1975. Rapid vegetative propagation of *Hyacinthus orientalis* L. in vitro. Sci. Hortic., 3(3): 293—297.
Pierik, R.L.M. and Ruibing, M.A., 1973. Regeneration of bulblets on bulb scale segments of hyacinth in vitro. Neth. J. Agric. Sci., 21: 129—138.
Pierik, R.L.M. and Steegmans, H.H.M., 1975b. *Freesia* plantlets from flowerbuds cultivated in vitro. Neth. J. Agric. Sci., 23: 334—337.
Pierik, R.L.M. and Steegmans, H.H.M., 1976a. Vegetative propagation of *Anthurium scherzerianum* Schott through callus cultures. Sci. Hortic., 4(3): 291—293.
Pierik, R.L.M. and Woets, J., 1971. Regeneration of isolated bulb scale segments of hyacinth. Acta Hortic., 23(2): 423—428.
Pierik, R.L.M., Steegmans, H.H.M. and Marelis, J.J., 1973. *Gerbera* plantlets from in vitro cultivated capitulum explants. Sci. Hortic., 1(1): 117—120.
Raju, M.V.S. and Hines, J.E., 1973. Regenerative studies on the detached leaves of *Echeveria elegans*. Regeneration and growth in leaves. Can. J. Bot., 51: 451—456.
Rao, P.S., Handro, W. and Harada, H., 1973. Hormonal control of differentiation of shoots and embryos in leaf and stem cultures of *Petunia inflata* and *Petunia hybrida*. Physiol. Plant., 28: 458—463.
Roest, S., 1976a. Flowering and Vegetative Propagation of Pyrethum (*Chrysanthemum cinerariaefolium* Vis.) in vivo and in vitro. Thesis; Agric. Res. Rep. 860, Wageningen, 104 pp.
Roest, S., 1976b. In vitro culture and mutation breeding. In: Proc. Workshop Eur. Comm. — Israel, The Use of Ionizing Radiation, Wageningen, 22—24 March. Assoc. Euratom—ITAL, Wageningen, pp. 427—433.
Roest, S. and Bokelmann, G.S., 1973. Vegetative propagation of *Chrysanthemum cinerariaefolium* in vitro. Sci. Hortic., 1: 120—122.
Roest, S. and Bokelmann, G.S., 1975. Vegetative propagation of *Chrysanthemum morifolium* Ram. in vitro. Sci. Hortic., 3: 317—330.
Roest, S. and Bokelmann, G.S., 1976. Vegetative propagation of *Solanum tuberosum* L. in vitro. Potato Res., 19: 173—178.
Ross, M.K. and Thorpe, T.A., 1973. Physiological gradients and shoot initiation in tobacco callus culture. Plant Cell Physiol., 14: 473—480.
Sagawa, Y., 1976. Potential of in vitro culture techniques for improvement of floricultural crops. In: K.C. Sink (Editor), Int. Symp., Floriculture, Plant Breeding and Genetics, East Lansing, Mich., 1976, pp. 61—66.
Saunders, J.W. and Bingham, E.T., 1972. Production of alfalfa plants from callus tissue. Crop Sci., 12: 804—808.
Scowcroft, 1975. The potential use of somatic cell genetics in plant improvement. Sabrao J., 7(2): 147—158.
Simonsen, J. and Hildebrandt, A.C., 1971. In vitro growth and differentiation of *Gladiolus* plants from callus cultures. Can. J. Bot., 49: 1817—1819.

Smith, H.H., 1974. Model systems for somatic cell plant genetics. BioScience, 24(5): 269—276.

Smith, H.H., Kao, K.N. and Combatti, N.C., 1976. Interspecific hybridization by protoplast fusion in *Nicotiana.* J. Hered., 67: 123—128.

Spiegel-Roy, P. and Kochba, J., 1976. Use of ionizing radiation for mutagenesis in tissue culture. In: Proc. Workshop Eur. Comm. — Israel, The Use of Ionizing Radiation, Wageningen, 22—24 March. Assoc. Euratom—ITAL, Wageningen, pp. 434—444.

Steward, F.C., Mapes, M.O. and Smith, J., 1958. Growth and organized development of cultured cells I: Growth and division of freely suspended cells. Am. J. Bot., 45: 693—703.

Steward, F.C. and Mapes, M.O., 1971a. Morphogenesis in aseptic cell cultures of *Cymbidium.* Bot. Gaz. (Chicago), 132: 65—70.

Steward, F.C. and Mapes, M.O., 1971b. Morphogenesis and plant propagation in aseptic cultures of *Asparagus.* Bot. Gaz. (Chicago), 132: 70—79.

Stichel, E., 1959. Gleichzeitige Induktion von Sprossen und Wurzeln an in vitro kultivierten Gewebestücke von *Cyclamen persicum.* Planta, 53: 293—317.

Syono, K. and Furuya, T., 1972. The differentiation of *Coptis* plants in vitro from callus cultures. Experientia, 28: 236.

Tran Thanh Van, M. and Drira, A., 1970. Definition of a simple experimental system of directed organogenesis de nuevo: organ neoformation from epidermal tissue of *Nautilocalyx lynchi.* In: Proc. 2nd Int. Plant Tissue Cult. Conf., Strassbourg, Colloq. Int. C.N.R.S., 193: 169—176.

Vasil, I.K. and Hildebrandt, A.C., 1966. Variations of morphogenetic behaviour in plant tissue cultures. Am. J. Bot., 53: 869—874.

Vasil, V. and Hildebrandt, A.C., 1965. Differentiation of tobacco plants from single isolated cells in microcultures. Science, 150: 889—892.

Venverloo, C.J., 1974. Exogenous regulation of the formation of roots or shoots in the same culture. In: H.E. Street (Editor), Third Int. Assoc. Plant Tissue and Cell Culture, Leicester. Abstr. No. 31.

Wilmar, C. and Hellendoorn, M., 1968. Growth and morphogenesis of *Asparagus* cells cultured in vitro. Nature (London), 217: 369—370.

Wirth, K., 1960. Experimentelle Beeinflüssung der Organbildung an in vitro kultivierten Blattstücke von *Begonia rex.* Planta, 54: 265—293.

Ziv, M., Halevy, A.H. and Shilo, R., 1970. Organs and plantlets regeneration of *Gladiolus* through tissue culture. Ann. Bot. (London), 34: 671—676.

Ziv, M., Kanterovitz, R. and Halevy, A.H., 1973. Vegetative propagation of *Alstroemeria* in vitro. Sci. Hortic., 1: 271—277.

Chapter 4

ROOT AND TUBER CROPS

CASSAVA

Cassava (*Manihot esculenta* Crantz, fam Euphorbiaceae), also known as manioc or tapioca, is an important source of food in the lowland tropics. Most of the world production, a total of approximately 62×10^9 metric tons, is consumed locally, with Africa as the most important producer. The crop, the root tubers forming the edible part, can grow reasonably well under poor conditions. The acreage used to grow cassava has increased considerably during the 20th century (Purseglove, 1974).

Breeding work in the tetraploid cassava has only quite recently started, and not much has been achieved yet. Selection programmes aim mainly at high yields, better protein quality, low hydrogen cyanide content and resistance to viruses.

Cassava and its wild relatives show an enormous genetic variation and there are a large number of cultivars. The crop is normally outcrossing but can also be propagated vegetatively. Interspecific hybridization is also used. According to Magoon (1968) many cultivars in India (and of course elsewhere in the world) are either chance seedlings or bud mutations. Bud mutations, in fact, have been reported to occur for many years, as has the use of colchicine in attempts to produce a higher level of polyploidy.

Moh (1963) was the first to use radiation on this crop. He irradiated nodes with up to 4 kR of γ-rays (400—500 R/min) subjecting 150 nodes to each treatment. The LD_{50}, i.e. the dose of radiation at which 50% of the treated material does not survive treatment, was at approximately 3 kR.

Vasudevan et al. (1967) have described the initiation of a mutation breeding programme in cassava, using a wide range of mutagens. The work started with the irradiation of selected varieties suffering from at least one serious defect. Morphological mutants were obtained after γ-irradiation of stem cuttings. One mutant showed an increase in starch content and a decrease in hydrogen cyanide content. A difference in the sprouting capacities of the two treated strains was observed after irradiation. Therefore, a dose between 4 and 7.5 kR was advised.

Abraham (1970) reported that maximum sprouting of buds from irradiated stem cuttings was obtained at doses of less than 1.5 kR, whereas at 5 kR, no sprouts were produced at all. Moh and Alan (1973), who support the findings of Abraham, mention that in general only low mutation frequencies could be obtained, but that some cultivars were relatively easy to mutate. Moreover, most mutants were chimeric in nature, two vegetative cycles being needed to obtain solid mutants. Mutations either concerned changes in the chlorophyll or were of a morphological nature. However, useful mutants were not found. Abraham (1970) also reported that some promising types had been obtained in a programme to induce mutations with a higher yield.

Moh and Alan (1973) compared the irradiation of pollen with that of seeds and stems. The advantage of pollen irradiation is that no chimeras are formed. Furthermore, it allows early evaluation of the mutants, while the heterozygotes obtained may also be of practical value. The authors claimed a reasonably high mutation rate following relatively low doses of γ-rays although, in our opinion, it seems hard to distinguish between the variability caused by outcrossing and that resulting from mutation induction. The authors looked for characters, such as early root maturity and low hydrogen cyanide content, both of which are rarely found in natural cultivars (Moh, 1976). Although some of the mutants appeared to be non-chimeric, the application of in vitro propagation (Kartha et al., 1974) might become a valuable technique to increase the efficiency of mutant production.

Finally, Nayar (1975) treated a high yielding tapioca cultivar with 0.5—5 kR γ-rays, as well as with ethyl methanesulphonate or colchicine to induce desirable characters. Several mutants and tetraploids, obtained from this and other cultivars, are under observation.

Literature cassava

Abraham, A., 1970. Breeding work on tapioca (cassava) and a few other tropical tuber crops. In: D.P. Plucknett (Editor), Tropical Root and Tuber Crops Tomorrow, Vol. 1. University of Hawaii, Honolulu, Hawaii, pp. 76—78.

Kartha, K.K., Gamborg, O.L., Constabel, F. and Shyluk, J.P., 1974. Regeneration of cassava plants from apical meristems. Plant Sci. Lett., 2: 107—113.

Magoon, M.L., 1968. Some immediate problems, possibilities and experimental approaches in relation to genetic improvement of cassava. Indian J. Genet., 28A: 109—125.

Moh, C.C., 1963. Radiosensibilidad de las especies de plantes tropicales: *Carica papaya, Manihot dulcis* y *Swietenia humilis.* Turrialba, 13: 180—181.

Moh, C.C., 1976. Screening for acyanogenic somatic mutations in cassava (*Manihot esculenta* Crantz). Mutat. Breed. Newsl., 8: 10—11.

Moh, C.C. and Alan, J.J., 1973. Methods for inducing mutations in cassava and the possible uses of the mutant. In: Induced Mutations in Vegetatively Propagated Plants. IAEA, Vienna, pp. 67—75.

Nayar, G.G., 1975. Improving tapioca by mutation breeding. J. Root Crops, 1(2): 55—58.

Purseglove, J.W., 1974. Tropical Crops: Dicotyledons. Longman, London, pp. 172—180.

Vasudevan, K.N., Nair, S.G., Jos, J.S. and Magoon, M.L., 1967. Radiation-induced mutations in cassava. Indian J. Hortic., 24(1—2): 95—98.

GARLIC

Garlic, *Allium sativum* L., family Liliaceae, and other food alliums like the onion, shallot and leek are among the most popular vegetables in many countries of the world. Modern cultivars of garlic are sterile; they differ in many of their characters, including age at maturity, bulb size, clove size and number, and scale colour, but whether this variation originated during the period when *A. sativum* reproduced sexually or has arisen through bud mutation since it became vegetatively propagated and sterile, is not known. Since the causes of the sterility are also unknown, the only way to obtain further genetic variability is by mutation, either spontaneous or artificially induced (McCollum, 1976).

To this end, cloves have been γ-irradiated or treated with chemicals, both treatments resulting in mutants, in vM_1, i.e. the first cycle of vegetative multiplication after mutation treatments (Sklyar, 1973). The optimum radiation dose seems to be (much) below 1 krad, though how much below depends upon the cultivar used (Zhila, 1975). So far, no commercial radiation-induced mutant has been reported. This is not surprising in view of the slow rate of vegetative propagation and the general difficulties related to chimera formation and subsequent selection for mutated characters that are not directly visible. It is, therefore, of interest to the mutation breeder that in vitro propagation techniques, such as meristem-tip culture, enabling him to multiply rapidly virus-free clones (Pena-Iglesias, 1974) have recently become available. Stem explants from *Allium porrum* L. have been used to produce adventitious plantlets (Debergh and Standaert-De Metsenaere, 1976). Consequently, large-scale production of adventitious plantlets on explants in vitro seems to be within the range of possibilities which would considerably speed up the production and selection of improved mutant cultivars.

Literature garlic

Debergh, P. and Standaert-De Metsenaere, R., 1976. Neoformation of bulbils in *Allium porrum* L. cultured in vitro. Sci. Hortic., 5: 11—12.

McCollum, G.D., 1976. Onion and allies. In: N.W. Simmonds (Editor), Evolution of Crop Plants. Longman, London, New York, N.Y., 339 pp.

Pena-Iglesias, A., 1974. Meristem-tip culture of garlic: Elimination of viruses from infected bulbs and rapid multiplication of virus-free clones. Proc. Int. Hortic. Congr., 19(1A): 62.

Sklyar, G.E., 1973. The induction of somatic chemical and radiation mutants in garlic. Biofiz. Issled. Rast. Geterozis, 1973: 90—95 (in Russian); Plant Breed. Abstr., 46: No. 2899.

Zhila, E.D., 1975. The radiosensitivity of aerial bulbils of garlic. Tsitol. Genet., 9(6): 501—504 (in Russian); Plant Breed. Abstr., 46: No. 8471.

POTATO

Introduction

The potato (*Solanum tuberosum* L.), an important food crop especially in the northern temperate zone, ranks fifth after wheat, maize, rice and barley in terms of total production of energy for human consumption. World production amounts to about 290×10^6 metric tons, approximately one half of which is used for direct human consumption, one third for fodder and 10% for seed production.

From a breeding point of view the potato is a difficult crop. Our knowledge of its genetics is very limited because it has tetrasomic inheritance ($2n = 4x = 48$) and small chromosomes. Many cultivars show poor flowering and a low fertility. Other stumbling blocks are the high degree of heterozygosity and the low heritability of many important characters.

Despite the large efforts being made, both by institutes and private breeders, results of cross-breeding programmes are often disappointing. Several cultivars of more than national importance, such as the Dutch cultivar Bintje, 'King Edward' in the U.K. and 'Russet Burbank' in the U.S.A., have been known for 50 years or more, and have not yet been replaced. Of course, one must take into account the fact that, in general, consumers who are accustomed to the taste and other characteristics of a certain cultivar, are reluctant to accept another cultivar with less palatability but better agronomic qualities, for example, with respect to yield or disease resistance.

Spontaneous mutations

Several cases of spontaneous mutation in potato, indicated by bud variations, were reported by Darwin (1868) and even earlier by Carrière (1865). As early as 1907 Cramer (1907) referred to some cases in which bud variations had led to the production of cultivars of practical value, for example, cv. White Fortyfold with its white tubers, obtained from cv. Purple Fortyfold.

During the first decades of this century, many scientists still doubted the occurrence of bud variations, bud sports or whatever they are called, let alone accepting such mutations as being of practical value to the breeder (Sutton, 1918; Salaman, 1926). A more optimistic view is presented by Dorst (1924) who, in an extensive review, presents data concerning the use of certain bud sports in practice. During the period 1919—1923 about 2—3% of the area planted with the cultivars Eigenheimer and Rode Star, was occupied by the bud sports Blauwe Eigenheimer and Bonte Rode Star in a certain part of The Netherlands.

According to Krantz (1951), 15% of the certified seed production of commercial varieties in the U.S.A. in 1951 came from bud sports. These sports displayed a change in tuber skin, i.e. either towards a russet structure or towards a different, more attractive colour. In 1959 the acreage of bud sports had risen to about 35% as Heiken et al. (1963) concluded from a survey by Turnquist (1960).

The frequency of spontaneous mutations appears to be very low. East (1917) mentioned only 12 clear cases on the 100 000 hills he examined.

Folsom (1923) found only five leaf mutants while examining more than 350 000 plants. Heiken (1960) found spontaneous aberration rates varying from 1.5×10^{-3} (for so-called bolters in certain cultivars) to 1.2×10^{-5} (for certain foliage mutants).

It seems that the registration of symptoms of virus attacks and bud sports must have occasionally been confused in early reports.

Induced mutations

Early reports. X-ray treatments of potatoes were performed for the first time by Jacobson (1923), Johnson (1928, 1937) and Sprague and Lenz (1929). Jacobson (1923) reported considerable increase in yield and larger tubers in two different cultivars. Johnson (1928), who administered a low unspecified dose of X-rays to tubers of cv. Early Ohio, observed increased formation of tubers but lower tuber weights. Sprague and Lenz (1929), on the other hand, treating tubers of the cvs Irish Cobbler and Green Mountain with 400—1 200 R of X-rays observed fewer but larger tubers and a somewhat higher total yield. No figures were given as to the number of tubers that had been treated. Mutations of tuber or foliage characters were not reported. In experiments with the so-called Colorado wild potato (*S. jamesii*), Johnson (1937) found increased tuber formation and increased weight per hill and per tuber after exposure to 1 500 R of X-rays.

The cases mentioned above give no definite proof that the effects were of a permanent genetic nature. The first reliable experiment in this respect is by Asseyeva and Blagovidova (1935). Four cultivars were X-irradiated with doses between 500 and 8 000 R; a total of 390 tubers per cultivar were treated. Altogether 23 foliage mutations were obtained, ten of which were in cv. Professor Wohltmann but none in cv. Epicure. According to Heiken (1960), Asseyeva and Blagovidova also tested the mutagenicity of several chemical agents, apparently without success. If one excludes work with colchicine and a report from Demidovic (1934), who claimed that storage of true potato seed for 5 years induces more mutations than any other method, no further mutation research on potato was performed till 1950. At that time, Sparrow and Christensen (1950) reported that X-irradiation had an inhibiting effect on the sprouting of potato tubers. Stanton and Sinclair (1951) treated their material with phosphorus-32 and described morphological changes in the leaves. They also reported that the apical area of shoots displayed a higher radiosensitivity than other plant parts.

Hagberg and Nybom (1954), who used phosphorus-32 and X-rays in small-scale experiments, obtained morphological changes which lasted through different cycles of vegetative propagation (indicated as vM_1, vM_2 etc.). With the work of Heiken in Sweden around the sixties, which will be discussed in a later section, the period of non-directional orientation work can be considered ended. After this, a true start was made on the application of mutation methods to potato breeding.

The starting material. Practically all mutagenic treatments of potato in the past have been performed on whole tubers or tuber halves. In the latter case one half is mostly used as a control. Tubers are usually divided longitudinally according to the method of Asseyeva (1927). Nayar and Dayal (1970), however, used "rose" and "heel" ends. Howard (1970) has repeatedly used rose ends only.

Dormant as well as sprouting material is treated. The use of dormant tubers normally leads to a lower proportion of chimeric plants, but sprouting tubers may produce higher mutation frequencies (Heiken, 1960).

There is a great deal of variation as regards the age of the treated tubers. Sometimes very small, premature tubers are used, sometimes tubers are irradiated immediately after harvesting. Experiments have occasionally been conducted with tubers that had been stored for several months. Gradually it was realized that the treatment of smaller structures, for example, young tubers, tuber parts, single-eye pieces, cuttings or adventitiously developed sprouts, is preferable. Single-eye pieces were used by Ferwerda (1965), Umaerus (1966), Van Harten (1970), Van Harten et al. (1973) and Upadhya and Purohit (1973) (Plate 4). In other experiments, Van Harten et al. (1972) started with eye-excised tubers which were sectioned into apical, middle and stolon parts in order to check the behaviour of the different sections with respect to mutation frequency and chimerism. They also compared the effects of irradiation on tuber eyes immersed in water to those on eyes irradiated under dry conditions (Van Harten and Bouter, 1976). Miedema (1973b) used sprouts which had developed adventitiously from irradiated rooted tuber slices. Rooted cuttings were used by McCrory and Grun (1969). Unrooted cuttings, from which the root area is inserted into a piece of lead during irradiation, have been used since 1974, at the institutes of both authors. This method is preferable, for example, if tuber production of a certain clone is low and if it takes too much time for the next tuber generation to be available. Another positive point of the method is the high uniformity of the starting material.

In addition to irradiation of vegetative parts, potato seed and pollen have also been treated for different purposes. This work will not be discussed here.

Practically all mutation work has been performed with the common, tetraploid potato. As early as 1964 Ferwerda (cited by Van Harten and Bouter, 1973) irradiated dihaploids and secondary diploids. Other preliminary work with potato material at various levels of ploidy has been reported by Gomez Cuervo and Nelson Estrada (1972) and by Moreno and Nelson Estrada (1973). An experiment involving 150 X-irradiated small dihaploid tubers of cv. Désirée was published by Van Harten and Bouter (1973). In this investigation all emerging sprouts were dissected ("milked") when about 3 cm long and planted individually over a period of 8 months. The majority of the latest developing sprouts especially were of adventitious origin and produced relatively high frequencies of non-chimeric mutants.

The use of sprouts and plantlets obtained from in vitro cultures of different kinds of potato material in mutation programmes will only be a question of time. Preparations at Wageningen (co-operation between the two authors' institutes) are now in their final stage (Roest and Bokelmann, 1976) (Plate 4). In Germany, callus growth and differentiation into plantlets as well as the production of single-cell cultures from dihaploid calli was reported (Behnke, 1975, 1976). Methods to produce callus from potato leaf strips have been described by Upadhya et al. (1976).

Mutagenic treatments. Over the years many different mutagenic agents have been applied. In the beginning mostly X-rays, ^{32}P and, to a smaller extent, γ-rays were used, in a wide range of doses, dose rates and concentrations. More recently neutrons have also been administered and, in a still increasing number of cases, different chemicals. Unfortunately, exact data are rarely given, especially in the early literature.

For X- and γ-rays, doses range from 400 to 10 000 R for vegetative parts, 2 000—3 000 R being the doses preferred by most workers. Doses of 8 000—10 000 R are usually lethal.

The effects of different dose rates have seldom been studied, although it appears that especially the rate of survival and the amount of chimerism may be affected in this way (McCrory and Grun, 1969; Van Harten et al., 1972). ^{60}Co or ^{137}Cs sources have been mainly used for acute γ-ray treatment in potato work. Essential differences in the effects of X-rays and γ-rays have not been reported.

Fast and thermal neutron irradiation has been applied, for example, by Solomko (1965a), Van Harten et al. (1972), Gomez Cuervo and Nelson Estrada (1972) and Miedema (1973b). Doses and dose rates used vary greatly and are dependent on the reactor used. High-energy protons, used recently by Tarasenko (1977), were found to be as effective as fast neutrons.

Reports concerning other physical agents such as different radioisotopes and ultraviolet radiation are relatively scarce (Jauhar and Swaminathan, 1967). Much work with chemicals has been done in Russia, for example by Korableva (1961), Tarasenko (1965), Solomko (1965a,b) and many others. Ferwerda (1965) and Umaerus (1966) were the first to apply ethyl methanesulphonate (EMS) solutions, which were dropped on eyes at concentrations of, for example, 15%. EMS has remained the most used chemical agent although several others, like ethylene imine (EI), diethyl methanesulphonate (DES), N-nitroso-methyl urethane (NMU), hydroxylamine, are mentioned on different occasions as well (Nayar and Chauhan, 1968; Jauhar, 1969a,b; Pavek, 1972; Upadhya and Purohit, 1973; Upadhya et al., 1974a,b). Again, a wide range of different concentrations, durations of treatment, temperatures, buffers etc. have been applied. As up till now results have been rather inconclusive, it seems premature to discuss these points in detail.

Types of induced mutations. Most cases in which the induction of mutations has been proved definitely, refer to genetic changes of morphological characteristics, for example, size, shape or colour of the foliage or of the subterranean parts.

The work of Heiken (1960) has confirmed that most mutations manifest themselves as periclinal chimeras (Plate 4) and often show pleiotropic effects. Most induced mutations are found to be "loss mutations". A first example of the opposite has been the induction of increased pigmentation in the corolla, again reported by Heiken (1960). This author has also made valuable observations and remarks with regard to the practical application of irradiation in potato breeding. He found that even drastic morphological mutations do not necessarily lead to inferior plants from a breeding point of view. Heiken did not observe differences between the spectra of spontaneous and induced mutations. Unfortunately, his work has not yielded mutations which were of direct practical value.

The induction of mutations for earliness, increased resistance to different diseases and increased starch content of the tubers was reported by Solomko (1962, 1965a). Kishore et al. (1963) found after X-irradiation of 55 plants 19 with increased resistance to *Phytophtora infestans*, but this material was tested during only one vegetative cycle. Some possibly induced resistance to leafroll virus was reported by Rudorf and Wöhrmann (1963) in vM_3. In general, these authors did not think much of mutation breeding as a method of practical value.

The induction of earliness and increased starch production was reported by Tarasenko (1965). Umaerus (1966) found tubers which showed less discolouration after cooking. A promising mutant with shortened stolons was obtained by Roer (1967). This mutant could not be utilized further because this positive change was accompanied by the occurrence of cracks in the tubers.

Kukimura and Takemata (1975) studied the induced variation in starch content, stem height and tuber yield after treating dormant tubers with cobalt-60 γ-rays. Doses of 4—16 kR were given. The plants were raised up to vM_5. The most important findings were that the genetic variation for starch content increased by the mutagenic treatment and that radiation damage was transmissible to generations later than vM_2.

Jauhar and Swaminathan (1967) induced shallow eyes and an improved tuber skin colour in two important Indian cultivars. Unfortunately, the clones were lost according to M.D. Upadhya (personal communication, 1976). "Hooded-eyes" were found by Udai Singh (1970). Day neutrality was induced into several Indian potato clones (Upadhya et al., 1974a; Kishore et al., 1975). Their efforts to induce resistance to bacterial wilt have remained abortive.

F.P. Ferwerda, whose work with several important Dutch cultivars, like Bintje, Désirée and Burmania, has unfortunately been buried mainly in internal reports (see Van Harten, 1970 and later), demonstrated that induced mutations may affect practically any plant character and may also work in opposing directions, for example, towards increased pigmentation of the tuber skin as well as towards paler types. Several characteristics of economic importance, like cooking quality, tuber flesh colour, eye depth, underwater weight, stolon length, resistance to *Phytophtora* and leaf roll virus, could be improved using mutagenic treatment. The experiments at our institute (IVP) were performed with the one intention of demonstrating the feasibility of the method and not in order to produce new cultivars. Resistance to *Phytophtora* was induced in a few clones of cv. Bintje, which remained significantly more resistant during 6 successive years of testing in infested field plots, without showing any aberration of the Bintje phenotype. This fact demonstrates that desirable mutations are not always accompanied by pleiotropic effects or, in any case, that such effects can be eliminated by further selection without losing the desired genetic change itself. At the Institute of Plant Breeding, Agricultural University, Wageningen, in other work, which is still in progress, some promising results have been found, especially with respect to radiation-induced higher levels of resistance to leaf roll. To find definite proof of such effects is a very tedious job. Many people do not realize that the considerable effects of soil and climatic conditions on field experiments, under conditions of either natural or artificial infection, make it absolutely necessary to repeat such experiments for 5 or more successive years. The same is true for yield trials, where contradictory results may be obtained in different years. So-called positive results which refer to only 1 or 2 years of experimental work are without value. This may also explain the contradictory results of the early reports by, for example, Johnson (1928, 1937) and Sprague and Lenz (1929).

The frequency of induced mutations. In the first publication in which frequencies of induced mutations are mentioned, Asseyeva and Blagovidova (1935) obtained ten morphological mutations after X-irradiation of 390 tubers of cv. Wohltmann. Another one, cv. Epicure yielded no visible mutation at all. The effect of the genetic constitution of the starting material on mutation frequency seems hereby to be sufficiently demonstrated. Another aspect that is normally neglected is that it is very difficult to compare the results of different authors, because the number of characters for which changes are scored differs from author to author. At a dose of 2 000—3 000 R of X-rays, which in most cases yields the best results, Heiken (1960) found 37% mutated plants in the most successful series. He considered the frequency of induced mutations in general high enough to justify further investigations. The reports of Solomko (1962, 1965a,b) are very optimistic; he claims that only 4—8% of the induced changes are of an adverse nature. In some well-known Russian cultivars like Early Rose, Epron and Sedov as many as 17% of induced mutants are reported to be useful. According to

Solomko, some of them could be used directly by the growers. Rudorf and Wöhrmann (1963), on the other hand, state that the frequency of induced mutations is too low for practical purposes. This is said to be a result of diplontic selection, i.e. competition between mutated and non-mutated cells (see Gaul, 1959).

Asseyeva and Yashina (1968) concluded that six of the 200 induced mutations they investigated had some practical significance. Only one of those six mutations was found in a cultivar of commercial value. These authors write rather pessimistically about their findings and add that the spectrum of artificially induced mutations is inferior to that of spontaneous ones.

Whether one agrees with the Russian view or not, one thing is sure: for practical mutation breeding in potatoes, high frequencies and a wide spectrum of mutations are necessary. If this can be achieved, the method seems at least as promising as starting from many thousands of potato seedlings from which, in many cases, only one new cultivar, or even none at all, is derived.

Efforts to induce such high frequencies, by overcoming diplontic selection, were described by Nayar (1969) and Nayar and Dayal (1970). Using a system in which the sprouts emerging from tuber halves were periodically removed and planted separately the frequency of vM_2 plants carrying mutations could be increased by up to 50%. This is 16 times higher than the frequency found by Heiken (1960) and three times higher than that reported by Nayar et al. (1965).

In an experiment with fast neutrons, Van Harten et al. (1972) obtained up to 28% visible mutants in vM_2. In another experiment with dihaploid potatoes (Van Harten and Bouter, 1973), up to 38% of plants obtained were visible mutants after treating whole tubers with 3 000 rad of X-rays at a dose rate of 1 000 rad/min.

Summing up, it appears from the results outlined above that the mutation frequency which can be induced is not really the factor limiting the practical application of mutation techniques to potato breeding.

Concluding remarks

After the important contributions of Heiken and others, it appeared that the release of new potato cultivars, obtained via the application of mutation techniques, would only be a question of time. At present, more than 15 years later, this expectation has not yet been realized — so far, only one commercial mutant is available (Table 6).

To explain this, the existence of certain technical problems is often cited. With respect to such (real or supposed) problems it can be concluded from the above literature review that high frequencies of mutations can be produced and, if experimental conditions are properly chosen, the mutations can be obtained practically chimera free. Up to now this has been a rather

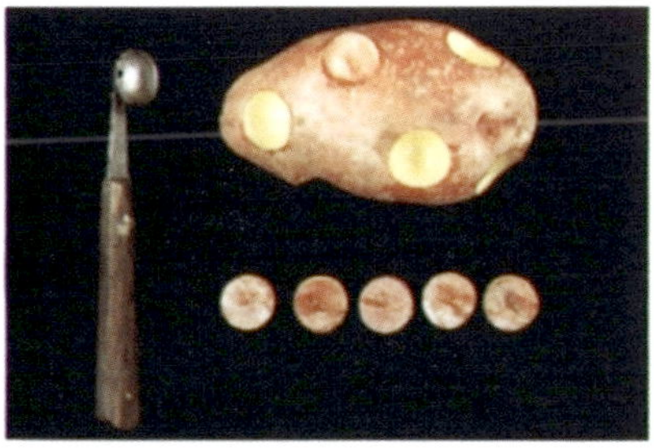

1

2

3

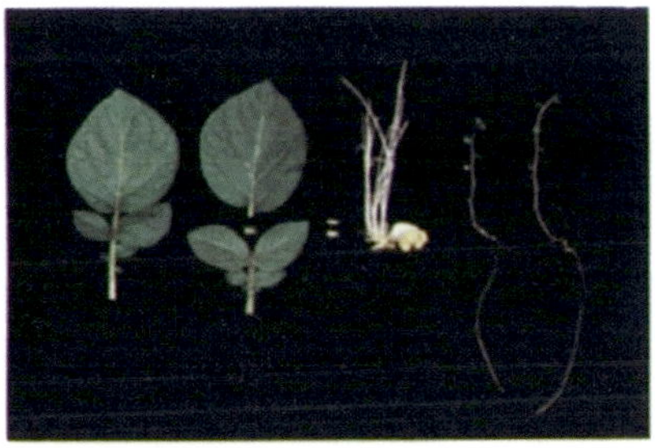

4

5

Plate 4. The in vivo single-eye vegetative propagation technique (1) in potato, *Solanum tuberosum*, cv. Desirée, results in "sectorial chimeras" (2) and, ultimately, in complete, periclinal chimeras (3), in this case a mutant with yellow—red splashed tubers (note also the reddish coloured stems), which indicates that only L-I is mutated from red to yellow. An in vitro adventitious bud technique (4) results in non-chimeral mutants i.e., with completely yellow tubers (5), indicating that all three cell layers are genetically yellow. (Slides, Institute of Plant Breeding, Agricultural University, Wageningen (1, 2, 3 and 5), and ITAL, Wageningen (4).)

TABLE 6

The commercial mutant of potato

Name of new cultivar	Place and date of release (or approval), and name of principal worker(s) and institute	Conditions and date of mutagenic treatment [Parent cultivar]	Main improved attributes of cultivar
Konkei, No. 45	Japan, 1973 Kazuo Asama and Ichiro Kaneka, Hokkaido Prefectural Konsen Agricultural Experiment Station	3 kR X-rays, 1962 [Benimaru]	Skin colour: red-yellowish white

Sources: lists prepared by the IAEA (1972, pp. 526—544) and *Mutation Breeding Newsletter*, 1 (1972) — 9 (1977).

laborious job and it seems desirable to look for better and quicker methods. Collection of additional data concerning the fate of a mutated cell in the apical area, or in other places from which regeneration takes place, is necessary. The behaviour of the irradiated apex itself also deserves further attention. Such investigations may also increase our insight into histogenic effects.

From a practical point of view, it would be very useful if whole plants could be grown from single mutated cells, not only to overcome chimerism, but also to make better use of the mutations induced in different parts or layers of the plant. In analogy with the positive results obtained for several ornamentals, the so-called adventitious bud methods using buds which often develop from a single epidermal cell seems to be very promising in potato as well. If this method could be applied, it would be a simple and rapid way of producing large numbers of uniform mutants.

In addition to the technical problems, there are a number of other reasons why, from a practical point of view, mutation breeding in potato has not been successful up to now. First of all, much of the research concerning mutations has never passed the stage of preliminary trials. Many experiments have been performed on an accidental basis, were conducted on far too limited a scale and for only very short periods of time. It is strange that plant breeders at scientific institutes as well as at private stations, used to starting more "classic" potato breeding programmes with many thousands of crosses, expect miracles from the irradiation of, let us say, 200 potato tubers or eye plugs.

An essential condition for successful mutation breeding work seems to be a considerable amount of realism with regard to the objectives as well as to the practical approach. As long as sufficient genetic variation can be obtained more easily or quickly in other ways, the mutation method should not be applied. However, the failure in The Netherlands to produce a suitable substitute for the famous cultivar Bintje in over 50 years, clearly demonstrates that even large amounts of conventional work do not always guarantee success.

In our opinion, mutation breeding should not be used to replace the production of initial potato crosses. The most promising approach appears to be to apply the method at an advanced stage of a breeding programme in order to cure certain, well-defined "defects" in otherwise promising breeding material and in generally accepted cultivars.

Literature potato

Asseyeva, T., 1927. Bud mutations in the potato and their chimerical nature. J. Genet., 19: 1—26.

Asseyeva, T. and Blagovidova, M., 1935. Artificial mutations in the potato. Bull. Appl. Bot. Genet. Plant Breed. (Leningrad). Ser. A, 15: 81—85 (in Russian).

Asseyeva, T.V. and Yashina, I.M., 1968. Somatic mutations in potato. Genetics (U.S.S.R.), 3: 145—164 (in Russian).

Behnke, M., 1975. Regeneration in Gewebekulturen einiger dihaploiden *Solanum tuberosum* — Klonen. Z. Pflanzenzücht., 75: 262—265.
Behnke, M., 1976. Kulturen isolierter Zellen von einigen dihaploiden *Solanum tuberosum* — Klonen und ihre Regeneration. Z. Pflanzenzücht., 78: 177—181.
Carrière, E.A., 1865. Production et Fixation des Variétés dans les Végétaux. Paris, 72 pp.
Cramer, P.J.S., 1907. Kritische Übersicht der bekannten Fälle von Knospenvariation, 4. Natuurkd. Verh. Holl. Maatsch. Wet., Haarlem, 474 pp.
Darwin, C., 1868. The Variation of Animals and Plants under Domestication, Vol. 1. Murray, London, 2nd edn, 473 pp.
Demidovic, A., 1934. A new possibility in breeding. Semenovodstvo (Seed Growing), 4: 8—10 (in Yugoslavian).
Dorst, J.C., 1924. Knopmutatie bij den aardappel en hare betekenis voor den landbouw. Genetica (The Hague), 6: 1—123.
Ferwerda, F.P., 1965. Mutagenic effects of X-rays and ethyl methane sulphonate on potato sprouts. In: The Use of Induced Mutations in Plant Breeding. FAO/IAEA Meeting, Rome (1964). Radiat. Bot., 5 (Suppl.): 687—690.
Folsom, D., 1923. Mutations of the potato: two somewhat unstable leaf-form sports of the Irish potato. J. Hered., 14: 45—48.
Gaul, H., 1959. Über die Chimärenbildung in Gerstenpflanzen nach Röntgenbestrahlung von Samen. Flora (Jena), Abt. B, 147: 209—241.
Gomez Cuervo, P.L. and Nelson Estrada, R., 1972. Artificial induction of mutants in the Criolla potato (*Solanum phureja* Juz. et Buk.). In: Proc. Study Group on Induced Mutations, Buenos Aires (1970). IAEA, Vienna, pp. 457—468.
Hagberg, A. and Nybom, N., 1954. Reaction of potatoes to X-irradiation and radiophosphorus. Acta Agric. Scand., 4: 578—584.
Heiken, A., 1960. Spontaneous and X-ray Induced Somatic Aberrations in *Solanum tuberosum* L. Almqvist and Wiksell, Stockholm, 125 pp.
Heiken, A., Ewertson, G. and Carlström, L., 1963. Studies on a somatic subdivided-leaf mutant in *Solanum tuberosum*. Radiat. Bot., 3: 145—153.
Howard, H.W., 1970. Genetics of the potato, *Solanum tuberosum*. Logos Press, London, 126 pp.
IAEA, 1972. Induced Mutations and Plant Improvement. IAEA, Vienna, 554 pp.
Jacobson, M., 1923. Die Wirkung der Röntgenstrahlen auf das Wachstum der Pflanzen. Beil. Rigaschen Rundsch., 54: 5.
Jauhar, P.P., 1969a. Induction of some rare somatic mutations in *Solanum tuberosum* L. by ionizing radiations and radiophosphorus. Eur. Potato J., 12: 8—12.
Jauhar, P.P., 1969b. Morphological and physiological effects of radiations and radioisotopes on potato, *Solanum tuberosum* L. Indian J. Agric. Sci., 39: 88—100.
Jauhar, P.P. and Swaminathan, M.S., 1967. Mutational rectification of specific defects in some potato varieties. Curr. Sci., 36: 340—342.
Johnson, E.L., 1928. Tuberization of potatoes increased by X-rays. Science, 68: 231.
Johnson, E.L., 1937. Tuberization of the Colorado wild potato as affected by X-radiation. Plant Physiol., 12: 547—551.
Kishore, H., Puskarnath and Singh, G., 1963. The effect of radiation on potato tubers. Indian Potato J., 5: 86—92.
Kishore, H., Das, B., Subramanyam, K.N., Chandra, R. and Upadhya, M.D., 1975. Use of induced mutations for potato improvement. In: Improvement of Vegetatively Propagated Plants through Induced Mutations, Tokai, Japan, 1974. IAEA, Vienna, pp. 77—82.
Korableva, N.P., 1961. Bestrahlungswirkungen auf anatomisch-physiologische Besonderheiten der Vegetationspunkte von Kartoffelklone. Dokl. Akad. Nauk. SSSR, 137: 454—457.

Krantz, F.A., 1951. Potato breeding in the United States. Z. Pflanzenzücht., 29: 388—393.
Kukimura, H. and Takemata, T., 1975. Induced quantitative variation by gamma-rays and ethylene-imine in tuber bearing plants. Gamma Field Symp., 14: 25—36.
McCrory, G. and Grun, P., 1969. Relationship between radiation, dose rate and lethality of diploid clones of *Solanum*. Radiat. Bot., 9: 27—32.
Miedema, P., 1973b. The use of adventitious buds to prevent chimerism in mutation breeding of potato. Euphytica, 22: 209—218.
Moreno, B.G. and Nelson Estrada, R., 1973. Induction of mutants in species of potatoes (*Solanum tuberosum* L.) by means of ionizing radiations. Rev. Inst. Colomb. Agropecu., 8: 117—129 (in Spanish).
Nayar, N.M., 1969. Considerations of overcoming intrasomatic selection during mutation breeding of vegetatively propagated plants. Theor. Appl. Genet., 39: 99—103.
Nayar, N.M. and Chauhan, H.S., 1968. A chemically induced systematic mutation in the potato (*Solanum tuberosum* L.) produced as a sectorial chimera. Jpn. J. Genet., 43: 389—392.
Nayar, N.M. and Dayal, T.R., 1970. A method of securing high frequencies of induced mutations in the potato. Z. Pflanzenzücht., 63: 155—160.
Nayar, N.M., Puskarnath and Hakim, K.L., 1965. Frequencies of certain types of gamma-ray induced mutations in potato. Indian Potato J., 7: 76—80.
Pavek, J.J., 1972. A preliminary trial with chemical mutagens applied to potato eyes. Am. Potato J., 49: 362 (abstract).
Roer, L., 1967. Mutations in potatoes induced by gamma irradiation. Euphytica, 16: 283—292.
Roest, S. and Bokelmann, G.S., 1976. Vegetative propagation of *Solanum tuberosum* L. in vitro. Potato Res., 19: 173—178.
Rudorf, W. and Wöhrmann, K., 1963. Versuche zur Auslösung von Mutationen durch Co^{60}-Bestrahlung auf vorgetriebene Augenkeime bei der Kartoffel. Z. Pflanzenzücht., 49: 397—414.
Salaman, R., 1926. Potato Varieties. Cambridge University Press, London, 377 pp.
Solomko, E.A., 1962. Ways of bringing induced somatic mutations in potato to light. Radiobiologiya, 2: 634—638 (in Russian; original not consulted).
Solomko, E.A., 1965a. Potato mutations induced by irradiation of vegetative parts of the plant. Radiobiologiya, 5: 547—551 (in Russian; original not consulted).
Solomko, E.A., 1965b. Useful artificial mutations in potatoes. Genetics (U.S.S.R.), 1: 233—239 (in Russian).
Sparrow, A.H. and Christensen, E., 1950. Effects of X-ray, neutron and chronic gamma irradiation on growth and yield of potato. Am. J. Bot., 37: 667 (abstract).
Sprague, H.B. and Lenz, M., 1929. The effect of X-rays on potato tubers for "seed". Science, 69: 606.
Stanton, W.R. and Sinclair, W.K., 1951. Effect of high concentration of phosphorus-32 on growth of potato. Nature (London), 167: 234—235.
Sutton, A., 1918. Do Potatoes Give Rise to New and Distinct Varieties by Bud Variation? Bull. No. 9, Messrs Sutton and Sons.
Tarasenko, N.D., 1965. Experimental somatic mutations in some potato cultivars. Genetics (U.S.S.R.), 5: 145—149 (in Russian).
Tarasenko, N.D., 1977. Mutagenic efficiency of high-energy protons in potato. Z. Pflanzenzücht., 79(1): 79—81.
Turnquist, O.C., 1960. Production of certified seed potatoes by varieties, 1959. In: J.C. Campbell (Editor), Potato Handbook, 1960. Potato Assoc. Am., New Brunswick, N.J., pp. 55—59.
Udai Singh, 1970. Radiation induced hooded eye mutants in potato. Sci. Cult., 36: 609—610.

Umaerus, M., 1966. Somatic aberrations in *Solanum tuberosum* induced by ethyl methane sulfonate and X-irradiation. Z. Pflanzenzücht., 55: 238—245.

Upadhya, M.D. and Purohit, A.N., 1973. Mutation induction and screening procedure for physiological efficiency in potato. In: Induced Mutations in Vegetatively Propagated Plants. Proc. Meeting FAO/IAEA, Vienna, 1972. IAEA, Vienna, pp. 61—66.

Upadhya, M.D., Anand, S.K. and Pandey, S., 1974a. Mutation induction for resistance to bacterial wilt in potato. In: Induced Mutations for Disease Resistance in Crop Plants. Proc. Meeting FAO/IAEA/SIDA, Novi Sad, 1973, pp. 172—173.

Upadhya, M.D., Dayal, T.R., Dev, B., Chandhra, V.P., Sharda, R.T. and Chandra, R., 1974b. Chemical mutagenesis for day-neutral mutations in potato. In: Polyploidy and Induced Mutations in Plant Breeding. Proc. Meeting FAO/IAEA, Bari, 1972. IAEA, Vienna, pp. 379—383.

Upadhya, M.D., Chandra, R. and Abraham, M.J., 1976. Mutagenic treatments towards increasing the frequency of day-neutral mutations and standardisation of procedures for tissue culture, in potato. In: Improvement of Vegetatively Propagated Crops and Tree Crops through Induced Mutations. Proc. Meeting, Wageningen, 1976. IAEA, Vienna, pp. 151—170.

Van Harten, A.M., 1970. Probleme und Perspektiven der Mutationszüchtung bei der Kartoffel. In: Bericht Arbeitstagung 1969, Gumpenstein (Österreich). Bundesversuchsanstalt für Alpenländische Landwirtschaft, Gumpenstein, pp. 116—133.

Van Harten, A.M. and Bouter, H., 1973. Dihaploid potatoes in mutation breeding: some preliminary results. Euphytica, 22: 1—7.

Van Harten, A.M. and Bouter, H., 1976. Differences in effects of dry and immersed irradiation in potato (*Solanum tuberosum* L.). In: Proc. Workshop Eur. Comm. — Israel, The Use of Ionizing Radiation, Wageningen, 22—24 March. Assoc. Euratom—ITAL, Wageningen, pp. 503—509.

Van Harten, A.M., Bouter, H. and Van Ommeren, A., 1972. Preventing chimerism in potato (*Solanum tuberosum* L.). Euphytica, 21: 11—21.

Van Harten, A.M., Bouter, H. and Schut, B., 1973. Ivy leaf of potato (*Solanum tuberosum*), a radiation-induced dominant mutation for leaf shape. Radiat. Bot., 13: 287—292.

SWEET POTATO

Ipomoea batatas (L.) Lam. or sweet potato (fam Convolvulaceae) is grown on a large scale in tropical areas, especially in Africa. Considerable quantities are also produced under more temperate conditions, especially in Japan and on a smaller scale in China, Taiwan the U.S.A. and New Zealand. World production is estimated to be more than 10^8 metric tons of tubers annually.

Tubers are used for human consumption, boiled or baked, but can also be used as fodder. Several products — flour, starch, glucose syrup and alcohol — can be manufactured from the tubers. Other parts of the plant are also utilized. Leaves are eaten as vegetables and fresh vines are given to cattle as fodder. Plants are normally propagated by means of stem cuttings 30—50 cm long. In temperate regions, sprouts or slips from small tubers, which are pulled from nursery beds and transplanted at a length of 20—30 cm, are often used. The cuttings are very easy to root.

Propagation by single-leaf cuttings is also possible (Purseglove, 1974). Wilson (1973) demonstrated that treatment of detached leaves of sweet potato with high nitrate-nitrogen levels produced up to six adventitious shoots. They were formed on roots, tubers, tuber stalks and, in a few cases, on the callus at the petiole base. It is not yet known whether or not the adventitious buds develop from single cells; this question is still being investigated.

From a breeding point of view, the sweet potato is a very difficult crop. Flowering is scarce, especially in temperate regions, and in all climates seed set is low. Both incompatibility and sterility barriers are known to exist and insight into these matters is only gradually being acquired. The hexaploid level of inheritance ($2n = 6x = 90$) makes genetic studies very complicated. It has been shown that many important characters inherit quantitatively (Jones, 1969; Jones et al., 1969).

Most cultivars of sweet potato are highly heterozygous and variable. In the past, improved cultivars were obtained by selection among existing clones in which spontaneous mutations occur quite frequently. Moreover, chance seedlings occur in nature. At present, generative breeding work is carried out with seedlings from open pollination as well as with controlled crossing. Breeding work in many countries such as Japan, Puerto Rico, Hawaii, the U.S.A. and New Zealand, aims mainly at producing high yielding, early, well-shaped, shallow rooting types with good storage and cooking qualities, palatability, and resistance to viruses and pests. Attractive skin colours, high vitamin content and frost resistance are also favoured (Pope, 1970). The complications met with in classical breeding work make the use of both spontaneous and induced mutations in sweet potato very attractive. As stated above, spontaneous mutations occur fairly often (Fujise, 1965).

From 1925 onwards, several instances of spontaneous mutation have been reported, for example, by Rosa (1926) and Miller (1930, 1935). Other useful references are given by Gustafsson and Gadd (1965a). Miller (1935) was also one of the first authors to work on X-ray induced mutations in this crop. Some differences in tuber colour and taste between treated and non-treated material were observed. More than 20 years later Cheng (1958a,b) reported that X-irradiation of tubers with 1 250 and 5 000 R produced 50% leaf-colour mutations and, more important, earlier maturing and higher yielding variants. At about the same time, Abraham (1957) published some preliminary work with X-rays in India. Mashimo and Sato (1959), in Japan, irradiated young shoots of cv. Norin-1 with 5—20 kR of X-rays. The material was cut into pieces and then studied until vM_7. Several mutations for tuber colour and morphological characters were obtained. The use of γ-rays and X-rays was mentioned in a short publication by Matsumura and Fujii (1958). Studies concerning radiobiological and physiological properties after irradiation with 1.2—7.5 kR of X-rays were carried out by Nishiyama et al. (1959) on leaves and leaf stalks. Irradiation of sweet potato roots with 10—40 kR was also mentioned by Osborne and Lunden (1961).

Hernandez et al. (1959) reported that some skin and flesh colour mutations had been induced with γ-rays (10—50 kR). In every case the colour changed from dark (e.g. orange) to light (white). In a later publication the same authors (Hernandez et al., 1964) compared mutation rates of three different cultivars treated with up to 50 kR of γ-rays and observed considerable differences between cultivars. Most commonly mutated "sectors" with a colour change, again from orange to white, were obtained.

The effect of irradiation on the yield of sweet potato was studied by Poole (1959) and Poole and Tanaka (1963). Marumine and Sakai (1961) reported studies on X-ray induced mutants, and Sakai (1966) studied the repeated application of phosphorus-32 to cuttings, leaves and tubers in order to obtain mutants with high starch contents. Increasing the starch content of sweet potato is given priority in breeding programmes in Japan. Sakai concluded that, in general, repeated treatment was an effective method for mutation induction as proved by stem diameter changes and colour mutations; he did not obtain increased dry matter content. He also referred to the work of Akita et al. (1960), who tried to breed practical cultivars by γ-irradiation of true seed, but this publication was not added to his list of references. Love (1969, 1972) noted mainly skin colour and flesh colour mutations after exposure of rooted cuttings to fast neutrons. Changes which darkened the skin were considerably less frequent than those resulting in lighter skin colours. Changes in flesh colour occurred more often than those in skin colour.

High yielding mutants with a high vitamin A content were obtained after irradiation by Miu et al. (1969). Kukimura (1971) achieved changes in tuber skin colour after using X-rays, γ-rays and the chemical mutagen, ethylene imine (EI). He claimed that (apparently) dominant mutations can be obtained at high frequencies. This statement, however, seems somewhat doubtful. His results could be explained by the occurrence of chimeric structures being uncovered by irradiation.

Soriano (1971, 1972) treated terminal vines, carrying between eight and ten axillary buds, with 800—3 200 R of γ-rays and obtained a number of variants with aberrant leaf characters. He ascribed the changes to chromosome aberrations.

Miu (1973) and Miu et al. (1973) reported the induction of improved tolerance in the important native cultivar Red-flesh-tail, from Taiwan, after treatment with 5 000 R of γ-rays. Plants with darker leaves and higher yields were also obtained.

Cuevas-Ruiz and Koo (1973) preferred lower doses (1 500—2 000 R) for mutation induction in sweet potato. Their studies revealed differences in response to the treatment between cultivars.

Kukimura and Takemata (1975) recently reported that mutants with increased as well as reduced sugar contents were obtained after treatment of shoots, dormant root tubers and seeds of sweet potato with cobalt-60 γ-rays (10—15 kR) or EI (at concentrations of 0.1 and 0.3%, and exposure for 3 h).

The above review clearly demonstrates that the induction of valuable traits in sweet potato by applying mutagenic treatment is quite feasible. In fact, this crop has very good prospects for further mutation work as it seems to be possible to use an adventitious bud technique (Wilson, 1973; J.E. Love, personal communication, 1973).

As it has also been demonstrated that mutations for most important characters can be induced relatively easily, one might expect practical results within the near future, provided the mutation breeding work is continued in a systematic fashion.

Literature sweet potato

Abraham, A., 1957. Breeding of tuber crops in India. In: Symp. Genetics and Plant Breeding, New Delhi. Indian J. Genet. and Plant Breed., 17: 212—217.

Cheng, C.F., 1958a. The effect of radiation on the first irradiated generation of sweet potato and peanut. In: Proc. 2nd UN Int. Conf. Peaceful Uses Atomic Energy. A/Conf. 15/P., 1131. UN, New York, N.Y., pp. 1—6.

Cheng, C.F., 1958b. Circumstances of the application of atomic power to the improvement of Taiwan Cropplant varieties. J. Agric. Assoc. China, 21: 65—68 (in Chinese).

Cuevas-Ruiz, J. and Koo, F.K.S., 1973. Effect of gamma irradiation on sweet potato stem cuttings. J. Agric. Univ. P.R., 57: 351—354.

Fujise, K., 1965. Spontaneous mutation and its use in sweet potato. The use of spontaneous mutations in plant breeding. Gamma Field Symp., 4: 43—53.

Gustafsson, Å. and Gadd, I., 1965a. Mutations and crop improvement III. *Ipomoea batatas* (L.) Poir. (Convolvulaceae). Hereditas, 53: 77—89.

Hernandez, T.P., Hernandez, T. and Miller, J.C., 1959. Effects of different rates of irradiation on sweet potato roots and tomato seeds. Proc. Am. Conv. Assoc. South. Agric. Work., Memphis, Tennessee, 56: 157—158.

Hernandez, T.P., Hernandez, T. and Miller, J.C., 1964. Frequency of somatic mutations in several sweet potato varieties. Proc. Am. Soc. Hortic. Sci., 85: 430—433.

Jones, A., 1969. Quantitative inheritance of ten vine traits in sweet potatoes. J. Am. Soc. Hortic. Sci., 94: 408—411.

Jones, A., Steinbauer, C.E. and Pope, D.T., 1969. Quantitative inheritance of ten root traits in sweet potato. J. Am. Soc. Hortic. Sci., 94: 271—275.

Kukimura, H., 1971. On the Artificial Induction of Skin Colour Mutation of Sweet Potato (*Ipomoea batatas* Lam) tuber. Inst. Radiat. Breed., Ohmiya, Ibaraki, Tech. News 8, 2 pp.

Kukimura, H. and Takemata, T., 1975. Induced quantitative variation by gamma-rays and ethylene-imine in tuber bearing plants. Gamma Field Symp., 14: 25—36.

Love, J.E., 1969. Mutation induction in the sweet potato by means of fast-neutron irradiation. In: Induced Mutations in Plants. Proc. Symp. Pullman, Wash., IAEA, Vienna, pp. 331—336.

Love, J.E., 1972. Somatic mutation induction in poinsettia and sweet potato. In: M.J. Constantin (Editor), Mutat. Breed. Workshop, Knoxville, University of Tennessee, Knoxville, Tenn.

Marumine, S. and Sakai, K., 1961. Studies on the induction of artificial mutation in sweet potato 1. X-ray induced mutations. Proc. Kyushu Crop Sci. Soc., Japan, 16: 4—6.

Mashimo, I. and Sato, H., 1959. X-ray induced mutations in sweet potato. Jpn. J. Breed., 8: 233—237 (in Japanese).

Matsumura, S. and Fujii, T., 1958. Induction of bud sports by X- and γ-rays. Annu. Rep. Nat. Inst. Genet., Japan, 8: 94—95.

Miller, J.C., 1930. A study of mutations of the Porto Rico sweet potato. Proc. Am. Soc. Hortic. Sci., 27: 343—346.
Miller, J.C., 1935. Further studies of mutations of the Porto Rico sweet potato. Proc. Am. Soc. Hortic. Sci., 32: 460—465.
Miu, T.S., 1973. An inquiry into indicators of cold tolerance among radiation-induced lines of sweet potato. Genetics, 74 (Suppl.): 183.
Miu, T.S., Lin, Y.T., Hsin, C.F. and Chang, C.S., 1969. The course of breeding of new mutant sweet potato lines R and R 19 resulting from irradiation treatment. Tai-wan Nung-yeh Chi K'an (Taiwan Agric. Q.), 5(1): 78—83 (in Chinese).
Miu, T.S., Yu, C.Y. and Chang, C.S., 1973. An inquiry into assessment of cold tolerance in groups of irradiation-induced lines of sweet potato. Tai-wan Nung-yeh Chi K'an (Taiwan Agric. Q.), 9(1): 28—40 (in Chinese).
Nishiyama, I., Okamoto, M. and Teramura, T., 1959. Radiobiological studies in plants. II. Effects of X-rays on the development of roots from leaf stalks of sweet potato. Rep. Kihara Inst. Biol. Res., 10: 33—36.
Osborne, T.S. and Lundén, A.O., 1961. The co-operative plant and seed irradiation program of the University of Tennessee. Int. J. Appl. Radiat. Isot., 10: 198—209.
Poole, C.F., 1959. Improvement in yield of sweet potato clones. Proc. Am. Soc. Hortic. Sci., 73: 445—452.
Poole, C.F. and Tanaka, J.S., 1963. Transmission of yield variation in irradiated sweet potato clones. Proc. Am. Soc. Hortic. Sci., 80: 488—496.
Pope, D.T., 1970. Recent progress and current needs of the sweet potato industry in the United States. In: D.P. Plucknett (Editor), Tropical Root and Tuber Crops Tomorrow, Vol. 1. University of Hawaii, Honolulu, Hawaii, pp. 7—11.
Purseglove, J.W., 1974. Tropical Crops: Dicotyledons. Longman, London, pp. 79—88.
Rosa, J.T., 1926. Mutations in the sweet potato. J. Hered., 17: 167—188.
Sakai, K., 1966. Induction of mutants in sweet potato by absorption of ^{32}P. Mutations induced by RI and chemicals. Gamma Field Symp., 5: 25—34.
Soriano, J.D., 1971. Induction of leaf mutations in irradiated vines of sweet potato. Sabrao Newsl., 3(2): 113—116.
Soriano, J.D., 1972. Effects of gamma radiation on buds of sweet potato (*Ipomoea batatas* (L.) Poir). Nat. Appl. Sci. Bull., 24(1—2): 41—53.
Wilson, L.A., 1973. Stimulation of adventitious bud production in detached sweet potato leaves by high levels of nitrogen supply. Euphytica, 22: 324—326.

YAM AND VARIOUS OTHER (TROPICAL) ROOT AND TUBER CROPS

Yams, *Dioscorea* spp., are mainly herbaceous climbing plants, grown for their edible tubers. They are especially important in West Africa, Malaysia, the Pacific and the Caribbean where they are staple carbohydrate food. The importance of yams is declining, although approximately 20×10^6 metric tons of tubers are still produced each year, two thirds of it in West Africa. Several species produce edible tubers, the two most common being *D. alata* and *D. rotundata* (Purseglove, 1975).

Very little scientific research has been done on yams and, for the most part, breeding work has simply been a case of selecting for high yield, shallow rooting, palatability and lack of toxicity. Not much is known about spontaneous bud sport occurrence and only two publications concerning the use of radiation on yams, were found. Abraham (1970) reported that by treating tubers with γ-rays high yielding mutants were obtained. Further details are not known to us. Koo and Cuevas Ruiz (1964b), however, irradiated the aerial tubers of several varieties after 1—2 months storage at room temperature. The dose used was 2 krad of γ-rays given at a dose rate of 1 krad/min. The first shoots to appear were removed, forcing new shoots to develop; these latter shoots were allowed to grow and observed. The results showed that solid (or apparently solid) mutant shoots appeared as early as the second generation.

Yams, which are normally propagated from tubers, can also be multiplied from detached leaves, which are far more attractive for mutation breeding. To improve the coarse flesh of large-tubered cultivars, leaves were irradiated with 2—4 kR γ-rays which, unfortunately, was too low a dose to have the desired effect (Dr K. Ramachadran, University of Kerala, personal communication, 1975).

So far, no commercially interesting mutants have been reported. Various other tropical tuber crops are propagated vegetatively and are, in principle, suitable material for mutagenic treatment that could improve yield, quality or other characteristics. Until recently, very little selection work has been done in such crops, using either conventional breeding methods or with mutation breeding.

Vasudevan et al. (1968) report the effects of γ-radiation on *Colocasia esculenta* (taro, tara, dalo, eddo, Poi), an important food crop with edible, starchy and tuberous rootstocks. A number of (chlorophyll) mutants were found which were of no commercial interest.

Pre-germinated buds of *Xanthosoma sagittifolium* Schott were treated with ethyl methanesulphonate to test the sensitivity of the buds to the chemical. No mutants were reported (Muñoz, 1975).

K. Ramachandran, University of Kerala, India (personal communication, 1975), irradiated rooted stem cuttings of the so-called Chinese potato, *Coleus parviflorus*. The apical bud was removed to stimulate axillary bud outgrowth. The aromatic tubers of this plant are used as a vegetable. Other programmes in progress at Kerala, include irradiation of ginger, *Zingiber officinale*, elephantfoot yam, *Amorphophallus campanulatus*, *Colocasia* and *Xanthosoma sagittifolium*. So far, no commercial results have been published.

Literature yam and various other (tropical) root and tuber crops

Abraham, A., 1970. Breeding work on tapioca (cassava) and few other tropical tuber-crops. In: D.P. Plucknett (Editor), Tropical Root and Tuber Crops Tomorrow, Vol. 1. University of Hawaii, Honolulu, Hawaii, pp. 76—78.

Koo, F.K.S. and Cuevas Ruiz, J., 1974b. Exploratory induction of solid mutations in yams by γ-irradiation. In: A.M. Srb (Editor), Genes, Enzymes and Populations. Plenum Publishing, New York, N.Y., pp. 331—336.

Muñoz, A.C., 1975. Algunos aspectos sobre la sensibilidad de la malanga (*Xanthosoma sagittifolium* Schott) a los agentes mutagenicos quimicos. Centr. Agric., Univ. Cent. Las Villas, 2(1): 51—62 (English summary).

Purseglove, J.W., 1975. Tropical Crops: Monocotyledons. Longman, London, pp. 97—117.

Vasudevan, K.N., Jos, J.S. and Magoon, M.L., 1968. Radiation induced mutations in *Colocasia esculenta* (L.) Schott. Indian J. Hortic., 25: 66—69.

Chapter 5

ORNAMENTAL CROPS

General

Ornamental crops include a large variety of different plants which, in some countries, make a substantial contribution to the value of the horticultural products of those countries. In The Netherlands, for instance, approximately 29% of the total value of agricultural and horticultural products is constituted of ornamental tuber and bulb crops, pot plants, cut flowers etc. (see Table 3 in the Introduction).

For various reasons mutation breeding has been especially successful in ornamentals. One of these is that the selection of directly perceptable characteristics, like colour, form or size, is generally not difficult. Another reason is that many cultivars are heterozygous and so have a comparatively high mutation frequency. In addition, in vivo or in vitro propagation methods are often available (Broertjes et al., 1968; Holdgate, 1977) which enable the successful production of solid mutants that can be recognized sooner and propagated more easily, without the complications attached to the production of mericlinal chimeras following the irradiation of multicellular apices.

Ornamental tuber and bulb crops contain a large, economically important variety of different plant species. As most of them do not breed true after seed propagation, the majority of these crops is propagated vegetatively. This, together with the generally low speed of propagation, is one of the main obstacles to breeding in general, and to mutation breeding in particular, since it may take years before a mutated sector appears and, ultimately, one or some of the vegetative progeny show the mutated character as a "complete" mutant.

This slow process cannot be speeded up since pruning, to promote the growth of axillary buds, is often impossible. It is therefore not surprising that irradiation at the right moment is recognized to be of great importance; that is, irradiation should be carried out at the earliest possible stage of development, when a mutated cell has the largest possible opportunity to make a substantial contribution to the formation of the new tuber, bulb or plant. In the *Dahlia*, for example, preference is given to freshly harvested tubers for irradiation. In tulip, the discussion about whether small or large bulbs should be irradiated early or late in the season, is partly connected with the problem of knowing which method will produce "solid" mutants soonest.

It is also not surprising that there is great interest in in vivo and in vitro adventitious bud techniques, not only to produce solid mutants, but also, and sometimes this is the main reason, for rapid vegetative propagation. The latter is important for breeders, since large-scale reproduction of interesting genotypes is much quicker by this method than by any other. Examples of the application of in vivo adventitious bud techniques include the following: hyacinth — irradiation just before scooping; *Muscari* — irradiation of parts of leaves, which after rooting produce bulblets; *Lilium* — irradiation just before or after scaling, and *Amaryllus* — irradiated bulbs are cut into pieces. It seems to be only a matter of time before use is made of explants for in vitro production of adventitious shoots, in view of the "explosive" progress in this field of research (Hussey and Hilton, 1975a). Their use will open up prospects for the more efficient application of mutation breeding to bulbous ornamentals.

Ornamental pot plants and cut flowers offer fewer difficulties in comparison to tuber and bulb crops, since many can be propagated readily by in vivo or in vitro adventitious bud techniques or using cuttings. Consequently, dozens of commercial mutants have been produced in such crops (e.g. *Achimenes* and other Gesneriaceae, *Begonia, Chrysanthemum*).

Woody ornamentals are not very attractive material for breeding since a great number of years are needed before selection can be carried out and it takes even longer for useful genotypes to be multiplied and ready to be put on the market. It may be that, here too, in vitro propagation techniques could speed up (mutation) breeding programmes.

Literature ornamental crops — general

Broertjes, C., Haccius, B. and Weidlich, S., 1968. Adventitious bud formation on isolated leaves and its significance for mutation breeding. Euphytica, 17: 321—344.

Holdgate, D.P., 1977. Propagation of ornamentals by tissue culture. In: J. Reinert and Y.P.S. Bajaj (Editors), Plant Cell, Tissue, and Organ Culture. Springer Verlag, Berlin, pp. 18—43.

Hussey, G. and Hilton, J., 1975a. Proliferation of in vitro monocotyledonous plantlets with cytokinin. John Innes Inst. Annu. Rep., 66: 53—54.

Tuber and bulb crops

AMARYLLUS (HIPPEASTRUM)

Hippeastrum, family Amaryllidaceae, mostly referred to as *Amaryllus*, has numerous commercial hybrid cultivars that are used as pot plants or for cut flowers. They propagate by seed or vegetatively, for example, by adventitious bulblet formation (Van Raalte, 1971). In vitro propagation using excised bulb scales is also possible (Mii et al., 1974).

Irradiation experiments have not yet been reported in the literature. In co-operation with the Association Euratom—ITAL at Wageningen, a few private breeders are carrying out mutation breeding experiments using irradiated bulbs which are either divided into bulb scales afterwards, or propagated via their bulblets. The objectives of the research are to investigate which of the two methods yields the higher percentage of complete periclinal chimeras or solid mutants and to find out whether flower colour mutants can be obtained from leading cultivars. Very recently, approximately 100 mutants have been obtained from artificial scales made from irradiated bulbs. All seemed to be solid but this still has to be confirmed by further vegetative propagation. One or two interesting flower colour mutants were found (M. van der Mey, personal communication, 1977).

The best radiation dose depends on the cultivar used and environmental conditions, but is between 0.5 and 1 krad.

Literature Amaryllus

Mii, M., Mori, T. and Iwase, N., 1974. Organ formation from the excised bulb scales of *Hippeastrum hybridum* in vitro. J. Hortic. Sci., 49: 241—244.

Van Raalte, D., 1971. Het Handboek voor de Bloemisterij. V. Snijbloemen. Born Periodieken, Assen, Amsterdam, pp. 480—499.

ANEMONE

Anemone, family Ranunculaceae, is propagated either by seed or vegetatively from tubers. It has never been reported in the literature as being a subject for mutation breeding. As part of a co-operative project between the Association Euratom—ITAL, Wageningen and the Bulb Research Centre at Lisse, both in The Netherlands, *Anemone coronaria* was irradiated on a small scale to determine its radiosensitivity. The optimum dose for seed is approximately 10 krad X-rays, whereas very small tubers were not even harmfully affected by doses up to 15 krad X-rays. The radiosensitivity of tubers, however, was dependent on the cultivar (Alkema, 1974). Since no results were obtained, the preliminary experiments were discontinued. Spontaneous mutations do occur, as was reported by Horovitz and Zohary (1966). In *Anemone coronaria* the scarlet flower colour seems to be recessive, as was also reported by these authors.

Literature Anemone

Alkema, H.Y., 1974. Mutatieveredeling bij Bolgewassen door Middel van Radioactieve Bestraling en andere Methoden. Rapport 24, Laboratorium voor Bloembollenonderzoek, Lisse, p. 15.

Horovitz, A. and Zohary, D., 1966. Spontaneous variegation for perianth color in wild *Anemone coronaria*. Heredity, 21(3): 513—515.

CHIONODOXA

Chionodoxa luciliae, family Liliaceae, has only been subjected to mutagenic treatment as part of a co-operative programme (see *Anemone*). The objective was to determine its radiosensitivity and to see whether genetic variability for flower colour and form could be obtained.

In one experiment all irradiated bulbs treated with doses up to 15 krad X-rays stayed alive, whereas in a second experiment all bulbs treated with 15—75 krad X-rays died (Alkema, 1974). At that point the experiments were discontinued because of reorganization in the breeding programme of the laboratory in question.

Literature Chionodoxa

Alkema, H.Y., 1974. Mutatieveredeling bij bolgewassen door middel van radioactieve bestraling en andere methoden. Rapport 24, Laboratorium voor Bloembollenonderzoek, Lisse, p. 15.

CROCUS

The only literature citation on the irradiation of *Crocus*, family Iridaceae, refers to that of Mitsukiri and Arai (1965), who treated the cultivar Mammoth Yellow with γ-rays. All kinds of morphological changes were studied, in relation to dose, but no mutations were reported.

From our own experience, acquired in co-operation with a commercial plant breeder, we know that the best X-ray dose is between 1 and 1.5 krad, if freshly harvested tubers are used. The cultivar Grote Gele ('Large Yellow'), several hundreds of tubers of which were irradiated, did not produce a single mutation.

Literature Crocus

Mitsukuri, Y. and Arai, K., 1965. The effect of radiation on *Crocus*. Nippon Univ. J. Radiat. Med. Biol., 1: 22—30 (in Japanese).

DAHLIA VARIABILIS

Much is known about the history and genetics of the garden dahlia, *Dahlia variabilis*, fam. Compositae, mainly due to the extensive genetical and cytological investigations of Lawrence (1931a,b, 1942; see also Crane and Lawrence, 1956).

The plant originates from Central America and two types were introduced into Europe in 1789. Until 1804, the variability was small but after the import of some seeds an explosive variation in form and flower colour appeared and within 10—15 years almost all colours and forms, which we know today, were available.

The great range of flower colours is explained by the assumption that the octoploid garden dahlia arose as a hybrid of two tetraploid species, either by the fusion of two unreduced gametes or by chromosome doubling. Either parent species happened to have two different flower colour pigments, so that in the garden dahlia four pigments are found, viz.:

deep anthocyanin	(B)
pale anthocyanin	(A)
yellow flavone	(Y) (later reported as being the chalcon butein)
ivory flavone	(I)

Most flower colour genes inherit in a tetrasomic way: Y and B are completely dominant in the simplex condition. A is cumulative from simplex to quadruplex and I is only partly cumulative. In addition, one or probably two tetrasomic inhibitors cumulatively inhibit the production of the yellow pigments (H). Y is epistatic to I. The main colours and genotypes are:

bayi = white	baYih = yellow
bayI = ivory	baYIh = yellow
bAyi = rosy-magenta	baYiH = cream to primrose
bAyI = bluish-magenta	bAYih = apricot
Bayi = rosy-purple	BaYih = scarlet
BayI = bluish-purple	BaYiH = crimson

Furthermore a very strong interaction occurs with regard to the production of the pigments, the yellow flavone suppresses the ivory flavone, while both flavones compete with the anthocyanins. The degree of interaction depends upon the total number and proportion of all flower colour genes present (Crane and Lawrence, 1956). The result is a very complicated genetical system for flower colour inheritance, the complexity of which is strongly enhanced by outcrossing caused by self-incompatibility.

For various reasons *Dahlia variabilis* is considered a promising species for mutation breeding. The high level of polyploidy and the great number of flower colour genes give a very wide variability after irradiation. Flower colour and other visible mutations can be observed in the material itself because of the high degree of heterozygosity. Moreover, one or a few characters in a leading cultivar obtained by cross breeding, could be changed or improved without changing the unique gene combination, as is the case in many asexually propagated plants. Cross breeding of a specific dahlia cultivar will never result in a genotype which is identical except for one desirable change, for instance flower colour. Therefore, mutation induction is the only means of improvement in those cultivars, which are prominent in the crop or in a specific group of it.

The choice of material is extremely difficult; the complicated genetics and the unknown genetic history of the modern cultivars make it impossible to choose cultivars with a known genetic constitution and, consequently, a predictable mutation spectrum. From the list on the previous page it is clear that white-flowering cultivars, based upon recessive genes only, are the least promising unless one counts upon chromosome aberrations with a dominant expression, which does not seem to be uncommon in polyploid material (Broertjes, 1976).

Therefore, the choice of material has to be made using other criteria, the overall quality of the cultivar in question being the main standard.

Dahlia is propagated from tubers which are harvested in the autumn; during the winter storage period eyes develop (Fig. 9a) and in early spring, in

Fig. 9. (A) Tuber of *Dahlia* in spring with well-developed buds (shoots). (B) Tuber of *Dahlia* immediately after lifting in autumn (buds invisible or even absent). (Photographs, ITAL, Wageningen.)

a heated greenhouse, several series of cuttings are made of the shoots formed at the crown of the tuber. Rooted cuttings are planted late in spring and their tubers are harvested in autumn.

Because of diplontic selection (e.g. the competition between mutated and normal cells) it is of vital importance to irradiate the eyes at the earliest possible stage of development, in order to give a mutated cell the best chance to take part in the formation of the shoot. Irradiation should therefore take place immediately after harvest, when no visible eyes can be detected on these so-called dormant tubers (Fig. 9b).

This method, viz., early irradiation combined with taking successive series of cuttings, has resulted in plants which are either completely normal or completely mutated. In some cases, half the plant was mutated and, only in a minority of cases, was a smaller part of the plant mutated.

It has not been ascertained whether the completely mutated plants obtained are the result of the propagation used or if they are due to a very small number of cells (one?) forming the eye. In this context, the question also arises, whether mutants are periclinal chimeras or solid mutants.

Although mutated characters are transferred via tubers to the following vegetative generation, a character will sometimes get lost, because the mutated sector did not take part in the formation of the new tubers. It could therefore be advisable to explore the possibilities of another way of vegetative propagation, viz., taking softwood cuttings after the detection of an interesting looking mutation. These rooted cuttings could undoubtedly be "hibernated" during the winter period in a heated greenhouse under long-day conditions and propagated in early spring. It is not yet known whether the tubers of such cuttings will develop eyes (as is the case in cuttings that are taken from the crown of the mother tuber). This subject needs further investigation.

Only very few authors report mutation breeding experiments with *Dahlia*, generally without (commercial) success (Das et al., 1974, 1975; De Loose, personal communication, 1976; Grabowska and Mynett, 1964b). Lantin and Decourtye (1970), who irradiated dormant tubers with γ-rays, obtained a number of mutants which have been released onto the market (L. Decourtye, pers. comm., 1977) (Table 7). A relatively large number of mutants was produced by Broertjes and Ballego (1967, 1968, 1969) and several of them have been introduced on the market. Similar results were obtained in other co-operative projects between Dahlia breeders and the Association Euratom—ITAL (see Table 7). These mutants were produced by the irradiation of freshly harvested tubers with 2—3 krad X-rays. Most mutants were improvements of the flower colour or the flower form of existing excellent cultivars.

Recent publications report a spontaneous deep rose magenta flower colour mutant in a white flowering cultivar (Singh et al., 1970).

An in vitro propagation technique has been developed by Asahira et al. (1975) which enables the breeder to regenerate plantlets from parts of a

TABLE 7

Commercial mutants of *Dahlia*

Name of new cultivar	Place and date of release (or approval) and name of principal worker(s) and Institute	Conditions and date of mutagenic treatment [Parent cultivar]	Main improved attributes of cultivar
Gracieuse	The Netherlands, 1966 Fa. Ballego en Zonen, Leiden *	Dormant tubers, 1—4 krad X-rays, 1963 [Salmon Rays]	Violet—mauve colour, spider-cactus type
Selection	The Netherlands, 1966 Fa. Ballego en Zonen, Leiden *	Dormant tubers, 1—4 krad X-rays, 1963 [Salmon Rays]	Larger blooms and longer stems than parent, but the same flower colour
Rotonde	The Netherlands, 1966 Fa. Ballego en Zonen, Leiden *	Dormant tubers, 1—4 krad X-rays, 1963 [Salmon Rays]	Vivid colour, true-pink flowering mutant with larger blooms than parent
Ornamental Rays	The Netherlands, 1966 Fa. Ballego en Zonen, Leiden *	Dormant tubers, 1—4 krad X-rays, 1963 [Salmon Rays]	Apricot coloured mutant with larger blooms than parent
Holland Jubilee	The Netherlands, 1967 Fa. Ballego en Zonen, Leiden *	Dormant tubers, 1—4 krad X-rays, 1963 [Arthur Godfrey]	Light orange throughout, bloom firmer and more regular than original cultivar
Progression	The Netherlands, 1967 Fa. Ballego en Zonen, Leiden *	Dormant tubers, 1—4 krad X-rays, 1963 [Arthur Godfrey]	Brick-red flower throughout
Rosy Mist	The Netherlands, 1967 Fa. Ballego en Zonen, Leiden *	Dormant tubers, 1-4 krad X-rays, 1963 [Arthur Godfrey]	Empire-rose flower throughout
Autumn Harmony	The Netherlands, 1967 Fa. Ballego en Zonen, Leiden *	Dormant tubers, 1—4 krad X-rays, 1963 [Arthur Godfrey]	Cadmium orange with scarlet centre
Explosion	The Netherlands, 1967 Fa. Ballego en Zonen, Leiden *	Dormant tubers, 1—4 krad X-rays, 1963 [Arthur Godfrey]	Blood red with bright yellow centre, massive bloom, full at centre with very fine habit

The Governor	The Netherlands, 1968 K. Maarse, Dzn. Jr, Aalsmeer *	Dormant tubers, 2 krad X-rays, 1963 [Authority]	Bloom 18 cm diam., plant 150 cm tall, deep copper-red flower colour
Dutch Visit	The Netherlands, 1968 Fa. Ballego en Zonen, Leiden *	Dormant tubers, 2 krad X-rays, 1963 [Arthur Godfrey]	Orange-red flower colour, bloom 30 cm diam. (rest of genotype unchanged)
Temptation	The Netherlands, 1968 Fa. Ballego en Zonen, Leiden *	Dormant tubers, 2 krad X-rays, 1963 [Arthur Godfrey]	Dark lacquer-red colour, giant flower
Wine Herald	The Netherlands, 1969 Fa. D. Bruidegom, Baarn *	Dormant tubers, 2 krad X-rays, 1967 [Holland Herald]	Improved flower colour (rest of genotype unchanged)
Raymond Smith	The Netherlands, 1970 Fa. Ballego en Zonen, Leiden *	Dormant tubers, 2 krad X-rays, 1965 [El Dorado]	Semi-cactus, light orange with large white tops
Motive	The Netherlands, 1971 Fa. Ballego en Zonen, Leiden *	Dormant tubers, 2 krad X-rays, 1963 [Arthur Godfrey]	Purple flowers
Maarse's Purple Wonder	The Netherlands, 1972 K. Maarse Dzn. Jr, Aalsmeer *	Dormant tubers, 1—2 krad X-rays, 1968 [Andries Wonder]	Purple flower colour
Maarse's Red Brown Wonder	The Netherlands, 1972 K. Maarse's Dzn. Jr, Aalsmeer *	Dormant tubers, 1—2 krad X-rays, 1968 [Andries Wonder]	Red-brown flower colour
Maarse's Golden Wonder	The Netherlands, 1972 K. Maarse Dzn. Jr, Aalsmeer *	Dormant tubers, 1—2 krad X-rays, 1968 [Andries Wonder]	Dark yellow flower colour
Adagio	France, 1970 B. Lantin and L. Decourtye, I.N.R.A. **, Etablissements TURC, Angers	Dormant tubers, 2—3 krad γ-rays, 1965 [Aztec]	Orange flower

TABLE 7 (continued)

Name of new cultivar	Place and date of release (or approval) and name of principal worker(s) and Institute	Conditions and date of mutagenic treatment [Parent cultivar]	Main improved attributes of cultivar
Allegro	France, 1970 B. Lantin and L. Decourtye, I.N.R.A., Etablissements TURC, Angers	Dormant tubers, 2—3 krad γ-rays, 1965 [Aztec]	Light purple flower
Altamira	France, 1970 B. Lantin and L. Decourtye, I.N.R.A., Etablissements TURC, Angers	Dormant tubers, 2—3 krad γ-rays, 1965 [Aztec]	Orange and red stripes (chimera)
Amalfi	France, 1970 B. Lantin and L. Decourtye, I.N.R.A., Etablissements TURC, Angers	Dormant tubers, 2—3 krad γ-rays, 1965 [Aztec]	Light and bright red flower
Annibal	France, 1970 B. Lantin and L. Decourtye, I.N.R.A., Etablissements TURC, Angers	Dormant tubers, 2—3 krad γ-rays, 1965 [Aztec]	Dark purple flower

* In co-operation with the Association Euratom—ITAL, Wageningen.
** I.N.R.A., Institut National de Recherches Agronomique.
Sources: lists prepared by the IAEA (1972, pp. 526—544) and *Mutation Breeding Newsletter*, 1 (1972) —9 (1977) and Association Euratom—ITAL, Wageningen.

mutated sector. For practical breeding, such sophisticated techniques do not seem necessary in *Dahlia* where the irradiation of 10 to 50 tubers and the subsequent production of a few hundred cuttings the next spring often produces sufficient genetic variation for successful selection.

Literature Dahlia variabilis

Asahira, T., Yamagata, H., Inagaki, M. and Osuga, S., 1975. Isolation and propagation in *Dahlia* by in vitro culture. In: Improvement of Vegetatively Propagated Plants through Induced Mutations, Tokai, 1974. IAEA, Vienna, pp. 1—23.

Broertjes, C., 1976. Mutation breeding of auto-tetraploid *Achimenes* cultivars. Euphytica, 25: 297—304.

Broertjes, C. and Ballego, J.M., 1967. Mutation breeding of *Dahlia variabilis*. Euphytica, 16: 171—176.

Broertjes, C. and Ballego, J.M., 1968. Addendum to: Mutation breeding of *Dahlia variabilis*. Euphytica, 17: 507.

Broertjes, C. and Ballego, J.M., 1969. Inconstances de mutations de Dahlia. Soc. Fr. Dahlia, Bull. Semest., 4: 6; 13—15.

Crane, M.B. and Lawrence, W.J.C., 1956. The Genetics of Garden Plants. McMillan and Co., London, pp. 86—90.

Das, P.K., Ghosh, P., Dube, S. and Dhua, S.P., 1974. Induction of somatic mutations in some vegetatively propagated ornamentals by gamma irradiation. Technology (Coimbatore, India), 11(2,3): 185—188.

Das, P.K., Dube, S., Ghosh, P. and Dhua, S.P., 1975. Mutation breeding in *Dahlia*. Indian J. Ornamental Hortic., 6(2): 3—8.

Grabowska, B. and Mynett, K., 1964b. The radiation of gamma Co^{60} rays on the dwarf *Dahlia*. Biul. Insto Hodowli Aklim. Rosl., 5—6: 17—26 (in Polish).

IAEA, 1972. Induced Mutations and Plant Improvement. IAEA, Vienna, pp. 526—544.

Lantin, B. and Decourtye, L., 1970. Obtention de types nouveaux chez le Dahlia par mutations provoquées. Lien Hortic., 7 (15 October). Rev. Fr. Maraîchage Hortic. Pépinière, No. 99: 5875—5878.

Lawrence, W.J.C., 1931a. The genetics and cytology of *Dahlia variabilis*. J. Genet., 24: 257—307.

Lawrence, W.J.C., 1931b. Mutation or segregation in the octoploid *Dahlia variabilis*. J. Genet., 24: 308—325.

Lawrence, W.J.C., 1942. The Origin of the Garden Dahlia, 1942. Hortus Printing Co., Burnley, London.

Singh, J.P., Arora, R.S., Dohare, S.R. and Sengupta, K., 1970. A spontaneous mutant for flower colour and shape in a white flowering Dahlia. Euphytica, 19: 261—262.

ENDYMION

Endymion hispanicus cultivar Excelsior (family Liliaceae) was used by one of the authors (Broertjes and Alkema, 1970) to study adventitious bud formation in vivo on detached leaf parts. The upper 10—15 cm of just mature, healthy leaves were either irradiated with X-rays or treated with a colchicine solution. The radiosensitivity was very high, 400 rad being an optimum dose, whereas colchicine sensitivity was very low, 1% during 7 h at room temperature not having any effect upon rooting, survival or bulblet production.

Since neither mutants nor tetraploids were found, it was impossible to decide whether the adventitious bulblets at the base of the leaf parts originate from a single (epidermal) cell. This method, however, would be a promising one if, in the future, better suited cultivars, for example, more heterozygous ones, were to be irradiated.

In co-operation with the Bulb Research Centre at Lisse, one other attempt was made — several cultivars were irradiated with a series of low doses, but no results were obtained (Alkema, 1974).

Literature Endymion

Alkema, H.Y., 1974. Mutatieveredeling bij bolgewassen door middel van radioactieve bestraling en andere methoden. Rapport 24, Laboratorium voor Bloembollenonderzoek, Lisse, p. 16.

Broertjes, C. and Alkema, H.Y., 1970. Mutation breeding in flower bulbs. In: First Int. Symp. on Flowerbulbs; Noordwijk/Lisse, Vol. II, pp. 407—411.

GLADIOLUS

During the centuries that *Gladiolus*, family Iridaceae, has been known, the interest in this plant has steadily increased, not only in Western Europe but also in some Mediterranean countries and in California, U.S.A. Its origin is hard to determine since large-scale cross breeding has been carried out among species and their hybrids, resulting in a large variety of flower forms and colours. Whether spontaneous sports are found in *Gladiolus* is not known to us, but on account of the apparently large heterozygosity, one would be inclined to expect them.

This is supported by several experiments in which it is demonstrated that mutations can easily be induced by radiation (Buiatti and Tesi, 1968; Buiatti et al., 1965b, 1967; Dryagina, 1964; Dryagina and Kazarinov, 1966; Dryagina et al., 1967; Grabowska and Mynett, 1970; Grabowska, 1972; Misra, 1976; Moës, 1966, 1969; Sheehan and Sagawa, 1959; and also Iba et al., 1964, 1965; Meshitsuka et al., 1963). In particular, dormant corms of the red-flowering tetraploid cultivar Hawaii was irradiated with doses of 2.5—15 kR of γ-rays, and a large variety of flower colour mutations (e.g. different shades of purple, crimson, scarlet, pink and white, as well as white edged petals, dark edged petals and white medium nerve on lower petals) was obtained. Ethylene methylsulphonate (EMS) treatment, on the other hand, did not produce mutations at all (Moës, 1966).

An interesting observation has been that the mutation frequency is higher in vM_2 (up to 60%) than in vM_1, whereas the rate of chimeras is much lower in vM_2 and completely mutated spikes are often observed. The phenomenon was studied extensively by Buiatti and Tesi (1968) and Buiatti et al. (1965b, 1967). They studied mutation frequency and mutation sector size in flower spikes in vM_1 plants grown from primary bulbs, in vM_2 plants from secondary bulbs and in vM_3 plants from the bulbils of secondary bulbs. A drastically larger sector size was found in vM_2 and vM_3, whereas the mutation frequency was lower than in vM_1. Furthermore, no correlation was found between mutations in these respective generations, though this is not surprising since mutations are induced independent of each other in the apices of the different buds. No correlation has been found in tulip either, for instance. From the results of their experiments using mainly dormant corms of triploid cultivars irradiated with 4—7.5 kR, they also conclude that the number of initial cells giving rise to second and third generation spikes is either very small or even one. On this basis, they conclude that the lower mutation frequency in vM_2 and vM_3 can be logically explained since a greater number of cells in vM_1 automatically allows a larger mutation frequency per spike. The results of Moës (1966, 1969) however, conflict with this conclusion.

From the articles mentioned above, it was not clear whether some of the mutations observed could be ascribed to uncovering existing chimeras (if they exist in *Gladiolus*) or whether the so-called wholly mutated plants were solid, non-chimeral mutants originating from one mutated cell or complete periclinal chimeras.

Various chemical mutagens have been used extensively by Dryagina (1975a) to stimulate germination, growth, flowering and vegetative reproduction. She treated corms with chemicals which included di-methyl sulphate (DMS), N-nitroso-N-methyl urethane (NMU) and nitroso di-methyl urea (NDMU). In addition, forms with improved flower and leaf colour or productivity, or resistance to lodging or diseases were produced by treating seeds. Whether these variants are mutants or recombinations, however, is not clear. One of them was put on the market, namely cv. Girl Friend (Dryagina, 1975b).

Some authors (Buiatti et al., 1969, 1970) have used somatic mutations in *Gladiolus* for studies on development. They found a decreasing mutation frequency with increasing flower order (from base to top) and, in addition, an increasing sector size which was also dependent on dose and environmental conditions. They consider their results as being in agreement with the selection theory of Gaul (see Chapter 2).

Our own experience is restricted to the irradiation of corms of *Gladiolus tubergenii*, cv. Fair Lady, with X-rays. The optimum dose was approx. 4 krad. No mutations have been reported yet.

In conclusion, it is clear that *Gladiolus* is a promising species for mutation breeding if the right genotype is irradiated and selection is carried out in the second or third vegetative generation of wholly mutated plants. Because of the desirability of using second or third generation plants, it could be of importance that the possibility of propagating *Gladiolus* in vitro (Hussey, 1975a, 1977; Hussey and Wyvill, 1972; Simonsen and Hildebrandt, 1971; Ziv et al., 1970) might speed up the whole process of selection and subsequent propagation.

Literature Gladiolus

Buiatti, M. and Tesi, R., 1968. *Gladiolus* improvement through radiation induced somatic mutation. Publ. No. 1, Laboratorio di Mutagenesi e differenziamento del C.N.R., Pisa, pp. 1–6.

Buiatti, M., Ragazzini, R. and Tognoni, F., 1965b. Effects of gamma irradiation on *Gladiolus*. Radiat. Bot., 5: 97–98.

Buiatti, M., Molino, M. and Tesi, R., 1967. Prove di miglioramento genetico mediante radiazioni ionizzanti in una pianta da fiore a propogazione vegetativa (gladiolo). Riv. Ortoflorofruttic. Ital., 51(6): 616–627.

Buiatti, M., Tesi, R. and Molino, M., 1969. A developmental study of induced somatic mutations in *Gladiolus*. Radiat. Bot., 9(1): 39–49.

Buiatti, M., Baroncelli, S., Tesi, R. and Boscariol, P., 1970. The effect of environment on diplontic selection in irradiated *Gladiolus* corms. Radiat. Bot., 10(6): 531–538.

Dryagina, I.V., 1964. On the effects of ionizing radiation on *Gladiolus*. Nucl. Sci. Abstr., 19 (1965), No. 38430.

Dryagina, I.V., 1975a. Forms of gladiolus produced by treatment with chemical mutagens. Moscow Univ. Biol. Sci. Bull., 30(1–2): 109–112.

Dryagina, I.V., 1975b. Gladioli forms obtained as a result of treatment with chemical mutagens. Vestn. Mosk. Univ., 6(2): 57–61.

Dryagina, I.V. and Kazarinov, G.E., 1966. Effects of ionizing radiation on gladiolus tubers and seeds. Nucl. Sci. Abstr., 20, No. 35290.

Dryagina, I.V., Kazarinov, G.E. and Kalinichenko, I.M., 1967. Mutations in fruit and ornamental plants induced by the action of neutrons and γ-rays Co^{60}. In: Symp., Mutational Process. Mechanics of Mutation and Induction Factors. G. Mendel Memorial Symp., Prague, 1965, pp. 131–135.

Grabowska, B., 1972. Attempts to induce mutations in gladiolus (*Gladiolus hybr.* hort.) by ^{60}Co gamma rays. Biul. Inst. Hodowli Aklim. Rosl., (5–6): 267–268 (in Polish with English summary).

Grabowska, B. and Mynett, K., 1970. Induction of changes in garden gladiolus (*Gladiolus hybr.* hort.) under the influence of gamma rays ^{60}Co. Biul. Inst. Hodowli Aklim. Rosl., (1–2): 85–88 (in Polish with English summary).

Hussey, G., 1975a. Totipotency in tissue explants and callus of some members of the Liliaceae, Iridaceae and Amaryllidaceae. J. Exp. Bot., 26: 253–262.

Hussey, G., 1977. In vitro propagation of *Gladiolus* by precocious axillary shoot formation. Sci. Hortic., 6: 287–296.

Hussey, G. and Wyvill, C., 1972. Propagation of bulbous species by tissue culture. Annu. Rep. John Innes Hortic. Inst., 68: 64–66.

Iba, S., Matsubara, H., Inokoshi, Y., Oka, M. and Meshitsuka, G., 1964. Effects of gamma irradiation on gladiolus. II. Tokyo Metrop. Isot. Res. Cent. Annu. Rep., 3: 109–113.

Iba, S., Matsubara, H., Oka, M., Meshitsuka, G., Matsumoto, H. and Kudo, T., 1965. Effects of gamma irradiation on gladiolus. III. Tokyo Metrop. Isot. Res. Cent. Annu. Rep., 4: 85–88.

Meshitsuka, G., Oka, M., Iba, S., Matsubara, H. and Inokoshi, Y., 1963. Effects of gamma irradiation on gladiolus. Tokyo Metrop. Isot. Cent. Annu. Rep., 2: 169–173.

Misra, R.L., 1976. A colour mutant in *Gladiolus* L. variety Picardy. Sci. Cult., 42(11): 569–571.

Moës, A., 1966. Mutations induites chez le glaïeul (*Gladiolus*). Bull. Rech. Agron. Gembloux, 1: 76–95.

Moës, A., 1969. Mutation breeding in tetraploid gladiolus (*Gladiolus gandavensis* V.H.). In: G.C. Chisa and G. Haussmann (Editors), Fifth Eucarpia Congress, Milan, 1968. Genet. Agrar., 23: 176–179.

Sheehan, T.J. and Sagawa, Y., 1959. The effects of gamma radiation on *Chrysanthemum* and *Gladiolus*. Proc. Fl. State Hortic. Soc., 72: 388–391.

Simonsen, J. and Hildebrandt, A.C., 1971. In vitro growth and differentiation of *Gladiolus* plants from callus cultures. Can. J. Bot., 49(10): 1817–1821.

Ziv, M., Halevy, A.H. and Shilo, R., 1970. Organs and plantlets regeneration of *Gladiolus* through tissue culture. Ann. Bot. (London), 34: 671–676.

HYACINTHUS

Hyacinthus, family Liliaceae, has a long history as an ornamental. Although there are few species, there is a large variety of single and double flowering cultivars with flower colours ranging over white, pink, red, blue and yellow.

Hyacinth is mainly propagated by scooping large bulbs, maintaining them at an appropriate temperature and air humidity and allowing them to produce a large number of bulblets at the base of the bulb scales. The existence of such an adventitious bud technique, and the fact that hyacinth exhibits polyploidy bode well for mutation breeding, as indicated by the occurrence of numerous sports.

One would therefore expect that mutation breeding has been contributing to the improvement of the crop for some time. This, for one reason or another, is not the case despite the relatively large number of experiments carried out, though they were to begin with only by one worker, namely De Mol (or De Mol van Oud Loosdrecht as he later called himself) (1926, 1931, 1933, 1934, 1937a,b, 1940, 1953). A complete list of his publications is given by Thamm (1956) in an introduction to an article by De Mol van Oud Loosdrecht (1956).

The main conclusions from his experiments, which unfortunately were sometimes carried out on too small a scale, are that flower colour mutates from blue, violet and red (to yellow) to white and the flower itself from single to double. Changes in colour intensity were mainly found in triploid cultivars. Light coloured cultivars easily mutated to white, whereas dark coloured ones produced the greatest variability in flower colour. The optimum dose seems to be around 300 rad of X-rays. No commercial cultivars are known to have resulted from this work.

In several co-operative programmes between the Association Euratom—ITAL and institutes or private plant breeders, large numbers of hyacinth bulbs have been irradiated, including the cultivars Amsterdam (100 bulbs; 250 rad X-rays), Anna Maria (100 bulbs; 250 rad X-rays), Blue Bird (530 bulbs; 350 rad X-rays), Blue Jacket (200 bulbs; 300 rad X-rays) and Pink Pearl (200 bulbs; 250 rad X-rays). So far, not a single mutation has been found.

In a co-operative project (Institute of Horticultural Plant Breeding, Wageningen, and the Association Euratom—ITAL, Wageningen) the cultivars Jan Bos, Ostara and Pink Pearl were irradiated with 300—500 rad X-rays or 100—

600 rad fast neutrons. During the experiments, it was also noticed that the radiosensitivity of the bulbs just before scooping was high, or, in other words, the regeneration capacity after scooping irradiated bulbs suffered very much, even from doses as low as 300 rad X-rays; whereas the radiosensitivity is much less if bulbs are irradiated a few weeks after scooping, when small bulblets have been formed (Fig. 10). However, even using the latter method no signs of mutations have been found, a fact which cannot be explained and is more or less unexpected.

One method which might be worth trying is the formation of adventitious bulblets at the base of leaf parts, the 10—15-cm upper part of healthy just mature leaves. Provided that one can effectively control diseases so that the rotting of the leaves can be prevented until rooting, bulblets are easily and

Fig. 10. View at the basal part of a hyacinth bulb which has developed adventitious bulblets after wounding and storing under the proper conditions. (Photograph, Bulb Research Centre, Lisse, The Netherlands.)

ery rapidly produced, as in *Muscari* and *Endymion* (Broertjes and Alkema, .970). In vitro propagation could also be tried, as soon as practically appli- able (Asjes et al., 1974; Hussey, 1975b; Pierik and Post, 1975; Pierik and ;teegmans, 1975d).

,iterature Hyacinthus

ısjes, C.J., Bunt, M.H. and Van Slogteren, D.H.M., 1974. Production of hyacinth mosaic virus-free hyacinths and lily symptomless virus-free lilies by meristem tip culture. Acta Hortic., 36: 223—228a.

›roertjes, C. and Alkema, H.Y., 1970. Mutation breeding in flower bulbs. In: First Int. Symp. on Flowerbulbs; Noordwijk/Lisse, Vol. II. pp. 407—411.

›e Mol, W.E., 1926. Heteroploidy and somatic variation in the Dutch flowering bulbs. Am. Nat., LX: 334—339.

›e Mol, W.E., 1931. Somatische Variation der Blumenfarbe der Hyazinthe durch Röntgenbestrahlung und andere äusseren Umstände. Z. Indukt. Abstamm. Vererbungsl., 59: 280—283.

›e Mol, W.E., 1933. Mutation sowohl als Modifikation durch Röntgenbestrahlung und die "Teilungshypothese". Cellule, 42: 149—162.

›e Mol, W.E., 1934. Drei aufeinanderfolgende Jahre der Mutation und Modifikation bei Hyazinthen, nach kaltem Aufbewahren während des Zellteilungsprozess zur Blütenbildung. Cellule, 43: 131—146.

›e Mol, W.E., 1937a. De invloed van Röntgenbestraling op bloembolgewassen (hyacinten en tulpen). Philips Tech. Tijdschr., 2(11): 321—328.

›e Mol, W.E., 1937b. Een vergelijking tussen het somatisch muteren der bloemkleur bij diploïde en tetraploïde variëteiten van *Hyacinthus orientalis*. Agricultura (Leuven), 40: 216—228.

›e Mol, W.E., 1940. Über Modifikationen, Mutationen und den Parallelismus dazwischen, im Zusammenhang mit Kältebehandlung von Hyazinthen. Genetica (The Hague), 22(1-2-3): 231—260.

›e Mol, W.E., 1953. X-raying of hyacinths and tulips from the beginning, before thirty years (1922) till today (1952). Jpn. J. Breed., 3(1): 1—8.

›e Mol van Oud Loosdrecht, W.E., 1956. Der Einfluss der Röntgenstrahlen auf die Entwicklung des Pollens und der Sprosse bei Tulpen. Bayerischer Landwirtschaftverlag, Bonn, München, Wien, pp. 5—128.

[ussey, G., 1975b. Propagation of hyacinths by tissue culture. Sci. Hortic., 3(1): 21—29.

ierik, R.L.M. and Post, A.J.M., 1975. Rapid vegetative propagation of *Hyacinthus orientalis* L. in vitro. Sci. Hortic., 3(3): 293—297.

ierik, R.L.M. and Steegmans, H.H.M., 1975d. Effect of auxins, cytokinins, gibberellins, abscisic acid and ethephon on growth of bulblets on excised bulb scale segments of hyacinth. Physiol. Plant., 34: 14—17.

'hamm, 1956. Rückblick auf die bisherige wissenschaftliche Arbeit des Amsterdamer Biologen Willem Eduard De Mol van Oud Loosdrecht. Bayerischer Landwirtschaftverlag, Bonn, Munich, Vienna, 128 pp.

RIS

Apart from a publication about the non-genetic effects of gamma radia- ion (Halevy and Shoub, 1965), not much has been published concerning nutation breeding in *Iris*, family Iridaceae, except for in a more general paper (Konzak and Randolph, 1956) and one by Hekstra and Broertjes (1968).

The latter irradiated dormant bulbs of the sterile cultivar Wedgwood with various doses of X-rays. The optimum moment of irradiation turned out to be directly after lifting the bulbs in August, when the growing point is at its youngest possible stage of development and radiosensitivity is lowest. The optimum dose at that time is approximately 1 krad.

Several flower colour and other mutations were obtained but none was better than some of the spontaneous mutants obtained earlier. This is partly due to the time-consuming stabilization of the mericlinal ("sectorial") mutants. This problem could be overcome by using in vitro propagation techniques, which would speed up the procedure and, moreover, may sometimes produce solid mutants which, in their turn, can be multiplied and rapidly judged (Fujino et al., 1972; Hussey, 1976; Hussey and Hilton, 1975b; Meyer, 1974; Meyer et al., 1975; Reuther, 1974).

Literature Iris

Fujino, M., Fujimura, T. and Hamada, K., 1972. Multiplication of Dutch Iris (*Iris hollandica*) by organ culture. J. Jpn Soc. Hortic. Sci., 4: 66—71.

Halevy, A.H. and Shoub, J., 1965. The effects of gamma-irradiation and storage temperature on the growth, flowering and bulb yield of Wedgwood Iris. Radiat. Bot., 5(1): 29—37.

Hekstra, G. and Broertjes, C., 1968. Mutation breeding in bulbous iris. Euphytica, 17: 345—351.

Hussey, G., 1976. Propagation of Dutch iris by tissue culture. Sci. Hortic., 4(2): 163—167.

Hussey, G. and Hilton, J., 1975b. Rapid propagation of Dutch iris by tissue culture. John Innes Inst. Annu. Rep., 66: 55—56.

Konzak, C.F. and Randolph, L.F., 1956. Radiation and iris breeding. Bull. Am. Iris Soc., 142: 68—76.

Meyer, Jr, M.M., 1974. Propagation of tall bearded *Iris* by in vitro techniques. Hortic. Sci., 9(3): 271 (abstract).

Meyer, Jr, M.M., Fuchigami, L.H. and Roberts, A.N., 1975. Propagation of tall bearded irises by tissue culture. Hortic. Sci., 10(5): 479—480.

Reuther, G., 1974. Die Anwendung verschiedener Methoden der Embryokultur bei der Iriszüchtung. Proc. Int. Hortic. Congr., 19(1A): 57.

LILIUM

The genus *Lilium* (fam. Liliaceae) is an important crop with numerous commercially grown cultivars. They originate from Europe, Asia and North and South America. By intercrossing many of the 87 known species, several hundreds of cultivars have been produced. Some are propagated by seed but most are propagated asexually, for instance, by using bulb scales and recently by in vitro methods as well.

Because of the unsurveyable variety of cultivars, a botanical classification was developed by the Lily Committee of the Royal Horticultural Society

Fig. 11. (A) Bulb scale of *Lilium* with rooting adventitious bulblets. (B) Well-developed and rooted bulblets on a bulb scale of *Lilium*. (Photographs, ITAL, Wageningen.)

TABLE 8

Commercial mutants of *Lilium*

Name of new cultivar	Place and date of release (or approval) and name of principal worker(s) and institute	Conditions and date of mutagenic treatment [Parent cultivar]	Main improved attributes of cultivar
Mies Bouwman	The Netherlands, 1977 A.J. Bischoff-Tulleken, Wieringerwerf *	X-rays, 1968 [Tabasco]	Orange flower colour, excellent forcing qualities
TX 68-1	The Netherlands, 1977 A.J. Bischoff-Tulleken, Wieringerwerf *	X-rays, 1968 [Tabasco]	Orange flower colour, excellent forcing qualities

* In co-operation with the Association Euratom—ITAL, Wageningen.

Sources: IAEA (1972, pp. 526—544) and *Mutation Breeding Newsletter*, 1 (1972) —9 (1977) and Association Euratom—ITAL, Wageningen.

and the American lily specialist Jan de Graaf, which classifies the cultivars and species (Van Raalte, 1967).

Most publications about *Lilium* deal with the fact that this plant has few, but large chromosomes, which makes it accessible for (cyto)genetic studies concerning, for example, radiosensitivity of meiotic stages, chromosome breakage, chiasmata and crossing over, dominant lethals, mutation induction and polyploidization (Bowen and Sparrow, 1954; Brown and Cave, 1953, 1954a,b; Brown and Zohary, 1953, 1955; Crouse, 1954; Emsweller, 1955; Iizuka and Ikeda, 1963, 1968; Loh and Cooper, 1966; Mitra, 1958).

Although the heterozygosity of most *Lilium* hybrids and the generally easy vegetative propagation (in vivo with bulb scales (Fig. 11) and in vitro) are two important positive circumstances for successful mutation breeding, few commercial results have (yet) been reported. As mentioned above (Broertjes, 1969a; Broertjes and Alkema, 1970), tens of thousands of bulb scales have been irradiated in programmes organized in co-operation between the Association Euratom—ITAL, Wageningen, and private plant breeders in The Netherlands. Several cultivars produced hundreds of solid mutants, a few of which were very promising. In addition, many complete tetraploids were produced by colchicine treatment of fresh bulb scales. (These results, complete mutants and polyploids, are in contradiction to those obtained by Emsweller and Stewart (1951) who report mixoploids after colchicine treatment. The moment of colchicine application, however, was not mentioned by these authors.)

So far, few commercial results can be reported, mainly because of virus problems (Table 8). Several promising mutants had to be made virus-free through in vitro propagation before more plants were obtained by propagation, again in vitro (Asjes et al., 1974; Zandbergen, 1975). They had to be tested before one of them, cv. Mies Bouwman, was put on the market (Van Nes, 1976; H.J. Bischoff-Tulleken, personal communication, 1977). More mutants can be expected by the combined use of fast in vitro propagation techniques, such as those described by Simmonds and Cumming (1976a,b) who have calculated that 1 g of callus can produce 6×10^{12} true to type plants, and the irradiation of virus-free material of excellent cultivars.

Literature Lilium

Asjes, C.J., Bunt, M.H. and Van Slogteren, D.H.M., 1974. Production of hyacinth mosaic virus-free hyacinths and lily symptomless virus-free lilies by meristem tip culture. Acta Hortic., 36: 223—228a.

Bowen, C.C. and Sparrow, A.H., 1954. Radiosensitivity of several meiotic stages of *Lilium*. Genetics, 39: 960.

Broertjes, C., 1969a. Mutation breeding of vegetatively propagated crops. In: G.C. Chisci and G. Haussmann (Editors), Proc. 5th Eucarpia Congress, Milan, 1968. Genet. Agrar., 23: 139—165.

Broertjes, C. and Alkema, H.Y., 1970. Mutation breeding in flower bulbs. In: First Int. Symp. on Flowerbulbs, Noordwijk/Lisse, Vol. II. pp. 407—411.
Brown, S.W. and Cave, M.S., 1953. Induced dominant lethality in *Lilium*. Proc. Natl Acad. Sci. U.S.A., 39: 97—102.
Brown, S.W. and Cave, M.S., 1954a. The detection and nature of dominant lethals in *Lilium*. I. Effects of X-rays on the heritable component and functional ability of the pollen grain. Am. J. Bot., 41: 455—469.
Brown, S.W. and Cave, M.S., 1954b. The detection and nature of dominant lethals in *Lilium*. II. Cytological abnormalities in ovules after pollen irradiation. Am. J. Bot., 41: 469—483.
Brown, S.W. and Zohary, D., 1953. Chiasmata and crossing-over in *Lilium formosanum*. Genetics, 38: 657.
Brown, S.W. and Zohary, D., 1955. The relationship of chiasmata and crossing-over in *Lilium formosanum*. Genetics, 40: 850—873.
Crouse, H.V., 1954. X-ray breakage of lily chromosomes at first meiotic metaphase. Science, 119: 485—487.
Emsweller, S.L., 1955. Improvement and propagation of lilies. Proc. Int. Hort. Congr., 14: 917—928.
Emsweller, S.L. and Stewart, R.N., 1951. Diploid and tetraploid pollen mother cells in lily chimeras. Proc. Am. Soc. Hortic. Sci., 57: 414—418.
IAEA, 1972. Induced Mutations and Plant Improvement. IAEA, Vienna, pp. 526—544.
Iizuka, M. and Ikeda, A., 1963. Effects of X-ray irradiation on *Lilium formosanum*. Proc. Am. Soc. Hortic. Sci., 82: 508—516.
Iizuka, M. and Ikeda, A., 1968. Induction of polyploidy in *Lilium formosanum* Wallace. Jpn. J. Genet., 43(2): 95—103.
Loh, P.T. and Cooper, D.C., 1966. Effects of gamma-irradiation on the pollen of *Lilium regale*. Can. J. Genet. Cytol., 8(1): 152—164.
Mitra, S., 1958. Effects of X-rays on chromosomes of *Lilium longiflorum* during meiosis. Genetics, 43(6): 771—789.
Simmonds, J.A. and Cumming, B.G., 1976a. Propagation of *Lilium* hybrids, I. Dependence of bulblet production on time of scale removal and growth substances. Sci. Hortic., 5: 77—83.
Simmonds, J.A. and Cumming, B.G., 1976b. Propagation of *Lilium* hybrids, II. Production of plantlets from bulb-scale callus cultures for increased propagation rates. Sci. Hortic., 5: 161—170.
Van Nes, C.R., 1976. Veel belangstelling voor sortiment snijlelies. Vakbl. Bloemisterij, 31(46): 24—25.
Van Raalte, D., 1967. Handboek der Bloemisterij, 2. Born Periodieken N.V., Assen, Amsterdam, 4th edn., 341 pp.
Zandbergen, J.K., 1975. In 1977 virusvrije lelies op de markt. Vakbl. Bloemisterij, 30(34): 26—27.

MUSCARI

Apart from our own work, no reports have been found concerning mutation breeding experiments with *Muscari*, family Liliaceae.

As was reported by Broertjes and Alkema (1970), *Muscari* can be propagated in vivo by leaf pieces which root readily and produce adventitious bulblets at the basal end of the rooted leaf (Fig. 12). This system has been used to study whether such adventitious bulblets originate from a single cell (C. Broertjes, unpublished data). To this end leaf pieces of *Muscari armenia-*

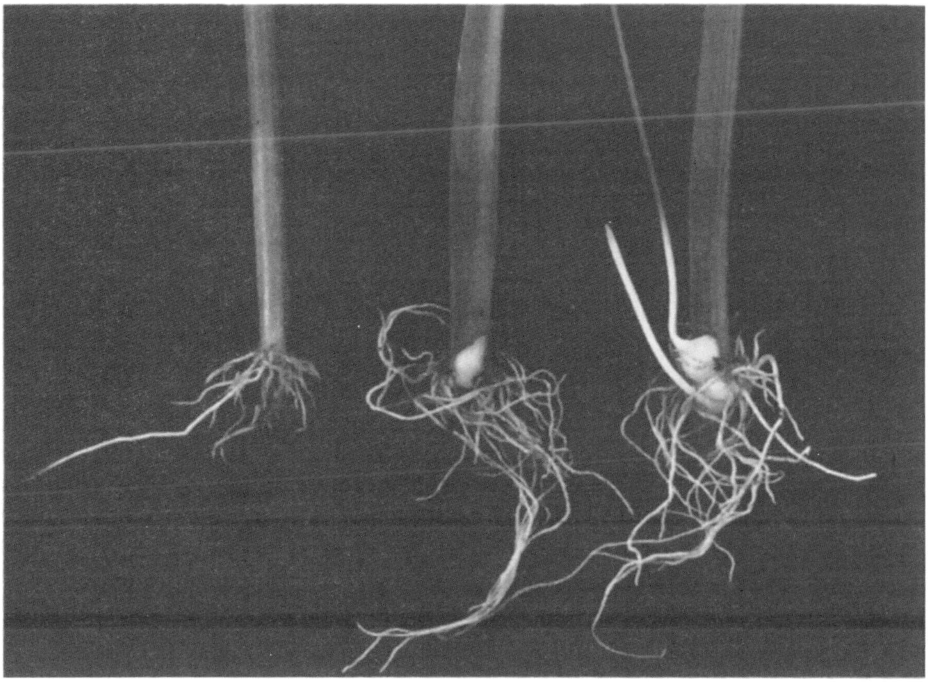

Fig. 12. Leaf segments of *Muscari,* 4, 8 and 12 weeks after detaching (from left to right). Adventitious bulblets develop, root and grow quickly. (Photograph, ITAL, Wageningen.)

cum were irradiated with a range of X-ray doses. The optimum dose turned out to be 1—1.5 krad of X-rays for maximal speed of rooting and greatest possible number of bulblets per leaf. Colchicine was also applied at a concentration of 0.2—0.4% for 7 h at room temperature.

Among the approximately 1 500 bulblets obtained from the various doses, 170 plants looked different to the control and 42 of them were selected for further propagation. No clear relationship between dose and mutation frequency could be observed. The mutations, which all had a solid, non-chimeral appearance, concerned time of flowering, size, form and colour of the inflorescence, growth habit, and leaf characters. After secondary selection on a clonal basis, one or two of the mutants, which grew true by normal vegetative propagation, seem to be of commercial interest and are probably autotetraploids resulting from the above-mentioned colchicine treatment. They have a very sturdy growth habit, an extremely large inflorescence and a thick pedicel which might make it suitable for early forcing.

In co-operation with the Bulb Research Centre, bulbs of *Muscari armeniacum* were irradiated with X-rays. The optimum dose lies between 3 and 5 krad. No mutants have been observed.

In vitro propagation of *Muscari* may also be a tool for future mutation induction and/or fast clonal propagation since Hussey and Wyvill (1972) succeeded in producing plantlets from callus.

Literature Muscari

Broertjes, C. and Alkema, H.Y., 1970. Mutation breeding in flower bulbs. First Int. Symp. on Flowerbulbs; Noordwijk/Lisse, Vol. II, pp. 407—411.

Hussey, G. and Wyvill, C., 1972. Propagation of bulbous species by tissue culture. Annu. Rep. John Innes Hortic. Inst., 63: 64—66.

NARCISSUS

Spontaneous somatic variation in cultivars of *Narcissus*, family Amaryllidaceae, have been reported by De Mol (1926) and by Zandbergen (1964) but the number of sports available commercially is restricted to very few or even none. This is surprising since daffodils have existed as a cultivated plant for a long time and have been subjected to cross-breeding programmes ever since. The heterozygosity, therefore, must be fairly large, whereas polyploidy seems to be restricted. Consequently, the low spontaneous mutation frequency cannot be explained and several attempts have been made to induce mutations, using a variety of cultivars (C. Broertjes, unpublished data). Amongst the collar daffodils, several hundreds of bulbs of the cultivars Chanterelle and Poesy were irradiated with the optimum dose of 700—800 rad of X-rays, but no mutations were observed. In the *Narcissus jonquille* cultivars Baby Moon and New Baby, with an optimum dose of 400—500 rad X-rays, no mutations were obtained either, except for a few variants with greener flowers.

Since the reproduction of daffodils is too slow and seems to be the bottleneck the development of an adventitious bud technique to speed up the process and to provide larger chances of survival for a mutated cell was attempted. The one-scale (Luyten, 1935) and the two-scale methods (Traub, 1935) were tried. Scales of cv. Golden Harvest, irradiated with 400 or 500 rad X-rays, produced plants in the second season which, when they flowered after 3 years, showed two aberrant plants (Alkema, 1974). This might be an indication that commercial mutation breeding becomes feasible as soon as a reliable and fast in vivo or in vitro propagation technique has been developed.

Literature Narcissus

Alkema, H.Y., 1974. Mutatie veredeling bij bolgewassen door middel van radioactieve bestraling en andere methoden. Rapport 24, Laboratorium voor Bloembollenonderzoek, Lisse, pp. 13—14.

De Mol, W.E., 1926. Heteroploidy and somatic variation in the Dutch flowering bulbs. Am. Nat., 40: 334—339.

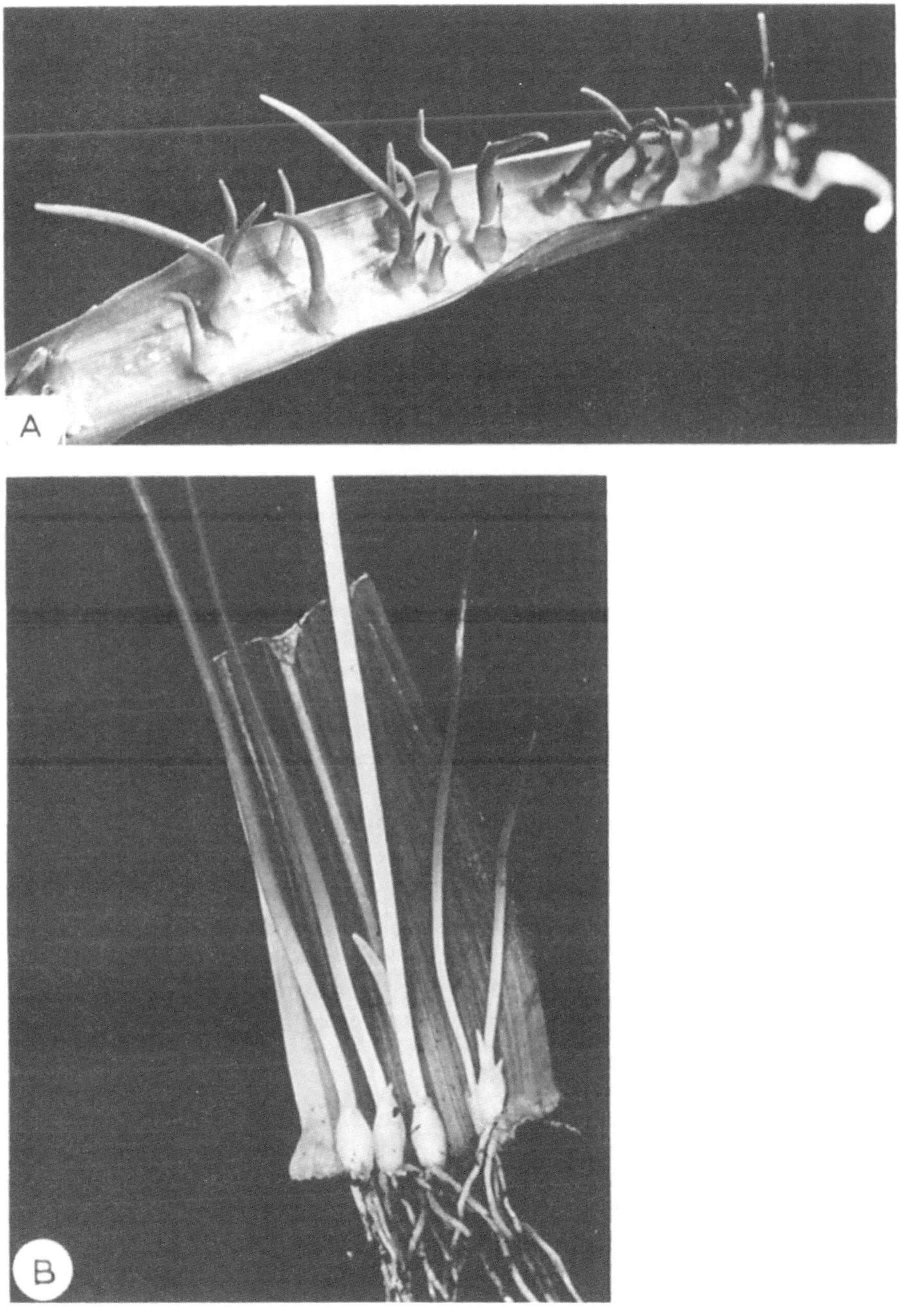

Fig. 13. (A) Leaf of *Ornithogalum thyrsoïdes* with (adventitious?) growing bulblets. (B) Leaf segment of a variegated radiation-induced mutant of *Ornithogalum thyrsoïdes* with well-developed rooted and growing adventitious bulblets, approximately 5 weeks after detaching. Note completely chlorophyll-less plantlet in the centre, which developed on a white sector of the leaf. (Photographs, ITAL, Wageningen.)

Luyten, I., 1935. Vegetative propagation of *Hippeastrum*. Yearb. Am. Amaryllus Soc., 2: 115—122.
Traub, H.P., 1935. Propagation of *Amaryllus* by stem cuttings. Yearb. Am. Amaryllus Soc., 2: 123—126.
Zandbergen, M., 1964. Why should daffodils not have a split cup? Daffodil Tulip Year Book, 29: 82—87.

ORNITHOGALUM

Concerning mutation breeding in *Ornithogalum*, family Liliaceae, no reports have been found in the literature. *O. thyrsoïdes* Jacq. has been used by ourselves to investigate whether adventitious bulblets formed at the base of rooted leaf parts originate from a single cell (C. Broertjes, unpublished data). Leaf parts readily produce bulblets within an extremely short period (Fig. 13). The number of bulblets per leaf part is dependent on the length of the leaf part (Table 5, page 36).

Leaf parts were irradiated with a series of X-ray doses (300—1 000 rad) or fast neutrons (160—500 rad). Colchicine treatment was also applied, at a concentration of 0.1—0.4% for 7 h at room temperature. The optimum doses for X-rays and fast neutrons were, respectively, 1 krad or somewhat higher and 330—400 rad. The production of bulblets at these doses was then about 50% as compared to the control.

Among the several thousands of plants observed only very few slightly aberrant types were found, including smaller plants, smaller flowers, different flower forms, flowers whose white colour had a greenish tinge and, in the colchicine group, a few larger plants. None of the mutants showed any sign of chimerism, even after clonal propagation. Since none of the mutants were of commercial interest, the experiments were discontinued.

In the future, in vitro propagation might be used as a tool to improve cultivars of *Ornithogalum* (Hussey and Wyvill, 1972).

Literature Ornithogalum

Hussey, G. and Wyvill, C., 1972. Propagation of bulbous species by tissue culture. Annu. Rep. John Innes Hortic. Inst., 63: 64—66.

POLYANTHES

Tuberose, *Polyanthes* (*Polianthes*) *tuberosa* L., family Amaryllidaceae, is a bulbous flower crop, popular in India. It is sterile and its natural variability is fairly restricted. Only white flowered, single and double types are known.

Gamma-ray treatments of bulbs with a series of doses revealed the optimum dose to be around 2 krad. No flower colour mutations have been found (Younis and Borham, 1975). A non-chimeric white flower mutant and a yellow margined leaf mutant were isolated after repeated vegetative propagation (Gupta et al., 1974).

Literature Polyanthes

Gupta, M.N., Sumiran, R. and Shukla, R., 1974. Mutation breeding of tuberose (*Polianthes tuberosa* L.). In: Symp., Use of Radiations and Radioisotopes in Studies of Plant Productivity, Pantnagar, pp. 169—179.
Younis, S.E.A. and Borham, J.H., 1975. The effects of gamma-irradiation on *Polianthes tuberosa*. Egypt. J. Bot., 18(1,3): 205—217 (not consulted).

RANUNCULUS

The only experience with *Ranunculus*, family Ranunculaceae, is the outcome of a single irradiation experiment with a series of X-ray doses of the cultivar Barbaroux. The LD_{50} dose is approximately 10 krad. The experiments were discontinued because the material died (Alkema, 1974).

New projects could perhaps profit from the fact that an in vitro propagation technique has been recently developed (Dorion et al., 1975).

Literature Ranunculus

Alkema, H.Y., 1974. Mutatieveredeling bij bolgewassen door middel van radioactieve bestraling en andere methoden. Report 24, Laboratorium voor Bloembollenonderzoek, Lisse, p. 15.
Dorion, N., Chupeau, Y. and Bourgin, J.P., 1975. Isolation, culture and regeneration into plants of *Ranunculus sceleratus* L. leaf protoplasts. Plant Sci. Lett., 5: 325—331.

SCILLA

No mutation breeding has been reported in *Scilla*, family Liliaceae, except preliminary studies with *Scilla sibirica*, in which bulbs were irradiated with a series of X-ray doses, before being scooped. The optimum dose seems to be between 500 and 1 000 rad. Since the bulbs did not form adventitious bulblets the experiments were discontinued (Alkema, 1974).

Scilla sibirica cv. Spring Beauty has been used to investigate the possibilities of an in vivo adventitious bud technique (C. Broertjes, unpublished data). Leaf parts were irradiated with X-rays or treated with colchicine. The optimum dose lies around 350 rad, if the number of adventitious bulblets at the basal end of the rooted leaf part is used as the criterion. Among the hundreds of adventitious plants neither mutants nor polyploids were observed.

Literature Scilla

Alkema, H.Y., 1974. Mutatieveredeling bij bolgewassen door middel van radioactieve bestraling en andere methoden. Report 24, Laboratorium voor Bloembollenonderzoek, Lisse, p. 16.

TULIP

The outstanding symbol of ornamental bulb crops is the genus *Tulipa*, family Liliaceae, a long and interesting history of which has provided mankind with an enormous variety of cultivars. The numbers are still increasing by the use of many botanical species in cross-breeding programmes.

Consequently, heterozygosity is large and it is not surprising that spontaneous mutations frequently occur in several groups of the tulip. Examples are the cultivar Murillo, a double and early flowering tulip, which produced a large number of sports. De Mol (1933), for instance, lists 38 flower colour sports of this cultivar, including rosa, violet, yellow, orange, brown and white. Of the so-called Darwin hybrids the cultivar Bartigon has produced many sports (De Mol, 1949).

Among the first and certainly the man who studied the use of radiation in tulip breeding most extensively in the past was De Mol or De Mol van Oud Loosdrecht as he later called himself. The number of his publications is too large for all of them to be mentioned here separately and, moreover, many of them are in Dutch. We, therefore, refer to an article by Thamm (1956) about the work of De Mol included in a publication of De Mol van Oud Loosdrecht (1956), in which a complete list of his articles is to be found.

It is impossible to summarize all these publications and, therefore, we restrict ourselves to a few remarks we think important in view of later work to be discussed. Some of the striking data are the high doses mentioned in his publications. This is probably due to the use of rather soft X-rays which have a restricted penetration depth and are therefore less able to destroy the apices of roots and bulblets even at high doses. Later on, the optimum doses mentioned were lower and the final advice was, that doses should not surpass 800 rad (which in our opinion is still a high or sometimes even lethal acute dose).

The size of the bulbs to be irradiated has been studied extensively and De Mol's conclusion was that small bulbs are to be preferred although they require more space. One should grow these bulbs for at least 4 years to allow the mutated cell to grow out and form (part of) a new bulb.

Many spontaneous and induced mutations in flower colour have led to the conclusion that pink seems to be the flower colour that gives rise to most other colours. The anthocyanin concentration can become higher, lower or even zero. Violet and pink can mutate into each other. Colour mutations for yellow flowers are rare. Fringed petals and parrot type are recessive, the latter being a periclinal chimera.

Several other authors have carried out large-scale mutation breeding research on tulips. Nybom (1961, p. 259) mentions the work of a Swedish plant breeder named Carlsson, but no other references to this work have been found. Some mutation breeding is carried out in Poland (Grabowska and Mynett, 1970), but most publications come from Japan (Matsubara et al., 1965; Matsuda, 1960; Myodo, 1942; Nezu, 1962, 1963a,b, 1964, 1965; Nezu and Obata, 1964a,b). A review article was published by Nezu in 1967 in which no less than 192 literature references are discussed. In The Netherlands commercial mutation breeding (C. Broertjes, unpublished data) as well as mutation breeding research (Eikelboom et al., 1975a,b) has also been carried out on a relatively large scale.

The experience obtained so far can be summarized as follows: The choice of cultivar is always difficult. It is completely dependent on the (generally unknown) genotype what kind of mutation one may expect. From the literature it appears that pink is a good starting point when flower colour mutations are desired, but other colours mutate too. Yellow mutations are rare. One should start with a healthy and good producer which is a new variety. The old cultivars have produced all possible mutants already, although there are examples of "identical" flower colour mutants differing, for instance, in productivity.

If one wants to know whether flower colour mutants are possible, irradiation of a number of bulbs directly after harvest is recommended. The next year one can observe whether or not mutations have been induced in the growing point which was in an ontogenetically very young developmental stage during the irradiation. At the same time, one could decide whether it seems promising to continue with these bulbs or whether the experiment should be discontinued. If one continues, it is of no use to pay special attention to plants which show mutations in the flower. There is no correlation between these mutations and the ones induced in the apices of the bulblets.

Mutations for other characters are, of course, also possible. Form, size, parrot, fringed and growth-habit mutations have been reported. They are generally unrecognizable as mericlinal ("sectorial") chimeras, so selection should not start until the third or fourth year after irradiation. This holds true for non-visible characters as well.

The second question is when to irradiate: directly after harvest when the bulblets are at the youngest possible stage of development or, during September, aiming at the cells in the bulblets which will later form the apices for the secondary bulblets or large bulbs with relatively more apices than smaller bulbs. Generally, large bulbs are preferred because more apices are irradiated in a given volume than when small bulbs are used. Moreover, small bulbs require more space when planted. Large bulbs are, in addition, less sensitive to radiation and can resist a higher dose (Custers et al., 1976).

The optimum dose, according to our experiments, lies around 500 rad X-rays for diploids (large bulbs) and somewhat higher for triploids. The best dose for small bulbs, irradiated directly after harvest, lies around 400 rad of X-rays.

The best time to irradiate has also been a subject of study. As has been stated above, irradiation directly after harvest is recommended when a cultivar is to be tested for its ability to mutate. For small bulbs, early irradia-

TABLE 9

Commercial mutants of the tulip

Name of new cultivar	Place and date of release (or approval) and name of principal worker(s) and institute	Conditions and date of mutagenic treatment [Parent cultivar]	Main improved attributes of cultivar
Faraday	The Netherlands, 1949 W.E. de Mol van Oud Loosdrecht	X-rays, 1936 [Fantasy]	Flower colour
Estella Rijnveld	The Netherlands, 1954 W.E. de Mol van Oud Loosdrecht	X-rays [Red Champion]	Flower colour

Sources: IAEA (1972, pp. 526—544) and *Mutation Breeding Newsletter*, 1 (1972) —9 (1977) and Association Euratom—ITAL, Wageningen.

tion and for large bulbs September irradiation seems to give the best results.

Finally, the selection should be continued for at least 3 or 4 years in order to give a mutated cell the chance to express itself in first or second generation bulblets. Continued selection can take place during further clonal propagation, especially for important characteristics like productivity.

Despite the enormous number of publications very few cultivars are the direct result of induced mutations. To our knowledge, De Mol has produced only two commercial tulip mutants, namely cv. Faraday, released in 1949 and resulting from the irradiation of cv. Fantasy in 1936 (it has a white, flushed salmon-pink flower colour), and cv. Estella Rijnveld, released in 1954 with a red flamed, white flower colour, a mutant of the cultivar Red Champion (Table 9). Van Eijk reports a few promising mutants which have already been released on a small scale for further testing (J.P. van Eijk, personal communication, 1977), such as a white flowering mutant of cv. Preludium, similar to 'Polar Ice' but a much better producer. Three mutants of cv. Lustige Witwe are of interest, namely a lilac-purple one with a large white edge, a pink one with a white edge and a fuchsin-purple one, darker than cv. Frederica. All mutants are, in addition, more productive, while the keeping quality and earliness are as good as the parental genotype. The bulbs were irradiated with 250—450 rad X-rays in 1965. Besides the above-mentioned promising mutants all sorts of other mutants were observed. Some of the mutants were released for commercial propagation (Custers et al., 1976).

Since in vitro propagation of monocotyledons has come within reach (Hussey and Hilton, 1975a), an attempt to speed up mutation breeding using this propagation method could be made.

Literature tulip

Custers, J.B.M., Van Eijk, J.P. and Sparnaay, L.D., 1976. New developments in mutation breeding of vegetatively propagated ornamental crops with special reference to quantitative characters. In: Proc. Workshop Europ. Comm. — Israel, The Use of Ionizing Radiation, Wageningen. Assoc. Euratom—ITAL, Wageningen, pp. 496—501.

Eikelboom, W., Van Eijk, J.P. and Broertjes, C., 1975a. Verbetering van het sortiment door kruisingen en mutatieveredeling (1). Bloembollencultuur, 85(44): 1013—1015.

Eijkelboom, W., Van Eijk, J.P. and Broertjes, C., 1975b. Verbetering van het sortiment door kruisingen en mutatieveredeling (2). Bloembollencultuur, 85(45): 1033—1036.

De Mol, W.E., 1933. Mutation sowohl als Modifikation durch Röntgenbestrahlung und die "Teilungshypothese". Cellule, 42: 149—160.

De Mol, W.E., 1949. Twenty-five years of tulip improvement by X-rays. Pap. Mich. Acad. Sci., Arts Lett., 35: 9—14.

De Mol van Oud Loosdrecht, W.E. 1956. Der Einfluss der Röntgenstrahlen auf die Entwicklung des Pollens und der Sprosse bei Tulpen. Mit Rückblick auf die wissenschaftliche Lebensarbeit des Verfassers, von K.J.J. Thamm. Bayerischer Landwirtschaftverlag, Bonn, 128 pp.

Grabowska, B. and Mynett, K., 1970. Induction of changes in garden tulips (*Tulipa hybr.* hort.) under the influence of gamma rays ^{60}Co. Biul. Inst. Hodowli Aklim. Rosl., 1—2: 89—92 (in Polish with English summary).

Hussey, G. and Hilton, J., 1975a. Proliferation of in vitro monocotyledonous plantlets with cytokinin. John Innes Inst. Annu. Rep., 66: 53—54.
IAEA, 1972. Induced Mutations and Plant Improvement. IAEA, Vienna, pp. 526—544.
Matsubara, H., Iba, S., Oka, M. and Meshitsuka, G., 1965. Effects of gamma-irradiation on tulip. II. Effects on various stages of development. Tokyo Metrop. Isot. Centr. Annu. Rep., 2(1963): 157—162.
Matsuda, G., 1960. Studies on the breeding of the special plants in the cold districts utilizing radioisotopes. I. Effects of gamma irradiation on tulip. Bull. Toyama Agric. Exp. Stn, 1.
Myodo, H., 1942. Effects of X-rays upon tulip plants when irradiated in different developmental stages of floral organs. J. Fac. Agric. Hokkaido Imp. Univ., 48: 359—382.
Nezu, M., 1962. The effect of radiation on tulip breeding. Gamma Field Symp., 1: 43—49.
Nezu, M., 1963a. Studies on the production of bud sports in tulips by ionizing radiation. III. Frequency and mechanism of color formation. Seiken Jihô, 15: 75—86.
Nezu, M., 1963b. Studies on the production of bud sports in tulips by ionizing radiation. IV. Comparison of color sports by γ-rays with similar virus disease phenomena. Seiken Jihô, 15: 87—97.
Nezu, M., 1964. Studies on the production of bud sports in tulips by ionizing radiation. V. Effects of total dose, fractionation and temperature on somatic mutation. Jpn. J. Genet., 39(6): 440—446.
Nezu, M., 1965. Studies on the production of bud sports in tulips by ionizing radiation. VI. Selection and observation of the mutant progeny. Jpn. J. Breed., 15(2): 113—118.
Nezu, M., 1967. Tulip breeding by bud sports induced by gamma rays. Toyama Agric. Exp. Stn, Spec. Rep., No. 7, pp. 1—74 (192 references).
Nezu, M. and Obata, S., 1964a. Studies on the induction of bud sports in tulips by ionizing radiation. I. Morphology and development of the bulbs. Idengaku Zasshi, 38: 386—391.
Nezu, M. and Obata, S., 1964b. Studies on the induction of bud sports in tulips by ionizing radiation. II. Acute and chronic radiation effects on plant growth and bulbil yield. Idengaku Zasshi, 38: 392—398.
Nybom, N., 1961. The use of induced mutations for the improvement of vegetatively propagated plants. In: Symp. Mutation and Plant Breeding, Cornell University, Ithaca, N.Y., NAS—NRC, Publ. 891: 252—294.
Thamm, K.J.J., 1956. Rückblick auf die bisherige wissenschaftliche Arbeit des Amsterdamer Biologen Willem Eduard De Mol van Oud Loosdrecht. Bayerischer Landwirtschaftverlag, Bonn, pp. 7—30.

ZEPHYRANTHUS

Zephyranthus, family Amaryllidaceae, is a bulbous plant with yellow flowers in many flushes. Only two publications which describe the effect of radiation upon this species are known. Spencer (1955) found variation in radiosensitivity among the different varieties and species and noticed early flowering in the treated population, which was not transmitted to the next generation. Banerjee (1967), who irradiated at a rather late stage of development, when buds were already well formed, did not find stimulation nor did he report mutations. The doses used were 1.2 and 5 kR of γ-rays.

Literature Zephyranthus

Banerjee, S.K., 1967. Effect of gamma radiation on some perennial ornamentals. Indian J. Genet. Plant Breed., 27(3): 417—422.
Spencer, J.L., 1955. The effects of X-radiation on the flowering of certain cultivated bulbs and corms. Am. J. Bot., 42: 13—29.

Pot plants

ANTHURIUM

Anthurium, family Araceae, has so far been propagated by seed. Recently, however, in vitro propagation techniques have been developed (Pierik, 1975d; 1976b; Pierik and Steegmans, 1975c, 1976a; Pierik et al., 1974a,b, 1975) opening up possibilities for the production of non-chimeral mutants as well as for a faster clonal propagation of selected superior strains. A mutation breeding project, in co-operation with the Association Euratom—ITAL, is in progress (R.L.M. Pierik, 1976; unpublished data).

Literature Anthurium

Pierik, R.L.M., 1975d. Callus multiplication of *Anthurium andraeanum* Lind. in liquid media. Neth. J. Agric. Sci., 23: 299—302.

Pierik, R.L.M., 1976b. *Anthurium andraeanum* plantlets produced from callus tissues cultivated in vitro. Physiol. Plant., 37: 80—82.

Pierik, R.L.M. and Steegmans, H.H.M., 1975c. Vegetatieve vermeerdering van *Anthurium scherzerianum* in vitro. Vakbl. Bloemisterij, 30(25): 21.

Pierik, R.L.M. and Steegmans, H.H.M., 1976a. Vegetative propagation of *Anthurium scherzerianum* Schott through callus cultures. Sci. Hortic., 4(3): 291—293.

Pierik, R.L.M., Steegmans, H.H.M. and Van Der Meys, J.A.J., 1974a. Plantlet formation in callus-tissues of *Anthurium andraeanum* Lind. Sci. Hortic., 2: 193—198.

Pierik, R.L.M., Van Der Meys, J.A.J. and Steegmans, H.H.M., 1974b. Vegetatieve vermeerdering van *Anthurium andraeanum* in kweekbuizen. Vakbl. Bloemisterij, 29(6): 12—17.

Pierik, R.L.M., Van Spronsen, J.C., Van Eyk-Bos, G., Groen, L.E. and Wiersma, P., 1975. Vegetatieve vermeerdering van *Anthurium andraeanum* in vitro. Vakbl. Bloemisterij, 30(27): 17.

BEGONIA

The genus *Begonia*, the only genus of the family Begoniaceae, is a very large one. More than a thousand species (Irmscher, 1960) originally from tropical areas in America, Asia and Africa are known. About 200 of them are cultivated and only a few of them have contributed to the present classification which is as follows (Van Raalte, 1969).

Group I	Winter flowering begonias
	(1) small flowers (Lorraine type)
	(2) large flowers (Elatior type);
Group II	Summer flowering begonias
	Semperflorens type;
Group III	Leaf begonias
	(1) Rex type
	(2) Diadema type;

Group IV Tuberous begonias
(1) large flowers
(a) Simplex type
(b) Duplex type
(c) Plena type
(2) small flowers (multiflora)
(3) fairly large flowers (multiflora maxima)
(4) hanging begonias;
Group V Bushy species

A great number of the types mentioned above are propagated asexually Groups I, III and IV-3), either by rooting shoots or by rooting (parts of) leaves on which adventitious plantlets generally develop easily (Heide, 1965; Karper, 1971; Mikkelsen, 1976a). Aseptic in vitro propagation is not difficult either (Arora et al., 1970; Bigot, 1974a; Fonnesbech, 1974; Hannings and Langhans, 1974; Thakur, 1973; Wirth, 1960). Some authors report the use of in vitro propagation techniques to produce solid mutants from radiation-induced chimeras (Matsubara et al., 1971; Shigematsu and Matsubara, 1972), which can also be achieved in vivo by repeated pruning (Matsubara et al., 1974).

In *Begonia* in vivo and in vitro propagation are easy, the degree of heterozygosity is probably high, there is a large range of chromosome numbers (Legro and Doorenbos, 1969, 1971; Legro and Haegemann, 1971) and, consequently, the chances of it being sterile as a result of triploidy, for instance, are relatively great (Zeilinga, 1962); these facts make *Begonia* a very suitable plant for mutation induction to improve cross-breeding products or attain specific objectives.

Despite the fact that numerous sports (spontaneous mutations) are known to exist (Lindeman, 1968), relatively few examples of purposeful production of artificially induced mutations have been reported.

Unpublished work on *Begonia* (C. Broertjes, unpublished data) in co-operation with private plant breeders has revealed that similar results can be obtained as those found after irradiation of *Saintpaulia* and other Gesneriaceae, that is, solid non-chimeral mutants are produced if freshly detached leaves are irradiated. Several hundreds of mutants were induced in cultivars like Marina, Weiszer Schnee and Schwabeland, which were either not useful or already existed as spontaneously occurring sports. The optimum dose for this material was 2—3 krad X-rays, at a dose rate between 50 and 200 rad/min. Similar observations were reported by Brown and Harney (1974).

Commercial results were obtained by Doorenbos and Karper (1975), who purposely produced (sterile) triploid hybrids between the tetraploid and tuberous *B. Bertinii compacta* 'Leuchtfeuer' or *B. Bertinii compacta* 'Sonnenschein' and the diploid winter-flowering *B. socotrana*. Leaves of a few of the best hybrids were irradiated to produce mutants with several different characters, including yellow-flowered mutants (yellow is recessive).

TABLE 10

Commercial mutants of *Begonia*

Name of new cultivar	Place and date of release (or approval) and name of principal worker(s) and institute	Conditions and date of mutagenic treatment [Parent cultivar]	Main improved attributes of cultivar
Begonia elatior hybrids			
Turo	The Netherlands, 1973 J. Doorenbos, Dep. Hortic., Wageningen	Leaf irradiation 1500—2500 rad X-rays [clone Le 1, from F_1 of cross *B. Bertinii compacta* 'Leucht-feuer' × *B. socotrana*]	Flower more vivid
Tiara	The Netherlands, 1975 J. Doorenbos and K. Karper, Dep. Hortic., Wageningen	X-rays, 1972 [So-1]	Yellow flower colour
Improved Aphrodite Rose (Enchantress)	U.S.A., 1974 Mikkelsens Inc., Ashtabula, Ohio	γ-rays, 1972 [Aphrodite Rose]	Ruffled flower, light rose red; propagates by leaf cuttings
Fantasy	U.S.A., 1975 Mikkelsens Inc., Ashtabula, Ohio	γ-rays, 1972 [Aphrodite Rose mutant]	Upright, compact slow growing; leaf propagated, deep rose-red colour
Aphrodite Joy	U.S.A., 1974 Mikkelsens Inc., Ashtabula, Ohio	γ-rays, 1972 [Aphrodite Rose]	Bright pink ruffled flowers; more vigorous and larger flowers than cv. Aphrodite Pink
Aphrodite Twinkles	U.S.A., 1974 Mikkelsens Inc., Ashtabula, Ohio	γ-rays, 1972 [Aphrodite Rose]	Dwarf, compact, slow growing, upright, pink flowers
Aphrodite Peach	U.S.A., 1974 Mikkelsens Inc., Ashtabula, Ohio	γ-rays, 1972 [Aphrodite Rose]	Peach coloured flowers, very floriferous, small foliage; short, compact; self branching; extremely high bud count on leaf cuttings

TABLE 10 (continued)

Name of new cultivar	Place and date of release (or approval) and name of principal worker(s) and institute	Conditions and date of mutagenic treatment [Parent cultivar]	Main improved attributes of cultivar
Elegance *	U.S.A., 1975 Mikkelsens Inc., Ashtabula, Ohio	γ-rays, 1972 [Aphrodite Rose mutant]	Very large double flowers, pink; very ruffled edges; propagates by leaf cuttings, except during summer
Red Elegance *	U.S.A., 1975 Mikkelsens Inc., Ashtabula, Ohio	γ-rays, 1972 [Aphrodite Rose mutant]	Large double flowers; red ruffled tepal edges; vigorous
Rose Elegance *	U.S.A., 1975 Mikkelsens Inc., Ashtabula, Ohio	γ-rays, 1972 [Aphrodite Rose mutant]	Large rose-red double flowers; ruffled tepals; vigorous
Mikkel Limelight	U.S.A., 1974 Mikkelsens Inc., Ashtabula, Ohio	Fast neutrons, 1973 [Aphrodite Rose mutant]	Very vigorous grower; large white flowers; propagates by leaf cuttings
Flambeau	U.S.A., 1976 Mikkelsens Inc., Ashtabula, Ohio	Fast neutrons, 1973 [Aphrodite Red]	Bright red double flower; propagates very well by leaf cuttings
Heirloom	U.S.A., 1975 Mikkelsens Inc., Ashtabula, Ohio	Fast neutrons, 1973 [Schwabenland Pink]	Deep bright pink coloured flowers; easier to propagate than parent; little serration on edges of leaves; more resistant to mildew
Northern Sunset	Canada, 1976 J.M. Molnar, Agric. Can. Res. Stn, Ottawa, Ont.	3 kR X-rays [Renaissance]	Semi-double rose coloured flower
Hoblanche	The Netherlands, 1977 W.J. Hofstede, Huissen **	2.5 krad X-rays, 1973 [pink sport of cv. Vuurgloed]	White flower colour
Begonia rex			
Gin-Sei	Japan, 1976 K. Shigematsu, H. Masubara and M. Oka, Tokyo Metrop. Isot. Res. Cent., Tokyo	Adventitious buds of cut leaves, 10 kR ^{60}Co γ-rays, 1966 [Winter Queen]	More showy leaf colour (changed from very volute leaves of silver—white to smooth green leaves with numerous silver—white spots) than original variety

Ryoku-Ha	Japan, 1976 H. Matsubara, K. Shigematsu and H. Suda, Tokyo Metrop. Isot. Res. Cent., Tokyo	Adventitious buds of cut leaves, 10 kR ^{60}Co γ-rays, 1966 [Winter Queen]	More fine wavy volute leaves (changed from very volute leaves of silver—white to green wavy volute leaves with a small number of silver—white spots) than original variety; ozone sensitive habit
Begonia masoniana			
Orange-Iron	Japan, 1976 H. Suda and H. Matsubara, Tokyo Metrop. Isot. Res. Cent., Tokyo	Adventitious buds of cut leaves, 0.5 kR ^{60}Co γ-rays, 1971 [Iron Cross]	More soft, warm colour (changed from large size and yellow leaves of wry heart form to middle size and orange—green symmetrical leaves) than original variety; very soft impressionable plants
Mini-Mini-Iron	Japan, 1976 H. Suda and H. Matsubara, Tokyo Metrop. Isot. Res. Cent., Tokyo	Adventitious buds of cut leaves, 1 kR ^{60}Co γ-rays, 1971 [Iron Cross]	Smaller plants (changed from large leaves with a purplich brown cross to small leaves with a red—brown stripe) than original variety; very miniature dwarfish plants, very pretty
Big-Cross	Japan, 1976 H. Suda and H. Matsubara, Tokyo Metrop. Isot. Res. Cent., Tokyo	Adventitious buds of cut leaves, 3 kR ^{60}Co γ-rays, 1971 [Iron Cross]	Larger leaves of wry heart form having larger bright purple—brown cross than original variety; very strong plants
Kaede-Iron	Japan, 1976 H. Suda and H. Matsubara, Tokyo Metrop. Isot. Res. Cent., Tokyo	Adventitious buds of cut leaves, 10 kR ^{60}Co γ-rays, 1972 [Iron Cross]	Larger leaves of irregular, pentagonal form having large subdued red—brown cross-shaped figures; very strong plants

* All three 'Elegance' cultivars occurred on the same plant, which itself was a mutation of 'Aphrodite Rose'.
** In co-operation with the Association Euratom—ITAL, Wageningen.
Sources: IAEA (1972, pp. 526—544) and *Mutation Breeding Newsletter*, 1 (1972)—9 (1977) and Assoc. Erautom—ITAL, Wageningen.

A much higher mutation frequency as well as a wider mutation spectrum was obtained by irradiation than is found in spontaneously occurring mutants. All or nearly all mutants were solid. One of the 96 yellow-flowering mutants was introduced to the market, under the name of 'Tiara' as was a more vividly flowering mutant (Table 10) (Doorenbos, 1973).

Similar results were obtained by Mikkelsen et al. (1975) who also produced hundreds of interesting mutants from several cultivars, including 'Aphrodite' and 'Schwabeland Red'. Over ten of the mutants were put on the market (Mikkelsen, 1976b) (Plate 5). In Canada, a mutant of Rieger *elatior* begonia, namely cv. Northern Sunset, was released (Harney, 1976; Molnar, 1976), and in Japan mutants of cv. Winter Queen were under trial in 1975 (Matsubara et al., 1975) and released in 1976 together with several other mutants (Anonymous, 1977).

In view of the rather high spontaneous mutation rate, it might appear that it is not at all necessary to use mutagens to provoke mutations in several (very heterozygous) begonia types. It is true that, in the course of time, all relevant (flower colour) mutations will occur. The advantage of artificial mutation induction is that, even before a new cultivar has been released a wider mutation spectrum can be produced within a short period of time.

Literature Begonia

Anonymous, 1977. List of mutant varieties. Mutat. Breed. Newsl., 9(1): 14—17.

Arora, Y.K., Nakao, S. and Nakajima, T., 1970. Perpetuation of *Begonia rex* by aseptic culture with micro-leaf cuttings under various conditions of auxin and cytokinin. Jpn. J. Breed., 20(5): 275—281.

Bigot, C., 1974a. Comparisons of bud forming potential in vitro of separated external and internal tissues. In: H.E. Street (Editor), Third Int. Assoc. Plant Tissue Cell Cult., Leicester. University of Leicester, Leicester, Abstr. No. 164.

Brown, J.L. and Harney, P.M., 1974. Induction of non-chimaeral mutants in *Begonia*. HortScience, 9(3): 276.

Doorenbos, J., 1973. 'Turo', een nieuwe Elatior-Begonia. Vakbl. Bloemisterij, 28(42): 25.

Doorenbos, J. and Karper, J.J., 1975. X-ray induced mutations in *Begonia* × *hiemalis*. Euphytica, 24(1): 13—19.

Fonnesbech, M., 1974. The influence of naphtyl acetic acid, benzyl adenine and temperature on plantlet development from petiole segments of *Begonia* × *cheimantha* grown in vitro. Proc. Int. Hortic. Congr., 19(1A): 66.

Hannings, D.W. and Langhans, R.W., 1974. In vitro propagation of the Rieger *elatior Begonia*. HortScience, 9(3): 271 (abstract).

Harney, P.M., 1976. Breeding of floricultural crops in Canada. Acta Hortic., 63: 19—26.

Heide, O.M., 1965. Interaction of temperature, auxins and kinins in the regeneration ability of *Begonia* leaf cuttings. Physiol. Plant., 18: 891—920.

IAEA, 1972. Induced Mutations and Plant Improvement. IAEA, Vienna, pp. 526—544.

Irmscher, E., 1960. Begoniaceae. In: Parey's Blumengärtnerei, Vol. 2. Paul Parey, Berlin, pp. 67—91.

Karper, J.J., 1971. Vermeerdering door bladstek in het bijzonder bij *Begonia*. Vakbl. Bloemisterij, 26(48): 9; 26(49): 8—9.

Legro, R.A.H. and Doorenbos, J., 1969. Chromosome numbers in *Begonia*. Neth. J. Agric. Sci., 17: 189—202.

1

2

3

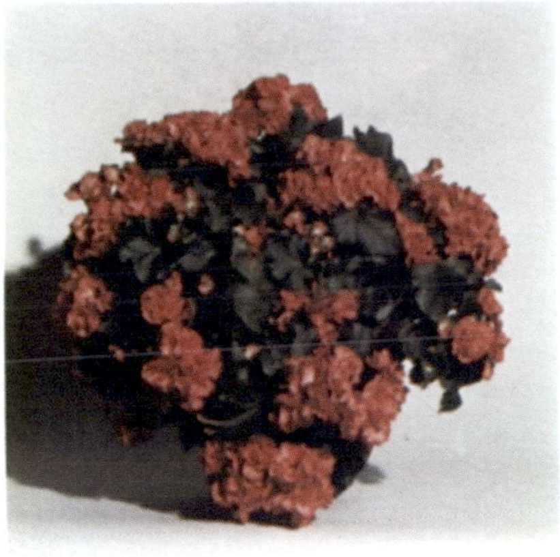

4

5

Plate 5. Mutants of *Begonia elatior* hybr. obtained by irradiation of detached leaves and the subsequent production of solid, non-chimeral mutants.
Top: cv. Peach (1) and cv. Joy (2) (both mutants of cv. Aphrodite Rose).
Centre: cv. Aphrodite Pink (3) (a mutant of cv. Aphrodite Rose partly mutated to light pink (white) that eventually became cv. Limelight). Bottom: cv. Fantasy (4) and cv. Elegance (5), both mutants of cv. Aphrodite Pink.
(Slides from J. Mikkelsen, Ashtabula, Ohio, who also sponsored this plate.)

Legro, R.A.H. and Doorenbos, J., 1971. Chromosome numbers in *Begonia*. 2. Neth. J. Agric. Sci., 19: 176—183.
Legro, R.A.H. and Haegeman, J.F.V., 1971. Chromosome numbers of hybrid tuberous begonias. Euphytica, 20(1): 1—14.
Lindeman, A., 1968. Mutationen bei Elatiorbegonien Rasse Rieger. Gartenwelt, 68: 266—267.
Matsubara, H., Shigematsu, K., Suda, H. and Hashimoto, S., 1971. The isolation of the mutated plants from sectorial chimera induced by irradiation in *Begonia* and *Chrysanthemum*. In: Japan Atomic Industrial Forum (Editor), Proc. 10th. Jpn. Conf. Radioisot., Tokyo. Tokyo Metrop. Res. Cent., Tokyo, pp. 374—376 (abstr. pap. 119 and 120).
Matsubara, H., Shigematsu, K. and Suda, H., 1974. The isolation and fixation of wholly mutant plant from sectorial chimera induced by gamma irradiation in *Begonia rex*. J. Jpn Soc. Hortic. Sci., 43(1): 63—68.
Matsubara, H., Suda, H., Sawada, T. and Nouchi, I., 1975. An ozone-sensitive strain from the mutants induced by gamma-ray irradiation of the *Begonia rex*, variety Winter Queen. Agric. Hortic., 50(6): 811—812 (in Japanese).
Mikkelsen, E.P., 1976a. Histology of Adventitious Shoot and Root Formation on Leaf-petiole Cuttings of *Begonia* × *hiemalis* Fotsch cv. Aphrodite Peach. Thesis, Michigan State University, East Lansing, Mich., 36 pp.
Mikkelsen, J.C., 1976b. Enlarging the scope of flower crop improvement: mutation breeding. In: K.C. Sink (Editor), Int. Symp. Floriculture, Plant Breeding and Genetics. East Lansing, Mich. Acta Hortic., 63: 197—202.
Mikkelsen, J.C., Ryan, J. and Constantin, M.J., 1975. Mutation breeding of Rieger's *elatior* begonias. Am. Hortic., 54(3): 18—21.
Molnar, J.M., 1976. Rieger *elatior* begonia cv. Northern Sunset. Can. J. Plant Sci., 56: 1003.
Shigematsu, K. and Matsubara, H., 1972. The isolation and propagation of the mutant plant from sectorial chimera induced by irradiation of *Begonia* rex. J. Jpn Soc. Hortic. Sci., 42(2): 196—200.
Thakur, S., 1973. In vitro foliar shoot bud formation in *Begonia semperflorens*. Curr. Sci., 42(12): 430—432.
Van Raalte, D., 1969. Het Handboek voor de Bloemisterij, IV. Born Periodieken N.V., Assen, Amsterdam, 4th edn, 468 pp.
Wirth, K., 1960. Experimentelle Beeinflüssung der Organbildung an in vitro kultivierten Blattstücken von *Begonia rex*. Planta, 54: 265—293.
Zeilinga, A.E., 1962. Cytological investigation of hybrid varieties of *Begonia semperflorens* LK. and Otto. Euphytica, 11(2): 126—136.

BROMELIACEAE

Several Bromeliaceae are propagated by seed but it is possible to propagate specific forms vegetatively, although this is a slow process since very few suckers are formed. In vitro propagation of *Aechmea fasciata* and *A. distichantha* is reported by Zimmer and Pieper (1975).

Mutation breeding has rarely been reported and the only worker who has carried out irradiation experiments is De Loose (1966, 1969b, 1973a). He irradiated *Guzmania paecockii* seeds of self-fertilized plants with 3.3 kR of γ-rays and obtained quite a high percentage of striped M_1 plants (6.87%).

TABLE 11

The commercial mutant of Bromeliaceae

Name of new cultivar	Place and date of release (or approval) and name of principal worker(s) and institute	Conditions and date of mutagenic treatment [Parent cultivar]	Main improved attributes of cultivar
Guzmania paecockii cv. Edith	Belgium, 1974 R. de Loose, Rijksstation voor Sierplanten-veredeling, Melle	Seeds; 3.3 krad ^{60}Co γ-rays, 1964 [*Guzmania paecockii*]	Striped leaves

Sources: IAEA (1972, pp. 526—544) and *Mutation Breeding Newsletter,* 1 (1972) —9 (1977) and Association Euratom—ITAL, Wageningen.

The chlorophyll-deficient stripes were either white or yellow and generally had vague outlines. One of the mutants is sold commercially under the name 'Edith' (Table 11).

Aechmea fasciata seedlings and seeds, and *Aechmea bromelifolia* seeds were irradiated with several doses of X-rays and treated with ethylene methanesulphonate (EMS) at concentrations of 0.5 and 1%. The optimum dose of X-rays is around 30—40 kR. As was found in *Guzmania*, rather high percentages of striped chlorophyll-deficient M_1 seedlings were obtained, some with the potential of being new cultivars.

Vriesea splendens, *V. saundersii* and *Canistrum roseum* seeds were treated as well, without, as yet, any commercial result.

Literature Bromeliaceae

De Loose, R., 1966. Quelques enseignements préliminaires resultant de l'application des rayonnements ionisants dans la mutagénèse chez *Rhododendron simsii* (*Azalea indica*) et *Guzmania paecockii*. In: Table Ronde sur l'Utilisation des Mutations Induites en Horticulture, Gembloux. Sémin. Hortic. Sci., 36—40.

De Loose, R., 1969b. Erfahrungen mit radioaktiven Bestrahlung von Zierpflanzen. Gartenwelt, 15: 348—350.

De Loose, R., 1973a. Bestralingen, mutatie- en isotopenonderzoek. Aktiviteitsverslag Rijksstation voor Sierplantenteelt, Melle, 1966—1971, VI, pp. 90—102.

IAEA, 1972. Induced Mutations and Plant Improvement. IAEA, Vienna, pp. 526—544.

Zimmer, K. and Pieper, W., 1975. Weitere Untersuchungen zur Kultur in vitro von *Aechmea*. Gartenbauwissenschaft, 40(3): 129—132.

CYCLAMEN

Although *Cyclamen*, family Primulaceae, is mainly propagated from seeds, there is an increasing interest in vegetative propagation, either in vivo or in vitro (Geier, 1977; Morel, 1975; Pierik, 1975c). Such vegetatively propagated clones will mainly be used for the production of F_1 hybrids, but the possibility of inducing solid, non-chimeral mutants during the process of inbreeding is also valuable. Solid mutants are easier to select and transfer to the next generation is guaranteed, in contrast to periclinal chimeras which are obtained after seed or tuber irradiation. Whether or not solid mutants are obtained if irradiated material is used for in vitro propagation remains to be seen.

Mutation induction was reported by Breider (1959) who irradiated young tubers and reported several mutations, one of which produced a mutant he called "rotundiflora". The radioresistance appears to be fairly high since doses of as high as 10 krad were not lethal. This was confirmed in a co-operative mutation breeding project with the Association Euratom—ITAL, in which the optimum dose for X-irradiation of seeds of a diploid cultivar was shown to be around 9—10 krad (C. Broertjes, unpublished data).

Literature Cyclamen

Breider, H., 1959. Röntgeninduzierte Mutationen bei *Vitis, Malus* und *Cyclamen*. Bayer. Landwirtsch. Jahrb., 36(4): 396–401.
Geier, T., 1977. Morphogenesis and plant regeneration from cultured organ fragments of *Cyclamen persicum*. In: Int. Symp. on Tissue Culture for Horticultural Purposes. University of Ghent, Ghent, Abstr. No. 20.
Morel, G., 1975. La multiplication végétative du cyclamen à partir de pétiole foliaire permettra-t-elle une nouvelle application de la culture 'in vitro' à l'horticulture. Pépinieristes Hortic. Maraichers, 158: 25–28.
Pierik, R.L.M., 1975c. Vegetatieve vermeerdering van *Cyclamen*. Vakbl. Bloemisterij, 30(28): 13.

EUPHORBIA

The family Euphorbiaceae contains over 1 600 species which grow under a great variety of conditions. Only a very limited number of species are commercially used. Large-scale (mutation) breeding is limited to *Euphorbia pulcherrima* Willd. ex Klotzsch (Poinsettia), whereas small-scale mutation breeding experiments are being carried out with *Euphorbia fulgens* Karw. and *Euphorbia milii* (*Euphorbia splendens*).

Euphorbia pulcherrima *Willd. ex Klotzsch (Poinsettia)*

Although this species has been known for some time (it was imported into the U.S.A. from Mexico in 1828 by the American ambassador to Mexico, Mr Poinsette), large-scale commercial production began after 1900. Several breeders have produced hybrids and many spontaneous sports have been found, either resulting from real mutational events or cell-layer rearrangements which produced different phenotypes. The latter phenomenon has been studied by several authors (Bergann, 1960, 1961, 1962, 1967b; Bergann and Bergann, 1960; Pötsch, 1966a) who also described similar effects in *Pelargonium*. In addition, a few publications have dealt with chimera formation and the pink bract colour, such as those by Robinson and Darrow (1929) and Stewart and Arisumi (1966). The conclusion of the latter authors was that pink flowers result from a recessive mutation from the red bract colour to white in L-I, while L-II and L-III remain genetically red. Stewart (1965) also used plastogene mutants of *E. pulcherrima* to study the morphology of development and the ontogeny of plant tissues.

Despite the considerable economic importance of this plant and the favourable conditions for mutant induction ("periclinal chimerism, differences in ploidy, genetic heterozygosity for bract colour and a natural tendency for chloroplasts to mutate" (Love, 1966, 1972)), Love is the only author to have published the results of mutation breeding experiments with Poinsettia. Irradiation of plants with some hundreds of rads of fast neutrons induced changes in flower colour, and size of bracts and leaves, as well as early maturity, branching characteristics and variations in plant height. The size of mutant sectors from adventitious buds was larger than those from terminal or lateral buds, and the mutation frequency was higher in adventitious buds.

The number of spontaneous sports, many of which are protected by plant breeders rights (economically important!), is very large and comprises almost all commercial cultivars. It is therefore surprising that, to our knowledge, no radiation-induced mutant cultivars have been reported. However, it is possible that several breeders who include mutagenic treatments in their breeding programme think it unnecessary to advertise the method by which their cultivars have been produced.

Euphorbia fulgens *Karw.*

Interest in this species is steadily growing and a number of spontaneous sports have recently been reported, commercialized and protected by plant breeders rights: for instance, the cultivars Orange, Salmonetta, Scarlet River and Quicksilver, which partly replace a few other sports that were previously sold commercially.

Irradiation of rooted cuttings with X-rays (optimum dose approximately 4 krad) resulted in a moderately large variation in flower colour and flower size (C. Broertjes, unpublished data). However, a problem is caused by the extremely small sector size which has made the recovery of the mutant sector almost impossible so far. An in vitro propagation method to regenerate plantlets from a mutated sector or to produce mutants seems to be a prerequisite for commercial mutation breeding.

One study of *Euphorbia splendens* (in fact *Euphorbia milii*) is reported in the literature (Koo and Cuevas-Ruiz, 1974a). Rooted cuttings were irradiated with approximately 2 krad of γ-rays, and a number of mutants with different leaf forms or colours were produced after several prunings.

Literature Euphorbia

Bergann, F., 1960. Über zwischenzellige Genwirkungen (Partnerinduktionen) bei der Pigmentbildung in der Brakteen der Periklinal-chimäre *Euphorbia pulcherrima* Willd. 'Eckes Rosa'. Ber. Dtsch. Bot. Ges., 73(11): 40—41.

Bergann, F., 1961. Eine weitere Trichimäre bei *Euphorbia pulcherrima* Willd. Biol. Zentralbl., 80(4): 403—412.

Bergann, F., 1962. Über den Nachweis zwischenzelliger Genwirkungen (Partnerinduktionen) bei der Pigmentbildung in den Brakteen der Periklinalchimäre *Euphorbia pulcherrima* Willd. 'Eckes Rosa'. Biol. Zentralbl., 81(4): 469—503.

Bergann, F., 1967b. Mutations-Chimären: Rohmaterial züchterischer Weiterbehandlung. Umsch. Wiss. Tech., 67(24): 791—797.

Bergann, F. and Bergann, L., 1960. Über die sogenannte Kräuselkrankheit der Poinsettia und die Beteiligung des "Dermatogens" bei der Mesophyllbilding. Flora (Jena), 149: 331—344.

Koo, F.K.S. and Cuevas-Ruiz, J., 1974a. Induction of somatic mutations in Jerusalem thorn (*Euphorbia splendens* Bojeri). Mutat. Breed. Newsl., 3: 13.

Love, J.E., 1966. Some effects of fast neutron irradiation on the somatic tissue of poinsettia. Proc. Am. Soc. Hortic. Sci., 89: 672—676.

Love, J.E., 1972. Somatic mutation induction in poinsettia and sweet potato. In: M.J. Constantin (Editor), Mutat. Breed. Workshop, Knoxville, Tenn. University of Tennessee, Knoxville, Tenn.

Pötsch, J., 1966a. Über die Auslösung extramutativer Strahlungseffekte an Klonsorten von *Euphorbia pulcherrima* Willd. Züchter, 36(1): 12—25.

Robinson, T.R. and Darrow, G.M., 1929. A pink poinsettia chimera. J. Hered., 20(7): 335—339.

Stewart, R.N., 1965. The origin and transmission of a series of plastogene mutants in *Dianthus* and *Euphorbia*. Genetics, 52: 925—947.

Stewart, R.N. and Arisumi, T., 1966. Genetic and histogenic determination of pink bract color in poinsettia. J. Hered., 57(6): 217—220.

FUCHSIA

There are a relatively large number of commercial cultivars of *Fuchsia*, family Oenotheraceae, which are easily propagated on a very large scale by shoot cuttings. In addition, many more cultivars have been and are being produced, mainly by amateur breeders who have obtained cultivars with beautiful flower colours and forms, and other useful characteristics.

Because of its economic importance, one would expect mutation breeding of *Fuchsia* to have been carried out in order to perfect the leading cultivars, but this is not the case. Only one small-scale experiment with a few cultivars has been made as a project in co-operation with the Association Euratom—ITAL. The cultivar Red Ribbons produced a mutant with a purple corolla. The optimum dose is about 2.5 krad X-rays. The cultivar Pink Calore was not shown to mutate (C. Broertjes, unpublished data).

It was a very small-scale experiment and, in our opinion, large(r)-scale irradiation of a few of the leading cultivars with a subsequent fast series of propagation with shoot cuttings would probably result in genetic variability which could be of commercial interest.

Methods of vegetative propagation other than rooting shoots are not available. In vivo adventitious shoot production on detached and rooted leaves is difficult and takes too long because of apparently necessary callus formation. In vitro propagation has, so far, not been reported.

GESNERIACEAE

General

The family Gesneriaceae is rather large and includes over a thousand, mostly tropical or subtropical species. Most are herbaceous with scaly rhizomes, tubers, stolons etc., but shrubs, vines and small trees also occur. There are two subfamilies, the Cyrtandroïdeae, including *Columnea*, *Episcia*, *Saintpaulia* and *Streptocarpus*, and the Gesnerioideae, including *Achimenes*, *Gesneria* and *Sinningia*, (*Gloxinia*) (Moore, 1957). Several of them contribute to the assortment of ornamental pot plants and mutation breeding has played a part in some genera, for example, *Achimenes*, *Saintpaulia* and *Streptocarpus* (Broertjes, 1969a). In fact, *Saintpaulia* (African violet) was the first species in which it was demonstrated that the apex of an adventitious bud on a detached leaf ultimately originates from a single epidermal cell (Naylor and Johnson, 1937; Sparrow et al., 1960). This avoids chimera formation, restricts diplontic selection and is in general advantageous for use in mutation breeding (Broertjes et al., 1968; Broertjes, 1972c). It has also been demonstrated in two other gesneriads, namely, *Achimenes* (Broertjes, 1972c) and *Streptocarpus* (Broertjes, 1969a), as well as in various other species.

Various Gesneriaceae can now be propagated in vitro which opens up possibilities for more rapid production of mutants and earlier screening (Grunewaldt, 1977; Raman, 1977).

Literature Gesneriaceae — General

Broertjes, C., 1969a. Mutation breeding of vegetatively propagated crops. In: G.C. Chisci and G. Hausmann (Editors), Proc. 5th Eucarpia Congr., Milan, 1968. Genet. Agrar., 23: 139—165.

Broertjes, C., 1972c. Use in Plant Breeding of Acute, Chronic or Fractionated Doses of X-rays or Fast Neutrons as Illustrated with Leaves of *Saintpaulia*. Thesis; Cent. Agric. Publ. Doc., Wageningen, Agric. Res. Rep. 776, 74 pp.

Broertjes, C., Haccius, B. and Weidlich, S., 1968. Adventitious bud formation on isolated leaves and its significance for mutation breeding. Euphytica, 17: 321—344.

Grunewaldt, J., 1977. Adventivknospenbildung und Pflanzenregeneration bei Gesneriaceae in vitro. Gartenbauwissenschaft, 42: 171—175.

Moore, Jr, H.E., 1957. African Violets, Gloxinias, and their Relatives. A Guide to the Cultivated Gesneriads. MacMillan, New York, N.Y., 323 pp.

Naylor, E.E. and Johnson, B., 1937. A histological study of vegetative reproduction in *Saintpaulia ionantha*. Am. J. Bot., 24: 673—676.

Raman, K., 1977. Rapid multiplication of *Streptocarpus* and gloxinia from in vitro cultured pedicel explants. Z. Pflanzenphysiol., 83(5): 411—418.

Sparrow, A.H., Sparrow, R.C. and Schairer, L.A., 1960. The use of X-rays to induce somatic mutations in *Saintpaulia*. Afr. Violet Mag., 13(4): 32—37.

Saintpaulia (*African violet*)

African violet originates from tropical East Africa, where it was discovered by Baron Walter von Saint Paul. Most present day cultivars are diploid having 2n = 28 chromosomes or 2n = 30 (Ehrlich, 1958) and are descendents of crosses between species and cultivars of *S. ionantha* H. Wendl. and *S. confusa*.

They readily reproduce from detached leaves which, after rooting, produce several or many plantlets at the bases of petioles of rooted detached leaves (Fig. 14). The number of plantlets depends on the cultivar, environ-

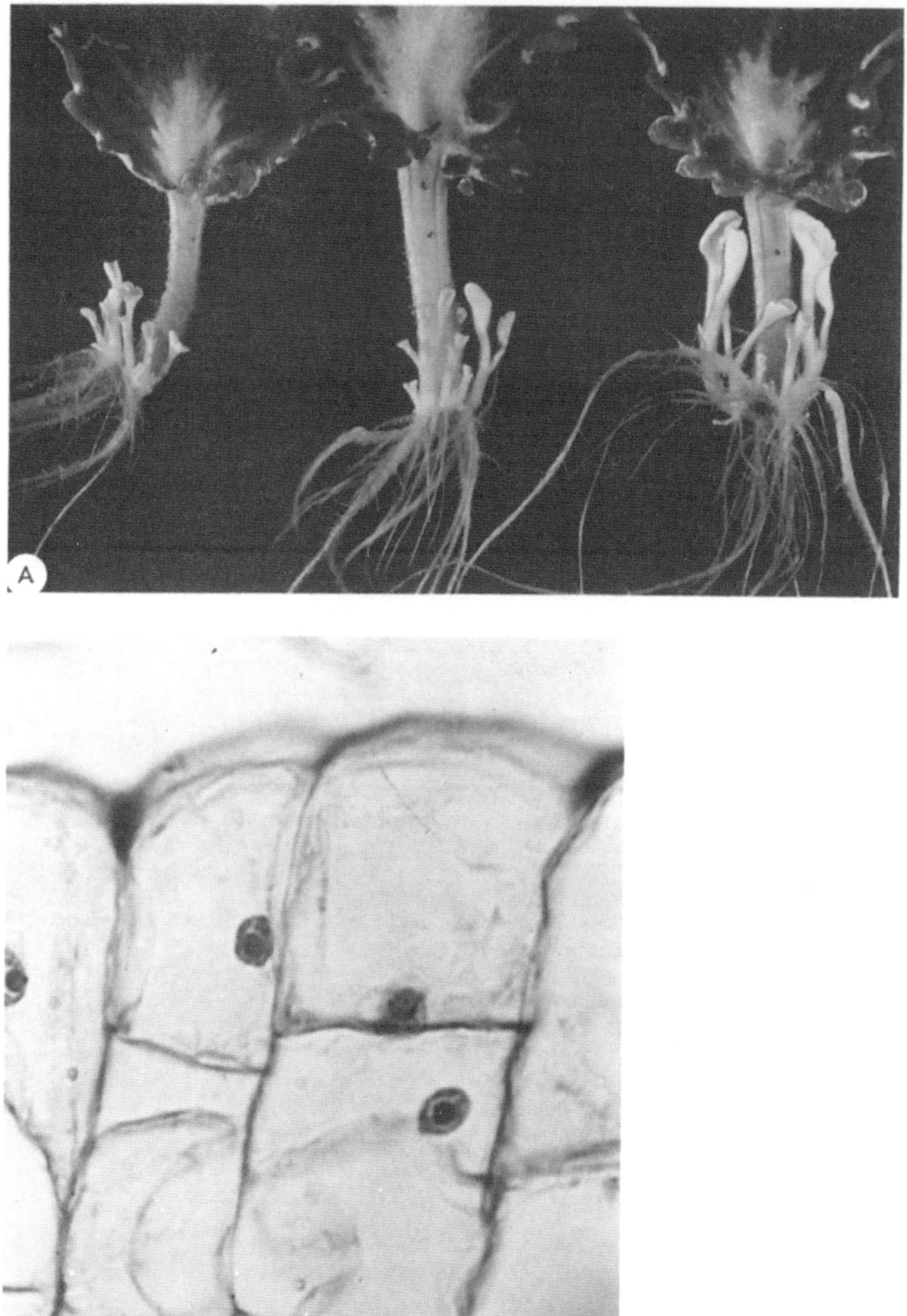
A
B

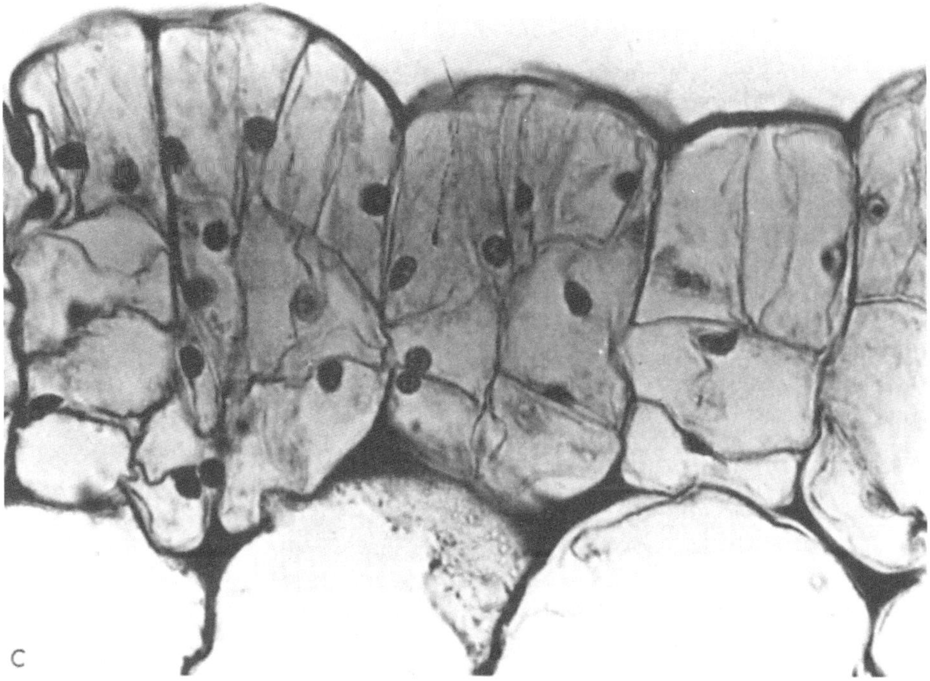

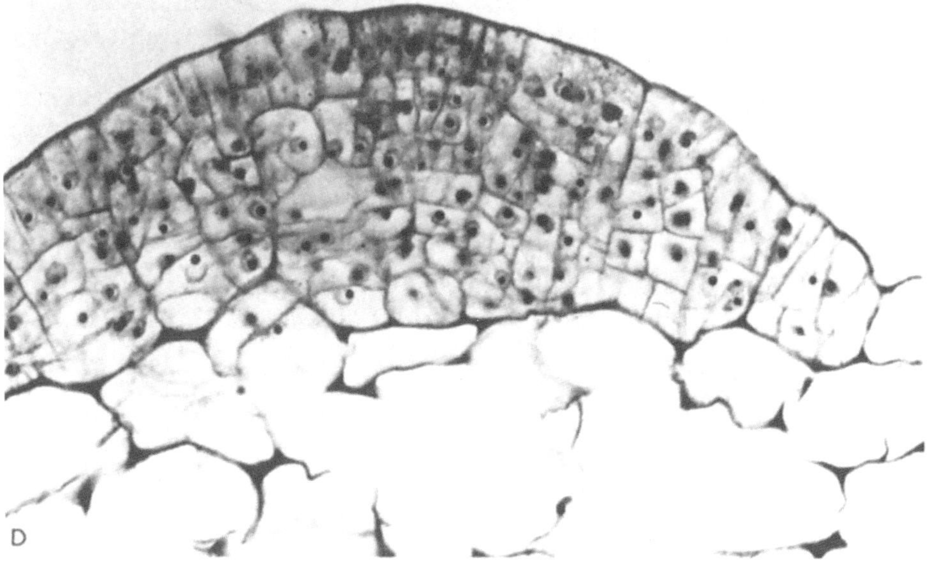

Fig. 14. Vegetative propagation of *Saintpaulia ionantha* by adventitious bud formation on detached leaves. (Photographs, ITAL, Wageningen.) (A) Rooted leaves with adventitious plantlets in different stages of development. (B) The first periclinal division cell wall in epidermal cells of the petiole, approximately 5 days after the leaf has been detached. (C) Epidermal cells are subdivided into several meristematic cells, approximately 1 week after detaching. (D) One of several meristems formed at the base of a petiole, approximately 10—14 days after detaching. This meristem will produce an apex and ultimately an

mental conditions and the application of growth regulators (Plummer and Leopold, 1957; Scott and Marston, 1967). Spontaneous mutations, rare in some clones but more frequent in others, have been reported since 1936, 50% being recessive and 50% being dominant (Reed, 1954). The first artificial mutations were probably induced by Sparrow et al. (1960), who confirmed that (almost) every new plantlet (adventitious plantlets at the base of the petiole of detached leaves) is derived from a single cell of the petiole, a fact first reported by Naylor and Johnson (1937). Broertjes (1972c, p. 14) used a slightly different definition: adventitious buds on detached leaves each ultimately originates from only one (epidermal) cell, which means that one, two or more vegetative daughter cells of one epidermal cell may have taken part in the formation of the apex of the adventitious plantlet.

The fact that solid tetraploids are obtained after colchicine treatment of detached leaves does not prove the single-cell origin of adventitious plantlets (Arisumi and Frazier, 1968), since more than one cell can become (auto)-tetraploid (Broertjes, 1974) whereas mutational events are unicellular.

Broertjes (1971b, 1972c) used the afore-mentioned type of adventitious bud formation on detached leaves to study the effect of acute, chronic or fractionated irradiation with X-rays or fast neutrons and found that, under certain conditions and with certain treatments, more mutants per 100 irradiated leaves were produced than was the case with the optimum acute dose (3 krad X-rays or 1.5 krad fast neutrons). Treatments which had this effect were a pre-treatment with 500 rad X-rays and a second dose of 6 krad separated by an 8-h interval (163%) and a similar treatment with fast neutrons (146%) (250 rad − 8 h − 1.7 krad). Other treatments (repeated irradiation, chronic irradiation) gave even higher percentages but are either impractical or too expensive.

In spite of the early available information of what is now recognized to be an important adventitious bud technique and the extensive experiments with ionizing radiation (Broertjes, 1971b, 1972c; Sparrow et al., 1960) or with mutagenic chemicals (Warfield, 1973) as far as we know, no one has ever produced a commercial mutant, either by radiation or by colchicine treatment. However, African violet does not escape the modern habit of mass production either, whereby high demands have to be fulfilled before a cultivar can meet the requirements of, for example, easy and fast reproduction, free flowering and non-breakable leaves. As soon as "ideal" cultivars have been produced mutation induction is used to create the necessary genetic variability within those outstanding cultivars. In fact, a number of such mutation breeding projects are at present in progress (C. Broertjes, personal communication, 1976).

A mutation project could simply consist of the administration of a series of doses, from 3 to 6 krad with steps of 0.5 krad, to every cultivar in question in order to determine the optimum dose for each. A larger number of leaves are then irradiated with the optimum dose, plantlets are raised and selection for desired traits among the solid mutants can start. It is important to irradiate detached leaves immediately after detachment, and an improvement on this is to cut off 0.5—1 cm of the petiole after irradiation and before planting. This is especially appropriate when leaves have to be shipped to and from an irradiation facility.

Recent developments concerning the in vitro propagation of *Saintpaulia* and other Gesneriaceae (Grunewaldt, 1976; Jungnickel, 1977) might be of interest for species which propagate less easily from leaves or for use in combination with chemical mutagens which are otherwise less effective than irradiation, probably because of inadequate penetration into the vegetative material.

Plastom mutations have been reported to produce variegated mutants after the use of N-nitroso-N-methylurea (Hentrich and Beger, 1974; Pohlheim, 1974a,b; Pohlheim and Beger, 1974; Pohlheim and Pohlheim, 1976).

Literature Saintpaulia

Arisumi, T. and Frazier, L.C., 1968. Cytological and morphological evidence for the single-cell origin of vegetatively propagated shoots in thirteen species of *Saintpaulia* treated with colchicine. Proc. Am. Soc. Hortic. Sci., 93: 679—685.

Broertjes, C., 1971b. Dose-fractionation studies and radiation-induced protection phenomena in African Violet. In: D.W. Benson and A.H. Sparrow (Editors), Survival of Food Crops and Livestock in the Event of a Nuclear War. Proc. Symp. Brookhaven Natl. Lab. Upton, 1970, U.S. Atomic Energy Commission, Oak Ridge, Tenn., pp. 325—342.

Broertjes, C., 1972c. Use in Plant Breeding of Acute, Chronic or Fractionated Doses of X-rays or Fast Neutrons as Illustrated with Leaves of *Saintpaulia*. Thesis; Centr. Agric. Publ. Doc., Wageningen, Agric. Res. Rep. 776, 74 pp.

Broertjes, C., 1974. The production of polyploids using the adventitious bud technique. In: FAO/IAEA/Eucarpia Conf., Mutations and Polyploidy, Bari, 1972. IAEA, Vienna, pp. 29—35.

Ehrlich, H.G., 1958. Cytological studies in *Saintpaulia* Wendl. (Gesneriaceae). Am. J. Bot., 45: 177—182.

Grunewaldt, J., 1976. Die in vitro Regeneration aus Blattstielquerschnitten von *Saintpaulia ionantha* H. Wendl. Gartenbauwissenschaft, 41(4): 145—148.

Hentrich, W. and Beger, B., 1974. Untersuchungen über die mutagene Effizienz von N-Nitroso-N-Methylharnstoff bei *Saintpaulia ionantha* H. Wendl. Arch. Züchtungsforsch., 4(1): 29—43.

Jungnickel, F., 1977. Induktion und Vermehrung von Mutanten bei *Saintpaulia ionantha* H. Wendl. in Steril-Kultur. Biol. Zentralbl., 96(3): 335—343.

Naylor, E.E. and Johnson, B., 1937. A histological study of vegetative reproduction in *Saintpaulia ionantha*. Am. J. Bot., 24: 673—676.

Plummer, T.H. and Leopold, A.C., 1957. Chemical treatment for bud formation in *Saintpaulia*. Proc. Am. Soc. Hortic. Sci., 70: 442—444.

Pohlheim, F., 1974a. Nachweiss von Mischzellen in variegaten Adventivsprossen von *Saintpaulia*, entstanden nach Behandlung isolierter Blätter mit N-Nitroso-N-Methylharnstoff. Biol. Zentralbl., 93: 141—148.

Pohlheim, F., 1974b. Induced plastom mutations in *Saintpaulia*. Proc. Int. Hortic. Congr., 19(1B): 888.

Pohlheim, F. and Beger, B., 1974. Erhöhung der Mutationsrate im Plastom bei *Saintpaulia* durch N-Nitroso-N-Methylharnstoff. Biol. Rundsch., 12(3): 204—206.
Pohlheim, F. and Pohlheim, E., 1976. Herstellung von Plastommutanten bei *Saintpaulia ionantha* H. Wendl. Biochem. Physiol. Pflanz., 169: 377—383.
Reed, S.C., 1954. African violet genetics. J. Hered., 45: 225—230.
Scott, M.A. and Marston, M.E., 1967. Effects of mist and basal temperature on the regeneration of *Saintpaulia ionantha* Wendl. from leaf cuttings. Hortic. Res., 7: 50—60.
Sparrow, A.H., Sparrow, R.C. and Schairer, L.A., 1960. The use of X-rays to induce somatic mutations in *Saintpaulia*. Afr. Violet Mag., 13: 32—37.
Warfield, D., 1973. Induction of mutations in African Violet (*Saintpaulia ionantha* Wendl.) by ethyl methane sulfonate. Hortic. Sci., 8: 29.

Streptocarpus

Streptocarpus is divided into two subgenera, the subgenus *Eustreptocarpus* (an acaulescent group with 2n = 32) from Southern Africa and the subgenus *Streptocarpella* (a caulescent group with 2n = 30) originating from Central Africa.

Until 1884, only five species in the acaulescent group were recognized, *S. rexii*, *S. polyanthus*, *S. gardeni* and *S. saundersii*, all with bluish flowers, and the white-flowering *S. parviflorus*. In 1884, *S. dunnii* was introduced into Europe from the Transvaal and has since been crossed with some of the acaulescent species, because of its bright brick-red flower colour. Further intercrossing and backcrossing has resulted in numerous hybrids and commercial cultivars (Brown, 1973, 1975a; Brown and Davies, 1970; Crane and Lawrence, 1956; Hilliard, 1966; Hilliard and Burtt, 1971; Lawrence, 1947, 1957, 1958; Lawrence and Sturgess, 1957; Lawrence et al., 1939).

The cultivar Constant Nymph was obtained by Lawrence from a cross between the hybrid 'Merton Blue' and *S. johannis*. It is an outstanding house plant which, under normal conditions, produces blue flowers from March to November. This cultivar revived interest in breeding *Streptocarpus*, by cross-breeding (Hammer, 1976) as well as mutation breeding (Broertjes, 1969b, 1970, 1973; Broertjes et al., 1969; Brown, 1971a, 1973; Brown and Davies, 1971; Davies, 1971; Zeven, 1972, 1973). The potentialities of mutation breeding are many since *Streptocarpus* reproduces easily from both small leaf segments (Doss and Kretchun, 1975) and leaf cuttings (Fig. 15). In the latter case hundreds of adventitious plantlets per leaf are produced if the proper technique (leaf halves) is used (Marston, 1964, 1964/1965), probably because the leaves contain a non-limiting level of endogenous cytokinin (Appelgren and Heide, 1972). By using this technique and irradiating leaves immediately after detaching, numerous complete, non-chimeral mutants have been produced and several are now sold commercially (Broertjes, 1969b), some being rather successful (Anonymous, 1975) (Table 12). Recently, a winter-flowering mutant of 'Constant Nymph', cv. Margaret, was introduced (Brown, 1974; Davies and Hedley, 1975; Van Raalte and Van Raalte-Wichers, 1974) which creates the possibility of year-round produc-

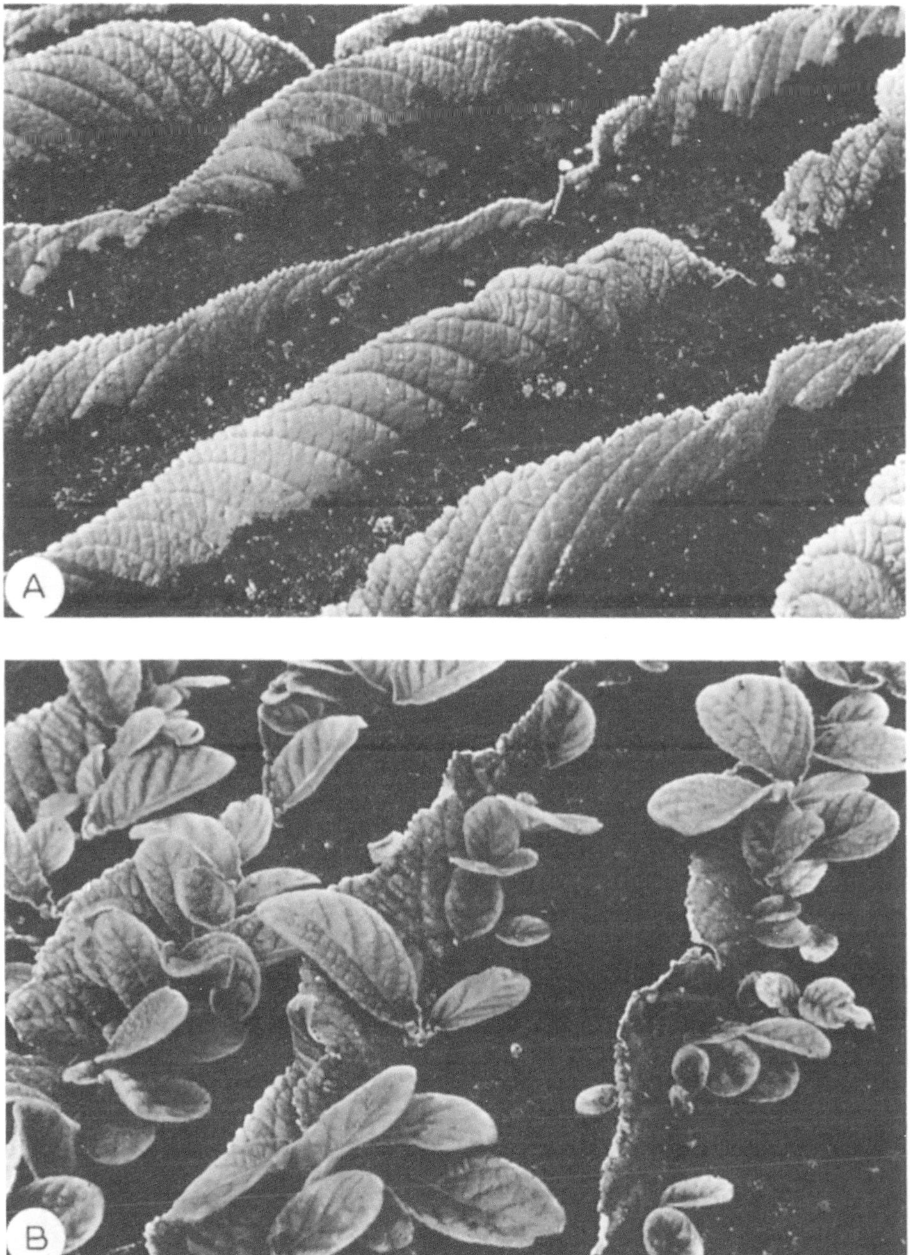

Fig. 15. Vegetative propagation of *Streptocarpus* by adventitious buds. (Photographs, ITAL, Wageningen.) The midrib of a leaf is taken out with a sharp knife. (A) The half leaves are planted in a peaty soil. (B) The half leaves root and each produces several tens of adventitious plantlets approximately 4 months after detachment.

TABLE 12

Commercial mutants of *Streptocarpus*

Name of new cultivar	Place and date of release (or approval) and name of principal worker(s) and institute	Conditions and date of mutagenic treatment [Parent cultivar]	Main improved attributes of cultivar
Purple Nymph	The Netherlands, 1969 C. Broertjes, Assoc. Euratom—ITAL, Wageningen	Leaves, X-rays, 1966 [Constant Nymph]	Larger and more purple coloured flower; plant more sturdy (rest of genotype unchanged)
Mini Nymph	The Netherlands, 1969 C. Broertjes, Assoc. Euratom—ITAL, Wageningen	Leaves, X-rays, 1966 [Constant Nymph]	Compact growth; very freely flowering (rest of genotype unchanged)
Blue Nymph	The Netherlands, 1969 C. Broertjes, Assoc. Euratom—ITAL, Wageningen	Leaves, X-rays, 1966 [Constant Nymph]	Light blue flower; finer plant growth (rest of genotype unchanged)
Netta Nymph	The Netherlands, 1969 C. Broertjes, Assoc. Euratom—ITAL, Wageningen	Leaves, X-rays, 1966 [Constant Nymph]	Dark blue netted and picotee flowers; extremely freely flowering
Cobalt Nymph	The Netherlands, 1969 C. Broertjes, Assoc. Euratom—ITAL, Wageningen	Combined colchicine and X-ray treatment, 1966 [Constant Nymph]	Compact plant, dark blue flowers, tetraploid
Margaret	United Kingdom, 1974 D.R. Davies, John Innes Institute, Norwich	Leaves, X-rays? 1972? [Constant Nymph]	Very freely flowering during winter
Snow-white	The Netherlands, 1973 Maassen, Lent *	Leaves, X-rays, 1971 [Maassen's White]	Dwarf habit; freely flowering; primula-like flowers
Albatros	The Netherlands, 1973 C. Broertjes, Assoc. Euratom—ITAL, Wageningen	Leaves, colchicine, 1971 [mutant 7111 of cv. Maassen's White]	Large white flowers; sturdy flower stalk; comparatively compact

* In co-operation with the Association Euratom—ITAL, Wageningen.

Sources: IAEA (1972, pp. 526—544) and *Mutation Breeding Newsletter*, 1 (1972) —9 (1977) and Association Euratom—ITAL, Wageningen.

tion of *Streptocarpus* cultivars. A chemical mutagen proved to be unsuccessful in *Streptocarpus* (Choudhary, 1976), probably partly due to penetration problems. These could be overcome if chemicals were used in combination with in vitro propagation (Handro and Rossini, 1974).

A very successful spontaneous mutant of cv. Constant Nymph, 'Maassen's White' (resulting from two simultaneous mutations of the same dominant gene), was used for genetic research (A.C. Zeven, personal communication, 1975) and the production of solid mutants and autotetraploids. One of the mutants, cv. Snow-white, was introduced to the market. It is a dwarf type with *Primula*-like white flowers. Another mutant, no. 7111 (flowering very freely with relatively long flower stalks), was used to produce autotetraploids with larger flowers and firm upright flower stems. Among the approximately 50 autotetraploid clones obtained, the best one was selected and sold commercially (Broertjes, 1973, 1974) under the name of 'Albatros'.

Various spontaneous mutants of *Streptocarpus*, such as the cultivars Boerenblauw, Blue Moon, Laura Nymph, and the most striking of all, the cultivar Maassen's White, are being marketed. This contrasts to the situation for *Saintpaulia* which has no known commercial spontaneous mutants.

Literature Streptocarpus

Anonymous, 1975. *Streptocarpus*-Versuche. Gartenbauschule Wien-Schönbrunn, Versuchsergebnisse 1975, Heft, 11. Gartenbauschule, Wien-Schönbrunn, pp. 113—115.

Appelgren, M. and Heide, O.M., 1972. Regeneration in *Streptocarpus* leaf discs and its regulation by temperature and growth substances. Phyiol. Plant, 27: 417—423.

Broertjes, C., 1969b. Mutation breeding of *Streptocarpus*. Euphytica, 18: 333—339.

Broertjes, C., 1970. Mutationszüchtung bei vegetativ vermehrbaren Zierpflanzen. Gartenwelt, 70(11): 266—268.

Broertjes, C., 1973. *Achimenes* en *Streptocarpus*. Nieuwe varieteiten door bestraling en colchicinebehandeling. Vakbl. Bloemisterij, 28(28): 11.

Broertjes, C., 1974. The production of polyploids using the adventitious bud technique. In: FAO/IAEA/EUCARPIA Conf., Mutations and Polyploidy, Bari, 1972. IAEA, Vienna, pp. 29—35.

Broertjes, C., Leffring, L. and Leuning, B., 1969. Nieuwe *Streptocarpus* varieteiten door bestraling. Vakbl. Bloemisterij, 24(22): 806—807.

Brown, A.G., 1971a. X-ray treatment of F_1-hybrid plants from *Streptocarpus johannis* × *S. hybridus*. John Innes Inst. Annu. Rep., 62: 49.

Brown, A.G., 1973. Hybrid *Streptocarpus*. J. R. Hortic. Soc., 98: 201—205.

Brown, A.G., 1974. *Streptocarpus* — the development of a new commercial pot plant. In: The Glasshouse Environment and Objectives in Breeding New Varieties of Ornamental Crops. Eucarpia Meeting on Ornamentals, Norwich. John Innes Inst., Norwich, pp. 59—66.

Brown, A.G., 1975a. *Streptocarpus* breeding. John Innes Inst. Annu. Rep., 66: 43.

Brown, A.G. and Davies, D.R., 1970. *Streptocarpus* breeding. John Innes Inst. Annu. Rep., 61: 37—38.

Brown, A.G. and Davies, D.R., 1971. The comparative roles of artificially induced mutations and hybridization in the breeding of *Streptocarpus*. In: Eucarpia Meet., Ornamentals, Wageningen. Inst. Hortic. Plant Breed., Wageningen, pp. 32—37.

Choudhary, D.K., 1976. Non-mutagenicity of sodium azide on *Streptocarpus hybridus*. Z. Pflanzenzücht., 76(2): 167—170.
Crane, M.B. and Lawrence, W.J.C., 1956. The Genetics of Garden plants. MacMillan, London, pp. 79—80.
Davies, D.R., 1971. Mutation breeding. Span, 14(2): 101—104.
Davies, D.R. and Hedley, C.L., 1975. The induction by mutation of all-year-round flowering in *Streptocarpus*. Euphytica, 24: 269—275.
Doss, P. and Kretchun, R.M., 1975. Propagation of cape Primrose (*Streptocarpus* 'Constant Nymph') from small leaf segments. Plant Propagator, 21(4): 8—9.
Hammer, P.A., 1976. Breeding *Streptocarpus* for pot plant use. Acta Hortic., 63: 99—100.
Handro, W. and Rossini, L.M.E., 1974. Induction of floral and vegetative buds in leaf discs of *Streptocarpus nobilis* cultured in vitro: physiological and structural aspects. In: H.E. Street (Editor), Third Int. Assoc. Plant Tissue Cell Cult., Leicester. University of Leicester, Leicester, Abstr. No. 163.
Hilliard, O.M., 1966. Studies in *Streptocarpus*. J.S. Afr. Bot., 32: 87—123.
Hilliard, O.M. and Burtt, B.L., 1971. *Streptocarpus*. An African Plant Study. University of Natal Press, Pietermaritzburg, 410 pp.
IAEA, 1972. Induced Mutations and Plant Improvement. IAEA, Vienna, pp. 526—544.
Lawrence, W.J.C., 1947. Studies on *Streptocarpus*. II. Complementary sublethal genes. J. Genet., 48: 16—30.
Lawrence, W.J.C., 1957. Studies on *Streptocarpus*. IV. Genetics of flower colour patterns. Heredity, 11(3): 337—538.
Lawrence, W.J.C., 1958. Studies on *Streptocarpus*. V. Speciation and gene systems. Heredity, 12(3): 333—356.
Lawrence, W.J.C. and Sturgess, V.C., 1957. Studies on *Streptocarpus*. III. Genetics and chemistry of flower colour in the garden forms, species and hybrids. Heredity, 11(3): 303—336.
Lawrence, W.J.C., Scott-Moncrieff, R. and Sturgess, V.C., 1939. Studies on *Streptocarpus*. I. Genetics and chemistry of flower colour in garden strains. J. Genet., 48: 16—30.
Marston, M.E., 1964. The propagation of plants from leaf-cuttings, with special reference to *Streptocarpus*. Proc. Int. Hortic. Congr., 16: 33—40.
Marston, M.E., 1964/1965. The morphology of a *Streptocarpus* hybrid and its regeneration from leaf cuttings. Sci. Hortic., XVII: 114—120.
Van Raalte, D. and Van Raalte-Wichers, D., 1974. Toekomst voor *Streptocarpus*. Vakbl. Bloemisterij, 29(23): 15.
Zeven, A.C., 1972. Inheritance of functional male sterility in *Streptocarpus* 'Constant Nymph' and its mutants. Euphytica, 21: 265—270.
Zeven, A.C., 1973. *Streptocarpus* 'Constant Nymph'. De oorzaak van de populariteit in Nederland. Vakbl. Bloemisterij, 28(35): 14—15.

Achimenes

Achimenes was originally found in Central and South America, from Mexico to Brazil and northern Argentina. One species, *A. coccinea*, occurs in Jamaica and was the first species to be introduced to Europe (in 1778) by William Forsythe. In later years, many of the other 25 species known today were introduced into Europe from South America and the United States.

During the 18th century many cultivars were produced by intercrossing available species and hybrids (Moore, 1957). Interest in the genus then decreased but, in 1945, it revived and many new cultivars were obtained by

Professor H.E. Moore (Cornell University) and Mr Paul Arnold, both working in the U.S.A.

In Germany K. Michelssen has recently introduced the Michelssen hybrids which are prolifically flowering cultivars with blooms of several different colours.

Achimenes is a low, generally hairy herb with toothed leaves. The leaves are opposing or in three's and have short petioles. Flowers occur singly or in pairs in the axils of leaves and vary in colour from white to yellow, pink, purple and blue. They reproduce from scaly rhizomes which are formed underground (Deutch, 1974), but propagation by rooting cuttings is also practised.

Achimenes has also been used to study the organogenesis of adventitious plantlets on detached leaves. As in *Saintpaulia* and *Streptocarpus* almost exclusively solid, non-chimeral mutants have been obtained after irradiation

Fig. 16. *Achimenes* cv. Paul Arnold (left) and one of the radiation-induced compact growing mutants, cv. Cupido. (Photograph, ITAL, Wageningen.)

TABLE 13

Commercial mutants of *Achimenes*

Name of new cultivar	Place and date of release (or approval) and name of principal worker(s) and institute	Conditions and date of mutagenic treatment [Parent cultivar]	Main improved attributes of cultivar
Early Arnold	The Netherlands, 1971 C. Broertjes, Assoc. Euratom—ITAL, Wageningen	Leaves, 2—4 krad X-rays or 1—2 krad fast neutrons, 1968 [Paul Arnold]	Flowers 1—2 weeks earlier; slightly different flower colour
Springtime	The Netherlands, 1971 C. Broertjes, Assoc. Euratom—ITAL, Wageningen	Leaves, 2—4 krad X-rays or 1—2 krad fast neutrons, 1968 [Paul Arnold]	Flowers 1—2 weeks earlier; slightly different flower colour
Compact Arnold	The Netherlands, 1971 C. Broertjes, Assoc. Euratom—ITAL, Wageningen	Leaves, 2—4 krad X-rays or 1—2 krad fast neutrons, 1968 [Paul Arnold]	More compact growth habit
Cupido	The Netherlands, 1973 C. Broertjes, Assoc. Euratom—ITAL, Wageningen	Leaves, 2—4 krad X-rays or 1—2 krad fast neutrons, 1968 [Paul Arnold]	Compact; sturdy; freely flowering
Orion	The Netherlands, 1973 C. Broertjes, Assoc. Euratom—ITAL, Wageningen	Leaves, 2—4 krad X-rays or 1—2 krad fast neutrons, 1968 [Paul Arnold]	Early flowering; large flower; sturdy but higher plant
Flamingo	The Netherlands, 1977 C. Broertjes, Assoc. Euratom—ITAL, Wageningen	Leaves, 3 krad X-rays, 1975 [Tango]	Good growth habit, somewhat more compact; flowers stand out better
Lollipop	The Netherlands, 1977 C. Broertjes, Assoc. Euratom—ITAL, Wageningen	Leaves, 1 krad fast neutrons, 1975 [Tango]	Compact; typical pink flower colour, somewhat fringed
Pink Attraction	The Netherlands, 1977 C. Broertjes, Assoc. Euratom—ITAL, Wageningen	Leaves, 2.5 krad X-rays, 1975 [autotetraploid form of cv. Repelsteeltje]	Compact, regular growth habit with dark green foliage; rather small flower with attractive pink colour

Sources: IAEA (1972, pp. 526—544) and *Mutation Breeding Newsletter*, 1 (1972) —9 (1977) and Assoc. Euratom—ITAL, Wageningen.

of freshly detached leaves, again demonstrating that adventitious plantlets originate from one (epidermal) cell (Broertjes, 1972b). It has also been shown that commercially useful mutants can be obtained within a very limited time period by using this technique in combination with radiation (Fig. 16) or colchicine (Broertjes, 1971a, 1972b, 1973, 1974, 1977b) (Table 13).

There is one complication which necessitates slightly unusual treatment during the vegetative propagation of mutants found in this species. Normally *Achimenes* is propagated by using the scaly rhizomes formed by cells at the basal part of an adventitious plantlet. Since this part may contain some cells that are genetically different from the rest of the mutant plantlet (probably caused at separation of the plantlet from the mother leaf), a certain percentage of the rhizomes formed are genetically different from the mutant. The complication can be eliminated in two ways:

(a) One or more (leaf) cuttings are made from the mutant plant and this material is used exclusively for the further clonal propagation of the mutant.

(b) Irradiation late(r) in the season, when only adventitious rhizomes are produced instead of plantlets. Since rhizomes appear to originate from one cell and the connective tissue dies away during storage, the above-mentioned complication is avoided.

Literature Achimenes

Broertjes, C., 1971a. *Achimenes.* Het verkrijgen van nieuwe varieteiten door bestraling. Vakbl. Bloemisterij, 26(34): 21.

Broertjes, C., 1972b. Mutation breeding of *Achimenes.* Euphytica, 21: 48—62.

Broertjes, C., 1973. *Achimenes* en *Streptocarpus.* Nieuwe varieteiten door bestraling en colchicine behandeling. Vakbl. Bloemisterij, 28(28): 11.

Broertjes, C., 1974. The production of polyploids using the adventitious bud technique. In: FAO/IAEA/EUCARPIA Conf., Mutations and Polyploidy, Bari, 1972. IAEA, Vienna, pp. 29—35.

Broertjes, C., 1977b. Nieuwe *Achimenes* cultivars verkregen door bestraling of door colchicinebehandeling. Vakbl. Bloemisterij, 32(43): 109.

Deutch, B., 1974. Bulblet formation in *Achimenes longiflora.* Physiol. Plant., 30(2): 113—118.

IAEA, 1972. Induced Mutations and Plant Improvement. IAEA, Vienna, pp. 526—544.

Moore, Jr, H.E., 1957. African Violets, Gloxinias and their Relatives. A Guide to the Cultivated Gesneriads. MacMillan, New York, N.Y., pp. 98—113.

Sinningia

The horticultural cultivars of *Sinningia,* often wrongly referred to as *Gloxinia,* are the results of intercrossing *S. perennis, S. caulescens, S. rubra, S. albiflora* and *S. macrophylla* (mostly imported at the beginning of the 19th century), during the second half of the 19th century by Belgian, German and English breeders. Later, because leaves of *S. perennis* are brittle, it was crossed with *S. regina* to introduce flexibility to the leaves (Van Raalte, 1969).

The modern assortment is propagated by seed and sometimes by tubers. Consequently, the cultivars are fairly homozygous for important (ornamental) characters and mutation is therefore not the most suitable breeding method for the genus. On the other hand, the plant can also be propagated by growing adventitious plantlets on detached leaves and this makes mutation breeding more attractive. It might also be effective if in vitro propagation techniques were used (Bigot, 1974b; Raman, 1977).

So far, no mutation breeding work has been published.

Cytology and interspecific hybridization was carried out with a few *Sinningia* and *Rechsteineria* species (Clayberg, 1970b). The same author also proposed the introduction of *Sinningia pusilla* as a laboratory plant (Clayberg, 1970a). The small plant can be propagated from seeds or vegetatively by rooting leaf cuttings, new shoots from a tuber, or sections cut from a tuber. Plants grow and flower almost continuously.

Literature Sinningia

Bigot, C., 1974b. In vitro organ culture of *Gloxinia hybrida* as a means of vegetative propagation. Proc. Int. Hort. Congr., 19(1A): 55.

Clayberg, C.D., 1970a. Genetics and culture of a new laboratory plant, *Sinningia pusilla*. J. Hered., 61: 11—14.

Clayberg, C.D., 1970b. Cytology of interspecific hybrids in *Sinningia* and *Rechsteineria* (Gesneriaceae). Can. J. Genet. Cytol., 12: 759—768.

Raman, K., 1977. Rapid multiplication of *Streptocarpus* and gloxinia from in vitro cultured pedicel explants. Z. Pflanzenphysiol., 83(3): 411—418.

Van Raalte, D., 1969. Het Handboek voor de Bloemisterij, IV. Born Periodieken N.V., Assen, Amsterdam, 4th edn, 468 pp.

Nautilocalyx

One other species in the family Gesneriaceae is to be mentioned, not because any (mutation) breeding work has been carried out with it, but because of its remarkable regeneration capacity, in vivo and in vitro. The superficial cell layers of *Nautilocalyx lynchei*, excised and cultured in vitro, have organo-genetic potentialities which can be controlled at will by using certain combinations of auxin, cytokinin and sucrose under controlled environmental conditions (light and temperature) (Tran Thanh Van, 1973; Tran Thanh Van and Drira, 1971). Leaf fragments readily produce buds and roots (Chouard, 1938; Venverloo, 1974).

Literature Nautilocalyx

Chouard, P., 1938. C.R. Acad. Sci., 207: 597—599.

Tran Thanh Van, M., 1973. In vitro control of de novo flower, bud, root and callus differentiation from excised epidermal tissues. Nature (London), 246: 44—45.

Tran Thanh Van, M. and Drira, A., 1971. Definition of a simple experimental system of directed organogenesis de nuevo: organ neoformation from epidermal tissue of *Nautilocalyx lynchei*. In: Les Cultures de Tissus des Plantes, Strassbourg, 1970. Colloq. Int. C.N.R.S., 193: 169—176.

Venverloo, C.J., 1974. Exogenous regulation of the formation of roots or shoots in the same culture. In: H.E. Street (Editor), Third Int. Assoc. Plant Tissue and Cell Culture, Leicester. University of Leicester, Leicester, Abstr. No. 31.

HIBISCUS

The Chinese Rose, *Hibiscus rosa-sinensis* L., family Malvaceae, contains a number of tropical cultivars which are sold as pot plants in moderate climates.

The only report about mutation breeding in *Hibiscus* is by Das et al. (1974) who obtained several mutations in flower form (from double to semi-double and single) and colour, such as from carmine red to deep red and from deep mauve to white. The optimum dose is around 10—20 kR at a rate of 30—50 R/h. Our own experience is that an acute dose of approximately 5 krad X-rays at a rate of 75 rad/min is the optimum treatment for rooted cuttings.

Literature Hibiscus

Das, P., Ghosh, P., Dube, S. and Dhua, S.P., 1974. Induction of somatic mutations in some vegetatively propagated ornamentals by gamma irradiation. Technology (Coimbatore, India), 11(2,3): 185—188.

KALANCHOË

Kalanchoë belongs to the fam. Crassulaceae and most species are found in tropical Africa, for example, *K. marmorata* Baker (2n = 34) in Ethiopia, *K. flammea* Stafp (2n = 34) in Somalia, and *K. blossfeldiana* Poelln. (2n = 34) and *K. tubiflora* (Harvey) Hamet. in Madagascar. *K. blossfeldiana* has played an important part in the development of most commercial cultivars. It was introduced in 1932 by Robert Blossfeld to Potsdam, Germany. Hybrids and sports were developed during the following decades.

Kalanchoë blossfeldiana reproduces by seeds, whereas all commercial cultivars reproduce from rooted cuttings, either stem cuttings or "leaf" cuttings (a leaf and part of the stem including at least one axillary bud).

Broertjes and Leffring (1972) decsribed a propagation technique, by which adventitious buds are produced on detached leaves (Fig. 17). Young leaves with their petioles produced the highest number of plantlets of the cultivars Annette and Josine. This method was used to investigate whether complete mutants could be produced from detached, irradiated leaves. All mutants found turned out to be solid and non-chimeral, while in addition, a few phenotypically and probably also genetically identical mutants were

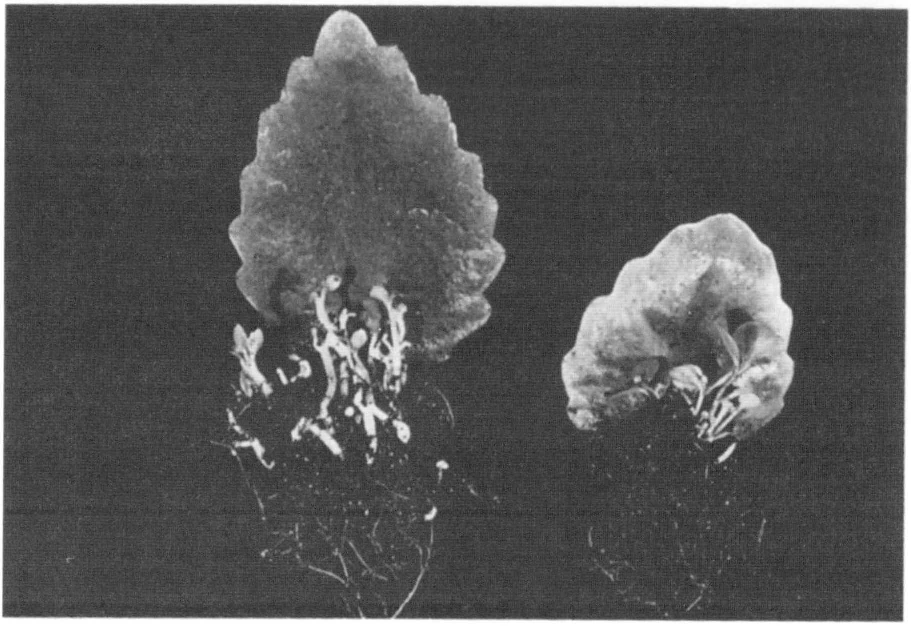

Fig. 17. Rooted leaves of *Kalanchoë* with adventitious plantlets developing on the basal part of the petiole and on callus formed at the base. (Photograph, ITAL, Wageningen.)

observed. This phenomenon was later observed in *Chrysanthemum* too (Broertjes et al., 1976).

A similar technique was used to produce mutants in *K. laciniata* by Nakornthap (1974) who reported mutations in leaf structure and habit. This was also reported by Shama Rao and Singh (1976) who, in addition, observed that radiation had a stimulative effect on bulbil formation in *K. daigremontiana*.

In vitro propagation of *K. pinnata* was described by Sebastian (1974). Since the in vivo adventitious bud technique cannot yet be applied to all commercial cultivars, such an in vitro propagation method could be important in mutation breeding.

The radiosensitivity of *Kalanchoë* depends on the species and the kind of plant material used. The optimum radiation dose for detached leaves is 2 krad X-rays or less (Broertjes and Leffring, 1972) or approximately 3—4 krad of γ-rays (Nakornthap, 1974). Exposure to similar doses has also been reported by Johnson (1948; small plants of *K. tubiflora*), Schwemmle and Röbbelen (1962; seed, pollen, egg cells, irradiated in situ) and Stein and Sparrow (1963, 1966). The latter irradiated seedlings of *Kalanchoë* cv. Brilliant Star with acute doses of 1.5 and 2 krad X-rays or chronically with γ-rays.

Commercial results have, to our knowledge, not been reported.

Literature Kalanchoë

Broertjes, C. and Leffring, L., 1972. Mutation breeding of *Kalanchoë*. Euphytica, 21: 415—423.
Broertjes, C., Roest, S. and Bokelmann, G.S., 1976. Mutation breeding of *Chrysanthemum morifolium* Ram. using in vivo and in vitro adventitious bud techniques. Euphytica, 25: 11—19.
Johnson, E.L., 1948. Response of *Kalanchoë tubiflora* to X-irradiation. Plant Physiol., 23: 544—556.
Nakornthap, A., 1974. Radiation-induced somatic mutations in *Kalanchoë* (*Kalanchoë laciniata*). Mut. Breed. Newsl., 3: 14.
Schwemmle, B. and Röbbelen, G., 1962. Haüfigkeit von Blattfarbmutationen nach Röntgenbestrahlung verschiedener ontogenetischen Stadien von *Kalanchoë*. Naturwissenschaften, 49: 64—65.
Sebastian, K.T., 1974. The effects of growth regulators on regeneration of *Kalanchoë pinnata* leaf fragments. In: H.E. Street (Editor), Third Int. Assoc. Plant Tissue and Cell Culture, Leicester. University of Leicester, Leicester, Abstr. No. 32.
Shama Rao, H.K. and Singh, Y., 1976. Gamma ray induced early bulbil formation and growth abnormalities in *Kalanchoë daigremontiana* H. and P. Indian J. Exp. Biol., 14(3): 358—360.
Stein, O.L. and Sparrow, A.H., 1963. The effects of chronic gamma irradiation on the growth of *Kalanchoë* cv. Brilliant Star. Radiat. Bot., 3: 207—222.
Stein, O.L. and Sparrow, A.H., 1966. The effect of acute irradiation in air, N_2 and CO_2 on the growth of the shoot apex and internodes of *Kalanchoë* cv. Brilliant Star. Radiat. Bot., 6: 187—203.

PELARGONIUM

Most of the approximately 250 species of the genus *Pelargonium* (fam. Geraniaceae) originate from South Africa. Some, such as *P. peltatum*, *P. zonale* and *P. graveolens*, were brought to Europe as early as 1609 by John Tradescant. Breeding, which started soon after introduction to Europe, led to the formation of several groups (e.g. *P. zonale* hybrids, *P. peltatum* hybrids, *P. grandiflora* hybrids) (Van Raalte, 1969).

Very little has been published about the ploidy of the different groups of cultivars. *P. zonale* cultivars seem to be diploid or tetraploid (2n = 17, 18, 35 and 36), whereas the chromosome number of *P. peltatum* is 36 (Darlington and Wylie, 1955) and that of *P. graveolens*, 72 (Skirvin and Janick, 1976a).

Most groups are propagated vegetatively, although there is a growing interest in the F_1 hybrids which are seed propagated. Recently, in vitro propagation has also been reported (Beauchesne, 1974; Skirvin and Janick, 1976a).

Spontaneous mutants (sports) have been reported, but the number found is small in comparison with that of commercial seedlings obtained via cross breeding. One of the reasons appears to be that sports are unstable and so, after a few generations, plants revert back to the original type.

Some of the literature concerning *Pelargonium* deals with experiments using radiation to induce uncovering and rearrangement in existing periclinal (tri-)chimeras, such as the work of Bergann with *P. zonale* cv. Madame Salleron (Bergann, 1967b; Bergann and Bergann, 1959) and a more general review by Pötsch (1964).

An interesting haploid (n = 9) *Pelargonium* cultivar is *P. zonale* cv. Kleiner Liebling (Daker, 1966). It is being used to study colchicine-induced diploids (Daker, 1967) as well as a system for testing mutagens (Pohlheim et al., 1972, 1976). The irradiation of plants with X-rays (optimum dose 1—1.25 krad) resulted in many mutations which affected chlorophyll content and growth rate of leaf cells.

The ontogeny of *Pelargonium* shoots was studied using mutant plastogenes in all possible chimeral combinations (Stewart et al., 1974).

Irradiation experiments at the Institute of Plant Breeding at Wageningen, The Netherlands, demonstrated that, starting from plants heterozygous for the three flower colour genes V, Sa and P (Craig, 1963), mutations in all three genes could be induced at rather low frequencies. Apart from changes in the shape of the petal, many orange—red or smaller flowers were observed. Leaf shape and leaf colour mosaics were also found (A.M. van Harten, unpublished data).

Commercial mutation breeding of *Pelargonium* has not been reported. Our own experience is restricted to a few investigations concerning an in vivo adventitious bud technique using cvs Salmon Beauty and Lärschen Müller, which occasionally produce shoots on rooted leaves. Small-scale irradiation with 0.5—1 krad X-rays, however, did not produce mutants (C. Broertjes, unpublished data). In vitro propagation techniques were more successful in producing variability (Skirvin and Janick, 1974, 1976a), even resulting in a calliclone that is now grown commercially, the *Pelargonium graveolens* cultivar Velvet Rose (2n = 144) (Skirvin and Janick, 1976b). Janick et al. (1977) compared in vitro and in vivo tissue culture systems in scented geranium and concluded, amongst other things, that adventitious shoots derived from single cells not belonging to the epidermal layer. Kameya (1975) used in vitro propagation of protoplasts to produce non-chimeral geranium plants, starting from a variegated one.

Literature Pelargonium

Beauchesne, G., 1974. Obtention de clone de *Pelargonium* en culture 'in vitro' à partir de méristèmes. Proc. Int. Hort. Congr., 19(1A): 64.

Bergann, F., 1967b. Mutations-Chimären: Rohmaterial züchterischer Weiterbehandlung. Umsch. Wiss. Tech., 67(24): 791—797.

Bergann, F. and Bergann, L., 1959. Über experimentell ausgelöste vegetative Spaltungen und Umlagerungen an chimärischen Klonen, zugleich als Beispiele erfolgreicher Staudenauslese. I. *Pelargonium zonale* Ait. 'Madame Salleron'. Züchter, 29(8): 361—374.

Craig, R., 1963. The Inheritance of Several Characters in the Geranium, *Pelargonium hortorum* Bailey. Thesis, Pennsylvania State Univ., University Park, Pa., 71 pp.

Daker, M.G., 1966. 'Kleiner Liebling', a haploid cultivar of *Pelargonium*. Nature (London), 211(5048): 549—550.

Daker, M.G., 1967. Cytological studies on a haploid cultivar of *Pelargonium*, and its colchicine-induced diploids. Chromosoma, 21(3): 250—271.

Darlington, C.D. and Wylie, A.P., 1955. Chromosome Atlas of Flowering Plants. Allen and Unwin, London, 519 pp.

Janick, J., Skirvin, R.M. and Janders, R.B., 1977. Comparison of in vitro and in vivo tissue culture systems in scented geranium. J. Hered., 68: 62—64.

Kameya, T., 1975. Culture of protoplasts from chimeral plant tissue of nature. Jpn. J. Genet., 50(5): 417—420.

Pohlheim, E., Pohlheim, F. and Günther, G., 1976. Mutagenicity testing of herbicides with a haploid *Pelargonium*. Eur. Environ. Mutagen Soc., Sixth Annu. Meet., Gernrode. P. 131.

Pohlheim, F., Pohlheim, E. and Günther, G., 1972. Die haploide *Pelargonium zonale* 'Kleiner Liebling' als Testsystem für Mutagene. Pädagog. Hochsch. Potsdam; Wiss. Z., 16: 65—70.

Pötsch, J., 1964. The Induction of Extramutative Radiation Effects on Periclinal Chimeras with Known Constitution of *Pelargonium zonale* Ait., *Euphorbia pulcherrima* Willd. and *Abutilon hybridum* Hort. Dissertation 64-1053, Potsdam, 214 pp.

Skirvin, R.M. and Janick, J., 1974. Calliclones in geranium. HortScience, 9(3): 270.

Skirvin, R.M. and Janick, J., 1976a. Tissue culture-induced variation in scented *Pelargonium* spp. J. Am. Soc. Hortic. Sci., 101(3): 281—290.

Skirvin, R.M. and Janick, J., 1976b. 'Velvet Rose' *Pelargonium*, a scented geranium. HortScience, 11(1): 61—62.

Stewart, R.N., Semeniuk, P. and Dermen, H., 1974. Competition and accommodation between apical layers and their derivatives in the ontogeny of chimeral shoots of *Pelargonium* ×*hortorum*. Am. J. Bot., 61(1): 54—67.

Van Raalte, D., 1969. Het Handboek voor de Bloemisterij, IV. Born Periodieken N.V., Assen, Amsterdam, 4th edn, 468 pp.

PEPEROMIA

Many *Peperomia* species (fam. Piperaceae) are botanical species or closely related to botanical species and therefore they are not very susceptible to mutation breeding. They can, however, easily be multiplied from adventitious buds growing on detached leaves or on leaf parts (Harris and Hart, 1964). It is not known whether the buds develop in the same way as those of Gesneriaceae. Pure green and pure yellow adventitious shoots on detached leaves from *Peperomia obtusifolia variegata* (Fig. 18) suggest that the shoots originate from one or a few cells only, as also indicated by a completely green (pigment-free) mutant of *P. resedaeflora* (C. Broertjes, unpublished data). The optimum dose for irradiation of *P. resedaeflora* and *P. polybotrya* leaves seems to be greater than 2—3 krad X-rays as this dose did not affect survival or plantlet production.

Literature Peperomia

Harris, G.P. and Hart, E.M.H., 1964. Regeneration from leaf squares of *Peperomia sandersii* A. DC: a relationship between rooting and budding. Ann. Bot. (London), 28: 509—526.

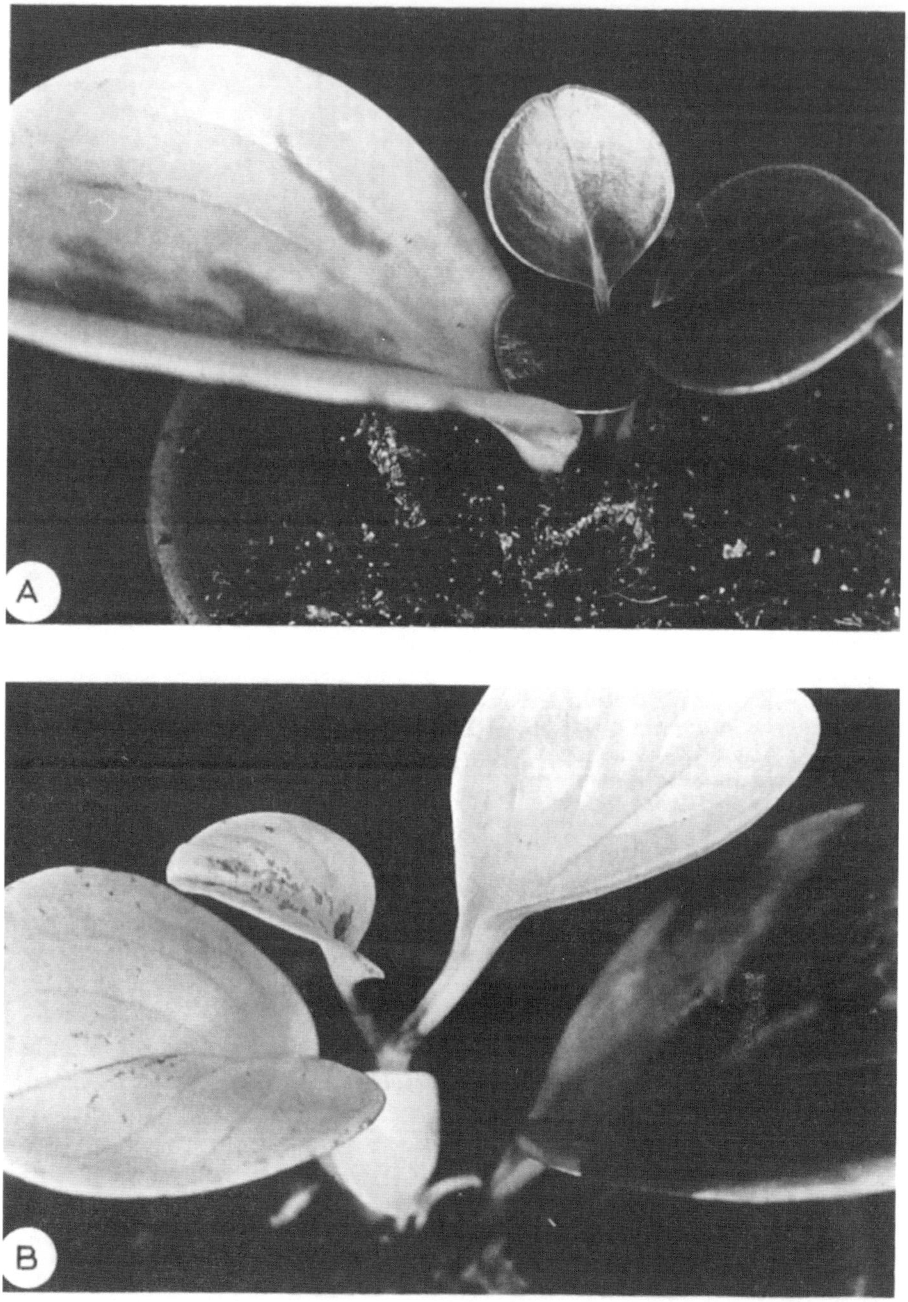

Fig. 18. Leaves of *Peperomia obtusifolia variegata*, a yellow-green variegated cultivar. The adventitious plantlets growing on detached leaves are either completely green (A) or almost completely yellow (B). They are thus very likely to have developed from few or only one cell. (Photographs, ITAL, Wageningen.)

RHODODENDRON (INCLUDING *AZALEA*)

The genus *Rhododendron* belongs to the family Ericaceae and is divided into four subgenera, the most important being (i) Eurhododendron: evergreen plants, commonly named *Rhododendron*, and (ii) Anthodendron: deciduous plants, usually called *Azalea*. The second group is divided into several species:

(1) *Rhododendron japonicum* Suringar, better known as *Azalea mollis*;
(2) *Rhododendron luteum* Sw. (*Azalea pontica* L.);
(3) *Rhododendron molle* G. Don. (*Azalea sinensis*);
(4) *Rhododendron obtusum* Planch., known as Japanese azalea;
(5) *Rhododendron simsii* Planch., the most important group, known as *Azalea indica*.

Furthermore, various hybrids among these groups are recognized, including *Azalea mollis* × *sinensis*, the *Azalea malvatica* hybrids and many others (Grootendorst, 1954; Van Raalte, 1969). In several of the species mentioned above, small-scale mutation breeding projects are in progress (co-operative programmes between the Experimental Station for Arboriculture, Boskoop, and the Association Euratom—ITAL, Wageningen) without commercial results so far. *Rhododendron simsii*, however, has recently been the subject of mutation breeding on a relatively large scale at Melle, Belgium (R. de Loose), and Dresden-Pillnitz, German Democratic Republic (H. Streitberg), resulting in commercial varieties (Table 14).

Eurhododendron (Rhododendron)

The most important *Rhododendron* cultivars belong to the section *Rhododendron*, for example, *Rhod. catawbiense* hybrids and *Rhod. ponticum* hybrids. Only a few species and cultivars such as *Rhod. impeditum*, *Rhod. ponticum* and *Rhod. catawbiense* 'Album' and Boursault', have been irradiated. Young, actively growing plants were generally used. Because of the high value of space and plants, the size of the plants and the long time between irradiation and flowering, however, the experiments were too small and no results were obtained. The radioresistance of *Rhododendron* is relatively high: a dose of 4—6 krad X-rays seems to be optimum; higher doses cause reduction in growth and abnormalities (C. Broertjes, personal communication, 1974).

Anthodendron (Azalea)

Rhododendron japonicum (Azalea mollis). Experience has been had of only one cultivar included in this group of *Rhododendron* by Grootendorst (1954), namely, *Rhod. altaclarensis* 'Christoffer Wren'. So far, no mutations have been observed. The optimum radiation dose for rooted cuttings or

TABLE 14

Commercial mutants of *Rhododendron simsii* (syn. *Azalea indica*)

Name of new cultivar	Place and date of release (or approval) and name of principal worker(s) and institute	Conditions and date of mutagenic treatment [Parent cultivar]	Main improved attributes of cultivar
Enzet-Rolko	German Democratic Republic, 1969 H. Streitberg, Institut für Obstbau, Dresden-Pillnitz	2 kR X-rays, 1959 [Ernst Thiers]	Smaller petals resulting in elegant flower; colour darker than original cultivar
Enzet-Rokola	German Democratic Republc, 1969 H. Streitberg, Institut für Obstbau, Dresden-Pillnitz	5.5 kR X-rays, 1959 [Mme John Haerens]	More intense flower colour (changed from pink to bright red); no change in flower colour when wilting
Cobalt	Japan, 1973 Masaru Akabane and Kanuma Branch, Tochigi Prefectural Agricultural Experiment Station	5 kR γ-rays, 1964 [Takasago]	Small petals and leaves due to dwarfish growth; more elegant than original cultivar
Adinda	Belgium, 1974 R. De Loose, Rijksstation voor Sierplantenteelt, Melle	Recurrent irradiation with ^{60}Co γ- and X-rays, 1965—1970 [Karl Glaser]	Yellow—red flower colour

Eroïca	Belgium, 1974 R. De Loose, Rijksstation voor Sierplantenteelt, Melle	Recurrent irradiation with ^{60}Co γ- and X-rays, 1965—1970 [Knut Erwen]	Orange flower colour
Mira	Belgium, 1974 R. De Loose, Rijksstation voor Sierplantenteelt, Melle	Recurrent irradiation with ^{60}Co γ- and X-rays, 1965—1970 [Euratom]	Orange flower colour
Mevr. R. de Loose	Belgium, 1974 R. De Loose, Rijksstation voor Sierplantenteelt, Melle	Recurrent irradiation with ^{60}Co γ- and X-rays, 1965—1970 [De Waele's Favorite]	Flower colour change from blue—red with white edge to yellow—red with white edge (flavonol synthesis blocked)
Pastorale	Belgium, 1974 R. De Loose, Rijksstation voor Sierplantenteelt, Melle	Recurrent irradiation with ^{60}Co γ- and X-rays, 1965—1970 [De Waele's Favorite]	Bluish red flower with narrow white edges
Saidjah	Belgium, 1974 R. De Loose, Rijksstation voor Sierplantenteelt, Melle	Recurrent irradiation with ^{60}Co γ- and X-rays, 1965—1970 [Euratom]	Yellow—red flower colour
Sierra Nevada	Belgium, 1974 R. De Loose, Rijksstation voor Sierplantenteelt, Melle	Recurrent irradiation with ^{60}Co γ- and X-rays, 1965—1970 [De Waele's Favorite]	Yellow—red flower with narrow white edges

Sources: IAEA (1972, pp. 526—544) and *Mutation Breeding Newsletter*, 1 (1972) —9 (1977).

young plants lies between 4 and 6 krad X-rays (C. Broertjes, personal communication, 1974).

Rhododendron obtusum (Japanese azalea). Most experiments using this species have been with the cultivar Silvester, a pink flowering pot plant, extremely suitable for very early forcing, and the cultivar Vuyk's Scarlet, one of the best Japanese azaleas with large flowers for outdoor growing. This species is much more radiosensitive than other *Rhododendron* species and the optimum dose is 1 500 rad X-rays, if recently rooted cuttings or young plants are irradiated.

Large numbers of plants, especially of cv. Silvester, have been irradiated, and various flower colour mutants were obtained (dark pink and orange, for instance), propagated and compared with each other (Miss Dr A. Heyting, Boskoop, personal communication, 1974). Several other cultivars were also subjected to irradiation. De Loose (1973a) reported several flower colour mutations in 'Kirin' (pink to white), and 'Toreador', 'Moederkensdag' and 'Brazier' (blue—red to yellow—red).

Rhododendron simsii (syn. *Azalea indica*). By far the largest amount of experience has been gained with this economically very important group of plants. At various places in Western Europe large-scale experiments and more or less practical mutation breeding is reported, such as that by R. de Loose (Melle, Belgium) and H. Streitberg (Dresden-Pillnitz, German Democratic Republic), while some work is going on in The Netherlands too, in co-operation with the Association Euratom—ITAL at Wageningen (C. Broertjes, personal communication, 1974).

Rhod. simsii (syn. *Azalea indica*) is an evergreen shrub that is propagated either by rooted cuttings or by grafting onto a rootstock. To ensure branching and the formation of a nicely shaped plant, repeated pinching of the shoots is recommended.

Irradiation with X- or γ-rays is generally carried out just before or after pinching. Either an acute dose of 4—6 krad is given or repeated irradiation with a generally lower dose, of 2 krad, for example, is given two, three, four or more times, usually at the time a plant is being pinched (De Loose, 1969b, 1970d, 1971a,b). Other workers, however, find (much) lower doses better: Streitberg (1965, 1966a,b, 1967a,b), for instance, obtained the greatest variation after a dose of 1 krad. In contrast to De Loose's procedure, irradiations were generally given during late summer. Radiosensitivity is highly dependent on the cultivar, some being (much) more sensitive than others. The doses are chosen to kill the main shoot meristem but to allow the axillary ones to remain alive and develop into new shoots.

The pigment composition in general and the nature of the flower colour mutants in particular have been studied extensively (De Loose, 1968, 1969a, 1970a,b, 1974d,e; Heursel, 1972, 1975) both being possible bases for cross breeding (De Loose, 1970c) and for mutation breeding of the species in general.

The main pigments are various anthocyanins (e.g. cyanidin, peonidin and malvidin) with various flavonols as co-pigments (e.g. azaleatin, quercetin, myricetin and kaempferol) (Heursel, 1972, 1975). The co-pigments are often lost when blue-red cultivars are irradiated, resulting in red and orange mutants. Further mutations can dilute the anthocyanins and result in pink and eventually white mutants. In addition, independent of all colours, mutations for white petal edges (of various widths) can be spontaneously obtained or artificially induced.

The spontaneous mutation rate for colour, form, size etc. is low and it may take decades before the full range is naturally obtained, despite the enormous number of plants that are produced and forced into bloom every year. Therefore, leading and relatively new commercial varieties, from which no — or only a restricted number — "sports" have developed, are generally selected for artificial mutation induction. It is important to get the full range of flower colour "sports" of such fast growing and easily forcible cultivars which also are generally early flowering, easily form flower buds, and have other good characteristics (De Loose, 1973a,b, 1974a; Streitberg, 1967a).

Numerous cultivars have been irradiated (De Loose, 1970d; Streitberg, 1966a,b) and hundreds of mutants were produced, several of which have been used in commerce since 1974 (De Loose, 1974b,c), as seen in Table 14.

Literature Rhododendron

De Loose, R., 1968. Kwalitatief onderzoek naar de bloemkleurpigmenten bij de Belgische hybriden van *Rhododendron simsii* Planch. (*Azalea indica* Linn.). Meded. Vlaam. Chem. Ver., 30(4): 99—123.

De Loose, R., 1969a. The flower pigments of the Belgian hybrids of *Rhododendron simsii* and other species and varieties from *Rhododendron* subseries *obtusum*. Phytochemistry, 8: 253—259.

De Loose, R., 1969b. Erfahrungen mit radioaktiven Bestrahlungen von Zierpflanzen. Gartenwelt, 15: 348—350.

De Loose, R., 1970a. Flower pigment composition of natural bud-variants among hybrid Chinese azaleas, *Rhododendron simsii* (Planch.). J. Hortic. Sci., 45: 265—274.

De Loose, R., 1970b. Flavonoid glycosides in the petals of some *Rhododendron* species and hybrids. Phytochemistry, 9: 875—879.

De Loose, R., 1970c. Perspektieven voor het bekomen van een geelbloeiende *Azalea indica*. Standpunt van een scheikundige. Belg. Tuinbouw, 51: 307—310.

De Loose, R., 1970d. Het bekomen van 'sporten' voor de bloemkleur bij de Belgische hybriden van *Rhododendron simsii* Planch. (= *Azalea indica*) door middel van ^{60}Co-gammastralen. Meded. Fac. Landbouwwet. Rijksuniv. Gent, 35(4): 1047—1074.

De Loose, R., 1971a. Mutatie-veredeling bij siergewassen met behulp van ioniserende stralingen. In: Grondslagen en Resultaten van de Mutatieveredeling. Rijkscentrum van Landbouwkundig Onderzoek, Ghent, 34—56.

De Loose, R., 1971b. Mutation research on the Belgian hybrids of *Rhododendron simsii* (Planch.). In: Eucarpia Meeting on Ornamentals, Wageningen. Inst. Hortic. Plant Breed., Wageningen, pp. 1—8.

De Loose, R., 1973a. Bestralingen, mutatie- en isotopenonderzoek. Aktiviteitsverslag Rijksstation voor Sierplantenteelt, Melle, 1966—1971, VI. Rijksstn Sierplantenteelt, Melle, pp. 91—102.

De Loose, R., 1973b. Het bekomen van 'sporten' voor de bloemkleur bij hybriden van *Rhododendron simsii* Planch. (*Azalea indica* L.) met behulp van ioniserende stralingen. Meded. Fac. Landbouwwet. Rijksuniv. Gent, 38(42): 2090—2109.

De Loose, R., 1974a. Nieuwe cultivar op komst van *Azalea indica* L.: een bonte sport van cv. De Waele's Favorite, bekomen door bestraling. Vakbondsnieuws Belg. Sierteelt, 18(4): 125—129.

De Loose, R., 1974b. Mutation breeding of the hybrids of *Rhododendron simsii* Planch. (*Azalea indica* L.). Mutat. Breed. Newsl., 3: 15.

De Loose, R., 1974c. Mutatie-veredelingsonderzoek op de *Azalea indica* L. Agricontact, Minist. Landbouw Belg., 34: 1—5.

De Loose, R., 1974d. Het chromatografie-patroon van de bloemkleurpigmenten bij de hybriden van *Rhododendron simsii* Planch. en *Rhododendron obtusum* Planch. (*Azalea indica* and *Azalea japonicum*). Meded. Fac. Landbouwwet. Rijksuniv. Gent, 39(1): 238—254.

De Loose, R., 1974e. Het chromatografie-patroon van de bloemkleurpigmenten bij de hybriden van *Rhododendron simsii* Planch. en *Rhododendron obtusum* Planch. (*Azalea indica* L. and *Azalea japonica* L.) Deel B. Bloemkleur en bloemkleurpigmenten bij spontane en kunstmatige sporten van *Azalea indica* L. en *Azalea japonica* L. Meded. Fac. Landbouwwet. Rijksuniv. Gent, 39(3): 1369—1384.

Grootendorst, H.J., 1954. Rhododendrons and Azaleas. Ver. Boskoop Cult., Boskoop Handelsdrukkerij, Boskoop, 216 pp.

Heursel, J., 1972. Studie van de Overerving en de Genotypische Variabiliteit van Kleur, Grootte en Vorm van de Bloemen bij *Rhododendron simsii* Planch. (*Azalea indica* L.). Rijkssn Sierplantenteelt, Melle, Meded. 28, 196 pp.

Heursel, J., 1975. Inheritance of the flavonols azaleatin and quercetin in *Rhododendron simsii* Planch. and *Rh. obtusum* Planch. Z. Pflanzenzücht., 74: 62—70.

IAEA, 1972. Induced Mutations and Plant Improvement. IAEA, Vienna, pp. 526—544.

Streitberg, H., 1965. Teilabschlussbericht zur Forschungsarbeit 'Züchterische Verbesserung volkswirtschaftlich wichtiger Zierpflanzen'. Teil a. Züchtung Azaleen unter Glas. Dtsch. Akad. Landwirtsch. Wiss., Berlin.

Streitberg, H., 1966a. Schaffung von Sprossvarianten bei Azaleen durch Behandlung mit Röntgenstrahlen. Z. Pflanzenzücht., 56(1): 70—87.

Streitberg, H., 1966b. Neue Rosen- und Azaleen-Sorte mit Hilfe der Röntgenbestrahlung. Dtsch. Gartenbau, 13: 267—268.

Streitberg, H., 1967a. Production of economically valuable variation in roses and azaleas by means of X-irradiation. In: H. Stubbe (Editor), Induced Mutations and their Utilization, Gatersleben, 1966. Abh. Dtsch. Akad. Wiss. Berlin Kl. Medizin, 2: 359—362.

Streitberg, H., 1967b. Schaffung von Sprossvarianten bei Azaleen durch Behandlung mit Röntgenstrahlen. Arch. Gartenbau, 15: 101—125.

Van Raalte, D., 1969. Het Handboek voor de Bloemisterij, IV. Born Periodieken N.V., Assen, Amsterdam, 4th edn, 468 pp.

1 2 3

4 5 6

Plate 6. *Alstroemeria* cv. Orchid fl. (1), surrounded by several radiation-induced mutants (2), some of which have been introduced to the market and are commercially very successful, such as cv. Canaria (3), cv. Zebra (4), cv. Yellow Tiger (5) and cv. White Wings (6). (Slides, ITAL, Wageningen. Plate sponsored by Fa. M.C. van Staaveren, Aalsmeer.)

Foliage plants

GENERAL

There is a large variety of vegetatively propagated pot plants that are ornamentally valuable because of the shape or colour of the plant and foliage. Examples are *Ampelopsis*, *Asparagus*, *Begonia rex*, *Cissus*, *Chlorophytum*, *Codiaeum*, *Coleus*, *Dieffenbachia*, *Dracaena*, *Fatshedera*, *Ficus*, *Hedera*, *Marantha*, *Monstera*, *Papyrus*, *Philodendron*, *Pilea*, *Rhoicissus*, *Sansevieria*, palm tree varieties and ferns.

Although spontaneous mutations must have played an important role in the development of the cultivars of many of them such as *Begonia rex* and *Hedera*, very little effort has been spent on artificially inducing mutations in such plants.

COLEUS

One of the species in which radiation has been used to induce mutations is *Coleus*. Non-genetic disturbances after the irradiation of cuttings with γ-radiation were observed in *Coleus blumei* (Love and Mullenax, 1964), while fast neutron irradiation of the cultivar Scarlet Red with 200—1 000 rad resulted in large numbers of variegated plants. One dwarf-type mutant was found (Love and Constantin, 1965, 1966). In a later publication it was shown that the differences between the mutants were due to variations in the amount of one of the anthocyanin pigments rather than differences in the structure of the pigment molecule itself. In other words, it was shown that fast neutrons are capable of modifying pigment synthesis in *Coleus* (Love and Malone, 1967).

Literature Coleus

Love, J.E. and Constantin, M.J., 1965. The response of some ornamental plants to fast neutrons. Tenn. Farm Home Sci., 56: 10—12.

Love, J.E. and Constantin, M.J., 1966. The induction of bud sports in *Coleus blumei* by fast neutrons. Proc. Am. Soc. Hortic. Sci., 88: 627—630.

Love, J.E. and Malone, B.B., 1967. Anthocyanin pigments in mutant and non-mutant *Coleus* plants. Radiat. Bot., 7: 549—552.

Love, J.E. and Mullenax, R.H., 1964. Exposure of coleus plants to gamma irradiation. Tenn. Farm and Home Sci., 50: 18—19.

FICUS

Rooted cuttings of *Ficus benjamina* and *F. diversifolia* were irradiated with a series of doses, varying in size up to 5 krad. One variegated mutant was obtained from *F. benjamina* which was not put on the market, however, because the variegated characteristic was insufficiently expressed in later generations (R. de Loose, personal communication, 1977).

HEDERA

Only one report concerning the irradiation of *Hedera helix* has been found: that by Knuth (1962). He irradiated several cultivars with various X-ray doses. The optimum acute dose was 4 krad. As far as we know, only one mutant was obtained, a mutant with more intense variegation than the mother cultivar, Glacier.

Literature Hedera

Knuth, M., 1962. Mutationen durch Röntgenbestrahlung auch bei Grünpflanzen möglich. Dtsch. Gartenbau, 9(6): 162—164.

FERNS

In ferns, such as *Adianthum*, *Asplenium*, *Platycerium* and *Pteris*, only a few preliminary irradiation experiments were carried out, but without success (C. Broertjes, unpublished data). Spores were irradiated with a series of doses to determine their radioresistance. Optimum X-ray doses were as follows: *Adianthum scutum*, 30 krad; *Asplenium nidus*, 10 krad; *Platycerium alcicorne*, over 50 krad, and *Pteris cretica albo lineata*, 30 krad.

It is questionable whether the irradiation of spores, as was also done by Howard and Haigh (1968), is the most appropriate way of obtaining mutants. A fairly strong tendency for "haplontic" selection between the mutated and the non-mutated (haploid) spores, as well as between the mutated and non-mutated (haploid) prothallia must exist. Moreover, the chance of a mutated spermatozoid (microgamete) fertilizing a mutated egg cell (macrogamete) seems to be restricted.

One wonders whether irradiation of prothallia or young fern plants growing from the diploid zygotes would result in ferns heterozygous for a given character and producing normal and mutated spores in great quantities. The offspring of one plant should be kept together so that the chance of mutated egg cells and spermatozoids produced on the mutant prothallium forming zygotes is greatly increased. In a sample of offspring from one irradiated plant, several mutant types should be found. (This idea was discussed with E.J. Klekowski Jr, Amherst, Mass., 1972.)

The combined use of in vitro techniques and mutagenic treatment could also be considered as a method to increase genetic variability in ferns. This seems to be possible since several authors have reported in vitro propagation of various ferns (Bristow, 1962; Caponetti, 1972a,b; Murashige, 1974 and

Peterson, 1970), while a high level of post-zygotic chromosome damage caused by a polluted environment was reported in *Osmunda regalis* (Klekowski, 1976).

Literature ferns

Bristow, J.M., 1962. The controlled in vitro differentiation of callus derived from a fern, *Pteris cretica* L., into gametophytic or sporophytic tissue. Dev. Biol., 4: 361—375.

Caponetti, J.D., 1972a. Morphogenetic studies on excised leaves of *Osmunda cinnamomea*: developmental capabilities of excised leaf primordial apices in sterile culture. Bot. Gaz. (Chicago), 133(3): 331—335.

Caponetti, J.D., 1972b. Morphogenetic studies on excised leaves of *Osmunda cinnamomea*: morphological and histological effects of sucrose in sterile nutrient culture. Bot. Gaz (Chicago), 133(4): 421—435.

Howard, A. and Haigh, M.V., 1968. Chloroplast aberrations in irradiated fern spores. Mutat. Res., 6(2): 263—280.

Klekowski, Jr, E.J., 1976. Mutational load in a fern population growing in a polluted environment. Am. J. Bot., 63(7): 1024—1030.

Murashige, T., 1974. Plant propagation through tissue culture. Annu. Rev. Plant Physiol., 25: 135—166.

Peterson, R.L., 1970. Bud development at the root apex of *Ophioglossum petiolatum*. Phytomorphology, 20(2): 183—189.

Cut flowers

ALSTROEMERIA

The genus *Alstroemeria*, a herb with leafy stems, is a member of the monocotyledonous family Amaryllidaceae. A Swedish botanist, Von Alstroemer, brought the first seeds of *A. pelegrina* from South America to Europe. Most, if not all, species originate from South America, mainly from central Chile and central Brazil, but also from Bolivia, Paraguay and Argentina (Koornneef, 1972; Van Raalte, 1971).

The species *A. aurantiaca*, *A. ligtu* and *A. pelegrina* are probably the parents of the first of the modern hybrids, which have larger flowers than the original species and are extremely long lasting as cut flowers (Oosthoek, 1967).

Many of the recent hybrids were developed by Goemans (1962) (Parigo Horticultural Co., Spalding, Great Britain). By crossing a sport of an unspecified species with a second species and subsequently with *A. aurantiaca* Goemans obtained the "Parigo hybrids", which include the pink-flowered 'Ballerina', 'Pink Attraction' and 'Pink Perfection', the red-flowered 'Carmen' and 'Pimpernel', the yellow-flowered 'Sussex Gold' and the pinky yellow-flowered cultivar Parigo's Charm, to mention just a few. A somewhat different type, *Alstroemeria* cv. Orchid fl (synonym cv. Walter Fleming), a white and yellow-flowered cultivar, originated much earlier, probably as a spontaneous interspecific hybrid. The cultivars Beauty, Regina and others are recent products of Van Staaveren's breeding programme.

The majority of these cultivars are sterile, probably either because they are interspecific hybrids with the diploid chromosome number (2n = 16) (cvs Beauty and Orchid fl) or because one of the parents (*A. aurantiaca*?) was an unknown spontaneous tetraploid resulting in triploid sterile seedlings (most of the Parigo hybrids). Triploids, such as cv. Regina, have also been observed as the results of crossing diploids, most likely because one of the gametes had an unreduced chromosome number. (So far, we have not checked the chromosome numbers but triploidy seems to be confirmed by (small) differences in radiosensitivity between the diploid and triploid categories.)

All these sterile hybrids are among the best of the present-day *Alstroemeria* cultivars. They are attracting a rapidly growing interest among cut flower growers because of the flower form, the increasing choice of flower colours, the favourable lasting quality and also because of the growing number of techniques for cropping (Anonymous, 1969; Van Dordt, 1969; Lelieveld, 1972a,b; Van Raalte, 1971, pp. 48—51; Verboom, 1972).

To our knowledge, nothing has been published about the genetics of *Alstroemeria*. Neither the number nor the nature of its flower colour genes, for example, is known, so that its suitability for mutation breeding has had to be investigated by trial and error. The same is true for other characters, like flower size and form, plant height and number or quality of the flowers. Spontaneous mutations occur only seldom.

One of the available methods of vegetative propagation involves dividing plants and making use of the fairly large number of fast-growing rhizomes that grow continuously and regularly develop shoot apices which later become flower-bearing shoots. Since a mutated cell has the largest chance of survival if induced at the youngest possible stage of development of meristem or apex, the best material to irradiate appears to be the actively growing rhizomes of young plants (Fig. 19).

Fig. 19. Actively growing rhizomes of a young *Alstroemeria* plant with buds in various stages of development (Broertjes and Verboom, 1974, Fig. 1). (Photograph, ITAL, Wageningen.)

TABLE 15

Commercial mutants of *Alstroemeria*

Name of new cultivar	Place and date of release (or approval) and name of principal worker(s) and institute	Conditions and date of mutagenic treatment [Parent cultivar]	Main improved attributes of cultivar
Yellow Tiger stavero	The Netherlands, 1970 M.C. van Staaveren, Aalsmeer *	±500 rad X-rays, 1967 [Orchid fl]	Yellow flower colour with striking black stripes (original cultivar white with some yellow)
Canaria stagula	The Netherlands, 1970 M.C. van Staaveren, Aalsmeer *	X-rays, 1967 [Orchid fl]	Yellow flower colour (original cultivar white with some yellow)
White Wings staretto	The Netherlands, 1972 M.C. van Staaveren, Aalsmeer *	X-rays, 1967 [Orchid fl]	White, except for yellow ears
Harmony stabrons	The Netherlands, 1972 M.C. van Staaveren, Aalsmeer *	X-rays, 1968 [Regina]	Bronze flower colour
Rosita stareza	The Netherlands, 1972 M.C. van Staaveren, Aalsmeer *	X-rays, 1968 [Regina]	Pink flower colour
Zebra stazeb	The Netherlands, 1975 M.C. van Staaveren, Aalsmeer *	X-rays, 1968 [Orchid fl]	Heavily striped flower
Harlequin	The Netherlands, 1973 A. Wülfinghoff, Rijswijk *	X-rays, 1970 [Paringo's Charm]	Orange—yellow flower colour
Rosali staliro	The Netherlands, 1975 M.C. van Staaveren, Aalsmeer *	X-rays, 1971 [Starosa]	Pink flower colour
Capitol	The Netherlands, 1977 A. Wülfinghoff, Rijswijk *	X-rays, 1972 [Carmen]	Salmon-pink flower colour
Fanfare	The Netherlands, 1977 A. Wülfinghoff, Rijswijk *	X-rays, 1972 [Carmen]	Red flower colour
Result	The Netherlands, 1977 A. Wülfinghoff, Rijswijk *	X-rays, 1972 [Carmen]	Bright red flower colour
Trident	The Netherlands, 1977 A. Wülfinghoff, Rijswijk *	X-rays, 1972 [Carmen]	Pink flower colour
Valiant	The Netherlands, 1977 A. Wülfinghoff, Rijswijk *	X-rays, 1972 [Carmen]	Light red flower colour
Zenith	The Netherlands, 1977 A. Wülfinghoff, Rijswijk *	X-rays, 1972 [Carmen]	Orange—red flower colour

* In co-operation with the Association Euratom—ITAL, Wageningen.

Sources: IAEA (1972, pp. 526—544) and *Mutation Breeding Newsletter*, 1 (1972) —9 (1977) and Association Euratom—ITAL, Wageningen.

According to Broertjes and Verboom (1974), the best time for irradiation is during March or April. First selections can then often be made in May or June and promising mutants can be isolated by cutting them from the rhizome. Following this, a second selection is carried out during the next flowering period. The optimum dose of X-rays is between 350 and 500 rad for diploid, and 500 and 700 rad for triploid cultivars.

Most experience was gained from the cultivars Orchid fl (synonym cv. Walter Fleming) and several other cultivars, including Regina, Beauty, Edison and Starosa. Many mutants were produced, especially of 'Orchid fl', and several of them have been put on the market (Plate 6) (Broertjes and Verboom, 1974). Numerous mutants have also been produced by irradiation of the cultivar Carmen, one of the triploid 'Parigo hybrids', several of these mutants too have been put on the market (Table 15).

An advantage of using young rhizomes for irradiation was the finding that nearly all mutants with altered directly visible characters seemed to be solid (non-chimeral). This phenomenon could not at the time be explained by the authors (Broertjes and Verboom, 1974). It was later suggested that a sympodial growth pattern could be responsible. Several species have a sympodial growth pattern, for example, *Galanthus* (Müller-Doblies, 1971) and a few other monocotyledons, including *Alstroemeria* (Buxbaum, 1951). In a sympodial growth system the chimeral problem is automatically solved, as found in tomato by Verkerk (1971), since axillary buds systematically become the main bud. In effect, this situation is comparable to the repeated pruning which is practised in crops like *Chrysanthemum* and leads to apparently solid mutants which are actually periclinal chimeras. It has not yet been investigated whether the apparently solid mutants in *Alstroemeria* are periclinal chimeras or completely mutated plants.

A second and fairly new method of propagation uses explants from inflorescence stems and the subsequent regeneration of adventitious plantlets in vitro (Ziv et al., 1973). This propagation technique might be an effective way of producing solid mutants if, as described in *Chrysanthemum* (Broertjes et al., 1976), the adventitious plantlets of *Alstroemeria* each develop from one original cell.

Literature Alstroemeria

Anonymous, 1969. These new hybrids could be winners under glass. Grower, 3832: 877—878.

Broertjes, C. and Verboom, H., 1974. Mutation breeding of *Alstroemeria*. Euphytica, 23: 39—44.

Broertjes, C., Roest, S. and Bokelmann, G.S., 1976. Mutation breeding of *Chrysanthemum morifolium* Ram. using in vivo and in vitro adventitious bud techniques. Euphytica, 25(1): 11—19.

Buxbaum, F., 1951. Die Grundachse von *Alstroemeria* und die Einheit ihres morphologischem Typus mit dem der echten Liliaceen. Phytomorphology, 170—184.

Goemans, J.A.M., 1962. Breeding of *Alstroemeria*. J.R. Hortic. Soc., 87(6): 282–284.
IAEA, 1972. Induced Mutations and Plant Improvement. IAEA, Vienna, pp. 526–544.
Koornneef, M., 1972. Karyotype analyse bij *Alstroemeria*. Verslag van het Ingenieurs-onderzoek voor Plantensystematiek, pp. 3–8.
Lelieveld, H.P.J., 1972a. De teelt van *Alstroemeria* (1). Vakbl. Bloemisterij, 27(27): 8–9.
Lelieveld, H.P.J., 1972b. De teelt van *Alstroemeria* (2). Vakbl. Bloemisterij, 27(28): 8–9.
Müller-Doblies, D., 1971. *Galanthus* ist doch sympodial gebaut! Ber. Dtsch. Bot. Ges., 84(11): 665–682.
Oosthoek, H., 1967. *Alstroemeria*-hybriden aantrekkelijk in meer dan één opzicht. Vakbl. Bloemisterij, 22: 1572–1573.
Van Dordt, H.W., 1969. Nieuwe *Alstroemeria*. Vakbl. Bloemisterij, 24(23): 846–847.
Van Raalte, D., 1971. In: Het Handboek voor de Bloemisterij. Snijbloemen. Born Periodieken N.V., Assen, Amsterdam, 4th edn, pp. 48–51.
Verboom, H., 1972. Het gebruik van kunstlicht bij de teelt van Vasa-*Alstroemeria*. Vakbl. Bloemisterij, 27(44): 20–21.
Verkerk, K., 1971. Chimerism of the tomato plant after seed irradiation with fast neutrons. Neth. J. Agric. Sci., 19: 197–203.
Ziv, M., Kanterovitz, R. and Halevy, A.H., 1973. Vegetative propagation of *Alstroemeria* in vitro. Sci. Hortic., 1: 271–277.

CARNATION

Of the various *Dianthus* species, belonging to the family of the Caryophyllaceae, *D. caryophyllus* L., the glasshouse carnation, is of greatest commercial interest. It gradually has become one of the most important cut flower crops since the development of the cultivar William Sim, in the U.S.A. in 1938, and particularly since an impressive number of (flower colour) sports and high quality strains were obtained from this cultivar by selection and stabilization within the genetic variation that arose spontaneously (Holly and Baker, 1963). Besides these strains, Chabaud carnations (spray types) have more recently attracted growing attention and, in 1975, already constituted approximately 50% of the total carnation production in The Netherlands. Most glasshouse carnations appear to be diploid (2n = 30) although a few tetraploids have been reported (Howard, 1968).

The first radiation-induced flower colour and single-flowered mutants in carnation were reported by Richter and Singleton (1955). They observed not only somatic mutations but also histological changes which were supposedly ascribed to the genetic replacement of the epidermal layer by deeper lying tissue. It has long been known that many sports of William Sim are periclinal chimeras, the somatic variants being genetically different from William Sim in the epidermal layer only (Heslot, 1964; Mehlquist et al., 1954). Both Heslot and Mehlquist have published a number of papers in which they studied the effect of radiation on morphological alterations in the shoot apex which lead to uncovering or rearrangement of periclinal chimeras (sports). It was found that high percentages of such tissue rearrangements could be obtained with a medium to a high radiation dose, compared to a relatively low mutation frequency with low doses (Sagawa, 1957; Sagawa and Mehlquist, 1956, 1957, 1959; Mehlquist and Sagawa, 1959, 1964). Although the commercial results are limited, these authors should be given credit for being among the first to stimulate interest in mutation breeding in general and its application to ornamentals in particular.

Comparable work has been carried out by several scientists in the years following this period. Induced chlorophyll deficiency and topographic analysis of partly mutated shoots were used to interpret meristem function (Bugnon et al., 1965; Dulieu, 1968, 1969; Gaufillier, 1965). Developmental morphology and the ontogeny of *Dianthus* tissues was also studied by Stewart (1965) using plastogene mutants.

The chimeral nature of most carnation sports has been investigated by several authors who used radiation as a tool for cell inactivation and tissue rearrangement (Farestveit and Klougart, 1966; Farestveit, 1969; Grabowska and Mynett, 1964a). Farestveit found that of 56 cultivars (most were mutants of 'Sim' or mutants of mutants), 21 were not chimeral, 29 were dichimeras and six were trichimeras, including eight, 24 and five Sim sports, respectively. The dichimeras generally sported back to the original red-flowering 'William Sim'. From these results it was concluded that the carnation has considerable ability to preserve stable chimeras.

The results discussed above were obtained by the irradiation of potted plants (topped to stimulate side shoot formation) with electrons of a linear accelerator which had a dose rate of 10^7—10^8 rad/sec. The optimum dose to provoke rearrangements was 5—6 krad. Mutations were rarely produced. Buiatti and others concluded from semi-acute γ-irradiation (6—12 kR; 600 R/h) that mutations frequency increased with increasing dose and was higher in the first main shoots than the laterals (Badr and Etman, 1977; Buiatti and Ragazzini, 1965; Buiatti et al., 1965a; D'Amato et al., 1964).

An interesting chimera in which to study the effect of radiation on tissue rearrangements was used by Péreau-Leroy (1969, 1970, 1974a,b, 1975). It was the orange flowering cultivar Jacqueline which has a genetically yellow L-I and genetically red L-II and L-III. By colchicine treatment, two types of cyto-chimeras were obtained, namely, 4n, 2n, 2n and 2n, 4n, 4n (L-I, L-II and L-III, respectively). From the colour and the size of the flower as well as the size of the cell nuclei, conclusions about the origin of shoots on irradiated plants could be drawn. The orange flower colour, for instance, results from the influence on the synthesis of pigments in the epidermal cells of the mesophyll which is carrying the red flower genotype, but is not pigmented itself. (This is not always the case; in white sports, for instance, L-I is white and L-II and L-III are red.) Another interesting phenomenon was Péreau-Leroy's inability to find the cyto-chimeras 2-4-2, 2-2-4, 4-4-2 and 4-2-4 using genetic techniques. In other words, L-II and L-III are apparently always genetically identical, both being either diploid (4-2-2) or tetraploid (2-4-4).

Dommergues and Gillot also used radiation to study chimeras. They found variations in the effect of ^{60}Co γ-rays (4, 6 or 8 krad) on the degree of uncovering observed, supposedly as the result of differences in dose rate (Dommergues and Gillot, 1965). The same authors later developed in vivo and in vitro adventitious bud techniques whereby they could produce genetically homogeneous plants, starting from chimeras. In the in vivo method, plants were topped and the wounded surface treated with 1% indolyl butyric acid to form callus from which shoots could ultimately grow. In the in vitro method, explants of the apex were used to produce callus from which adventitious shoots developed. When a white sport was used, approximately 70% of the shoots in vivo and 100% of the shoots in vitro were non-chimeral white flowering plants (23 of 32 plants, and nine of nine plants, respectively) (Dommergues and Gillot, 1973). Therefore, in carnation the shoots are often formed by epidermal cells; in other plants, for example, *Euphorbia*, similar in vivo methods lead to the formation of shoots from L-III tissue.

Several different types of explants have been used to produce adventitious shoots in vitro: parts of the apex, the leaves and the internodes (Hauzinska, 1974; Spinsky et al., 1974). Capacity for shoot formation depended very much on the kind, as well as on the age, of the plant parts.

The most important work for (commercial) mutation breeding, however, was carried out by Sparnaay and co-workers (Sparnaay, 1974; Sparnaay and Demmink, 1970, 1971; Sparnaay et al., 1974a,b). They usually irradiated unrooted cuttings, the lower 2—3 cm of which was protected by lead to ensure maximum root formation. The dose used to increase the variability of quantitative characters such as yield was 2.5—5 krad X-rays, whereas a higher dose was necessary for flower colour mutations.

The main problem for mutation breeders is to find a compromise between the necessary size of an experiment, statistically speaking, and its practical possibilities, especially when improvement of characters that are not directly visible is required. Because of the chimeral nature of the material, both directly and some time after the irradiation, the amount of material to be observed rapidly increases by continuous vegetative propagation. Therefore, it is of the utmost importance that pre-selection methods which reduce the amount at an early stage were developed by Sparnaay's research group. If selecting for increase in yield, for instance, probable mutants are pre-selected on the basis of clonal yield data.

Sparnaay also claims that irradiation can only be a means to increase variability significantly in such characteristics as flower production, time of flowering and higher yield if "old", long existing cultivars are used. In young, new cultivars, the existing natural variability in the material is large enough, and is not significantly increased by additional irradiation. Usual selection procedures in untreated material are sufficient to improve several quantitative characters (Custers et al., 1976).

At first sight, it does not seem to be logical to choose the difficult way of improving characters that are not directly visible, rather than to try to

TABLE 16

Commercial mutants of carnation

Name of new cultivar	Place and date of release (or approval) and name of principal worker(s) and institute	Conditions and date of mutagenic treatment [Parent cultivar]	Main improved attributes of cultivar
Sim Feu Follet	France, 1972 P. Péreau-Leroy, Cent. Etud. Nucl., Cadarache	5 krad γ-rays, 1965 [Sim Jacqueline]	Large yellow flower with broad red stripes
Uconn White Sim No. 1	U.S.A., 1962 G.A.L. Mehlquist, Conn. Agric. Exp. Stn, New Haven, Conn.	Rooted cuttings, γ-rays [White Sim]	Fewer ray flowers, ''holds'' longer after cutting

Sources: IAEA (1972, pp. 526—544) and *Mutation Breeding Newsletter*, 1 (1972) —9 (1977).

improve the flower colour of productive clones or seedling cultivars. Since low doses are relatively ineffective in inducing flower colour changes, even in large populations, a high dose is necessary. The flower colour mutations thus obtained, however, are often, if not always, accompanied by so many unfavourable mutations affecting yield, quality etc., that such an approach is felt to be unpracticable (L.D. Sparnaay, personal communication, 1975).

The use of adventitious bud techniques, either in vivo by decapitation of plant shoots or in vitro, to obtain solid mutants on a large commercial scale, has not yet been reported.

So far, only a very limited number of commercial mutants has been produced or reported (Table 16), despite the fact that carnation was one of the first objects of radiation research studies.

Literature carnation

Badr, M. and Etman, M., 1977. Gamma-radiation induced effects on the X_1-generation in carnation (*Dianthus caryophyllus,* L). Egypt. J. Genet. Cytol., 6: 32—43.

Bugnon, F., Dulieu, H. and Gaufillier, J., 1965. Sur les variations topographiques, le long de la pousse, entre les deux constituants d'une chimère chlorophyllienne sectoriale où méricline, chez l'Oeillet (*Dianthus caryophyllus* L.); essai d'interprétation ontogénique. C.R. Acad. Sci., 260: 211—213.

Buiatti, M. and Ragazzini, R., 1965. Gamma-ray induced changes in the carnation, *Dianthus caryophyllus* L. Radiat. Bot., 5(2): 99—105.

Buiatti, M., Ragazzini, R. and D'Amato, F., 1965a. Somatic mutations in the carnation induced by gamma radiation. In: The Use of Induced Mutations in Plant Breeding. FAO/IAEA/Eucarpia Meeting, Rome, 1964. Radiat. Bot., 5 (Suppl.): 719—723.

Custers, J.B.M., Van Eijk, J.P. and Sparnaay, L.D., 1976. New developments in mutation breeding of vegetatively propagated ornamental crops with special reference to qualitative characters. In: Proc. Workshop Eur. Comm. — Israel, The Use of Ionizing Radiation, Wageningen, 22—24 March. Assoc. Euratom—ITAL, Wageningen, pp. 496—501.

D'Amato, F., Moschini, E. and Pacini, L., 1964. Mutations somatiques de l'oeillet induites par les radiations gamma. Caryologia, 17(1): 93—101.

Dommergues, P. and Gillot, J., 1965. Variation de la réaction des boutures d'oeillet à l'irradiation gamma. In: The Use of Induced Mutations in Plant Breeding. FAO/IAEA/ Eucarpia Meeting, Rome, 1964. Radiat. Bot., 5 (Suppl.): 713—718.

Dommergues, P. and Gillot, J., 1973. Obtention de clones génétiquement homogènes dans toutes leurs couches ontogéniques à partir d'une chimère d'oeillet américain. Ann. Amélior. Plant., 23(2): 83—93.

Dulieu, H., 1968. Emploi des chimères chlorophylliennes pour l'étude de l'ontogénie foliaire. Bull. Sci. Bourgogne, XXV: 1—60.

Dulieu, H., 1969. Mutations Somatiques Chlorophylliennes Induites et Ontogénie Caulinaire. Thesis, Faculté des Sciences, Dijon, C.N.R.S. AO3016.

Farestveit, B., 1969. Flower colour chimeras in glasshouse carnations, *Dianthus caryophyllus* L. Yearb. R. Vet. Agric. Coll. Copenhagen, 1969: 19—33.

Farestveit, B. and Klougart, A., 1966. Bestråling af Nelliker med accellererede elektroner. Horticultura, 20: 9—12.

Gaufillier, J., 1965. Recherches sur l'Utilisation des Déficiences Chlorophylliennes Induites pour l'Interprétation du Fonctionnement des Méristèmes Terminaux de Tiges Feuillets: Cas de l'Oeillet (*Dianthus caryophyllus* L.) Faculté des Sciences, Dijon, 34 pp.

Grabowska, B. and Mynett, K., 1964a. The effect of gamma Co^{60} rays on the carnations of glasshouse races. Biul. Inst. Hodowli Aklim. Rosl., 5-6: 13—15 (in Polish).

Hauzinska, E., 1974. L'organogénèse dans le tissu de cal de l'oeillet (*Dianthus caryophyllus* L.) dans les conditions de cultures in vitro. Proc. Int. Hortic. Congr., 19(1A): 60.

Heslot, H., 1964. L'induction expérimentale de mutations chez les plantes florales. In: P.V. Séance 16 décembre, Acad. Agric. France. Paris, pp. 1281—1308.

Holly, W.D. and Baker, R., 1963. Carnation Production Including the History, Breeding, Culture and Marketing of Carnations. William C. Brown, Dubuqne, Iowa, 142 pp.

Howard, G.S., 1968. Polyploid carnations: their identification and breeding behaviour. J. Hered., 59: 77—80.

Mehlquist, G.A.L. and Sagawa, Y., 1959. The effect of gamma radiation on carnations. Proc. Int. Bot. Congr., 9(II): 258.

Mehlquist, G.A.L. and Sagawa, Y., 1964. The effect of gamma radiation on carnations. Proc. Int. Hortic. Congr., 16(4): 10—18.

Mehlquist, G.A.L., Ober, D.E. and Sagawa, Y., 1954. Somatic mutations in the carnations (*Dianthus caryophyllus* L.) Proc. Natl Acad. Sci. U.S.A., 40: 432—436.

Péreau-Leroy, P., 1969. Effet de l'irradiation gamma sur une chimère complexe d'oeillet Sim. In: Induced Mutations in Plants; Pullman, Wash., 1969. IAEA, Vienna, pp. 337—344.

Péreau-Leroy, P., 1970. Détection par irradiation gamma de l'influence de la structure en chimère péricline d'une variété d'oeillet sur la coloration de la fleur. C.R. Acad. Sci., Sér. D., 270(6): 810—812.

Péreau-Leroy, P., 1974a. Comparaison des effets physiologiques et génétiques provoqués par des irradiations aux neutrons et aux rayons gamma de boutures d'oeillets, *Dianthus caryophyllus* L. In: Biological Effects of Neutron Irradiation, Munich—Neuherberg, 1973. IAEA, Vienna, pp. 441—449.

Péreau-Leroy, P., 1974b. Genetic interaction between the tissues of carnation petals as periclinal chimeras. Radiat. Bot., 14: 109—116.

Péreau-Leroy, P., 1975. Recherche Radiobiologiques sur des Chimères d'Oeillet, *Dianthus caryophyllus* L. Thesis, Université de Clermont-Ferrand, Sér. E, No. 199, 169 pp.

Richter, A. and Singleton, W.R., 1955. The effect of chronic gamma radiation on the production of somatic mutations in carnations. Proc. Natl Acad. Sci. U.S.A., 41(5): 295—300.

Sagawa, Y., 1957. Nitrogen as a modifying factor in the growth and flowering responses of carnations to radiation. Plant Physiol., 32 (Suppl. L), 1 p.

Sagawa, Y. and Mehlquist, G.A.L., 1956. Radiation — A new tool in carnation breeding. Carnation Craft, 37: 3—5.

Sagawa, Y. and Mehlquist, G.A.L., 1957. The mechanism responsible for some X-ray induced changes in flower color of the carnation, *Dianthus caryophyllus*. Am. J. Bot., 44(5): 397—403.

Sagawa, Y. and Mehlquist, G.A.L., 1959. Some X-ray induced mutants in the carnation. J. Hered., 50(2): 78—80.

Sparnaay, L.D., 1974. Progress report on clonal selection in carnations after irradiation. In: C. Broertjes (Editor), Meeting of the Mutation Breeding Contact Group, Wageningen, October 1974. External Rep. No. 23, Assoc. Euratom—ITAL, Wageningen, pp. 28—31.

Sparnaay, L.D. and Demmink, J.F., 1970. Yield and quality of the glasshouse carnation (*Dianthus caryophyllus* L.) after mutagenic radiation with X-rays. Euphytica, 19(3): 310—317.

Sparnaay, L.D. and Demmink, J.F., 1971. Mutation breeding as a means of improving productivity in commercial carnation cultivars. In: Eucarpia Meeting on Ornamentals, Wageningen. Inst. Hortic. Plant Breed., Wageningen, pp. 38—44.

Sparnaay, L.D., Demmink. J.F. and Garretsen, F., 1974a. Sélection clonale chez les oeillets. In: Eucarpia Meeting on Ornamentals, Fréjus. Inst. Hortic. Plant Breed., Wageningen, pp. 51–62.

Sparnaay, L.D., Demmink, J.F. and Garretsen, F., 1974b. Clonal selection in carnation. In: Eucarpia Meeting on Ornamentals, Fréjus. Inst. Hortic. Plant Breed., Wageningen, pp. 39–50.

Spinsky, P., Beck, G.E. and McCown, B.H., 1974. Callus cultures of *Dianthus* species. HortScience, 9(3): 270.

Stewart, R.N., 1965. The origin and transmission of a series of plastogene mutants in *Dianthus* and *Euphorbia*. Genetics, 52: 925–947.

CHRYSANTHEMUM

Chrysanthemum, a genus of the Compositae (Asteraceae) family, contains annual and perennial plants commercially grown in greenhouses as pot plants or for cut flower production.

Chrysanthemum morifolium is a species with an extremely long history as a cultivated plant; it was grown in China during the time of Confucius, approximately 500 B.C. Between 725 and 750 A.D. a few cultivars were imported into Japan where, in later years, the flower became sacred. At the end of the 18th century and during the 19th century a number of forms were introduced to France and Great Britain. Nowadays many breeders in Western Europe, the U.S.A., Japan and other countries carry out breeding programmes to improve such characteristics as colour, form, size, production, quality, disease resistance and reaction to environment, mainly by cross breeding but increasingly by inducing mutations as well.

Sports, the results of spontaneous mutations, have been reported by various authors (Anderson, 1935; Dowrick, 1951, 1958; Jank, 1955). Some cultivars have given rise to a great number of sports such as 'Sweetheart', 'The Favourite' (Anonymous, 1961, Bowen et al., 1962) and 'Indianapolis' (Yoder Bros, Inc., 1967). Wasscher (1956) reported that approximately 30% of the cultivars originated as sports; the area occupied by sports, however, is often even larger than the acreage occupied by the original seedlings.

Most mutation breeding has been carried out with the hexaploid *Chrysanthemum morifolium* Ram. (6n = 54), also referred to as *C.* × *hortorum* or *C. indicum*. Some attention has been paid to the economically much less important diploid seed-propagated annual species (2n = 18), such as *C. carinatum* (p. 172).

Radiation-induced sports have been reported by many authors (Bowen et al., 1962; Broertjes, 1966a,b; Cawse, 1965, 1966; Chan, 1966; Crandell et al., 1966; Das et al., 1974; Fujii and Matsumura, 1967; Gupta, 1966; Gupta and Shukla, 1971c; Matsumura et al., 1961; Nakajima and Kawara, 1967; Nybom, 1961; Ruprecht, 1961; Satory, 1975; Shcherbakov, 1965; Sheenan and Sagawa, 1959; Simon, 1967; Van Hoeck, 1962; Yamaguchi and Takato, 1970). The significance of artificial sports for commercial chrysanthemum growing has increased considerably. Large-scale commercial (year-round) production of pot plants and cut flowers require cultivars which react uniformly to environmental conditions (light, temperature, nutrition, chemical growth regulation etc.). It is vital that the full ranges of colour and form are available at the same time (Machin, 1971). Artificial induction of mutations has therefore become an important tool to speed up the generation of "sports" in general, and particularly that of the outstanding products of cross breeding, which is a relatively simple affair with present-day knowledge.

Mutation breeding of the hexaploid and heterozygous *Chrysanthemum morifolium* has become an everyday practice of many a breeder and the best strategy seems to be to perfect the main products of cross breeding (health, favourable reaction to environment, productivity etc.) and to produce a family of sports as quickly as possible.

Since most of the commercial breeders do not advertise the origin of their products, we will never know which part of the assortment was produced by using artificially-induced mutations, but it already exceeds several hundreds of cultivars at the moment. The details of a number of commercial mutants, mostly recent ones, are presented in Table 17, and some recent commercially very successful mutants of excellent cultivars are shown in Plates 1 (p. 9) and 7 (p. 181).

In this section the important factors to consider in the development of a mutation breeding programme in *Chrysanthemum* will be discussed (cf. Broertjes, 1977a). After irradiation, one may expect a sometimes enormous variability for almost any character because of the generally large heterozygosity and the high ploidy level. Since, so far, no practical in vivo propagation method, in which a plant originates from a single cell, is available, one must reckon with the consequences of chimera formation. This implies that one of the first things that should be considered when starting a mutation breeding programme is the improvement of directly visible characters. Even then, it is often very hard to recognize minor but still important changes in flower colour, size and form, particularly during the "sectorial chimera stage".

The ploidy and the heterozygosity cause complex heredity of most characters, so that the determination of the best starting material is generally a matter of experience. Statistical analysis of spontaneous sports supplemented, if possible, with information from (specifically planned) experiments, evidently makes one's choice much easier and more justified.

An example is presented by the work of Jank (1957a,b), who demonstrated by statistical analysis as well as by experimental mutation induction that the pink flower colour is the best starting point for the induction of other flower colours, followed by white, bronze, red, purple, yellow, salmon, gold orange, yellow—bronze, yellow with red, and brown. This has been confirmed by Broertjes (1966a,b) who also obtained numerous flower colour mutants from various pink-flowered cultivars.

TABLE 17

Commercial mutants of *Chrysanthemum*

Name of new cultivar	Place and date of release (or approval) and name of principal worker(s) and institute	Conditions and date of mutagenic treatment [Parent cultivar]	Main improved attributes of cultivar
Dr X	U.S.A., 1966 P.C. Crandall, W.J. Clore and R.A. Nilan, Wash. State Agric. Exp. Stn, Vancouver, Prosser, and Pullman, Wash.	Rooted cuttings, 1.2 krad X-rays, 1963 [Dr Dave]	Darker purple—red flower colour
Izetka Köpenicker Bronze Vogue	German Democratic Republic, 1962 H. Jank, Inst. Zierpflanzenbau, Humboldt-Universität, Berlin	1—2.5 kR X-rays, 1956 [Vogue]	Red—bronze colour, inward-curved type, 15-cm flower diam.
Izetka Filmstar Bronze	German Democratic Republic, 1966 H. Jank, Inst. Zierpflanzenbau, Humboldt-Universität, Berlin	1—2.5 kR X-rays, 1958 [Filmstar]	Bronze colour, 6—8 flowers per stem; sturdy; dark green foliage
Izetka Marienhain Dunkelrosa	German Democratic Republic, 1966 M. Knuth, Inst. Zierpflanzenbau, Humboldt-Universität, Berlin	1—2.5 kR X-rays, 1957 [Izetka Marienhain]	Dark pink, 20-cm flower diam., 6—8 flowers per stem; weather resistant
Izetka Marienhain Cremeweiss	German Democratic Republic, 1966 M. Knuth, Inst. Zierpflanzenbau, Humboldt-Universität, Berlin	1—2.5 kR X-rays, 1957 [Izetka Marienhain]	Cream flower, other attributes as above
Izetka Marienhain Hellgelb	German Democratic Republic, 1966 M. Knuth, Inst. Zierpflanzenbau, Humboldt-Universität, Berlin	1—2.5 kR X-rays, 1957 [Izetka Marienhain]	Bright yellow, other attributes as above
Izetka Herbstgold	German Democratic Republic, 1964 M. Knuth, Inst. Zierpflanzenbau, Humboldt-Universität, Berlin	1—2.5 kR X-rays, 1959 [Izetka Köpenicker Rayonnante]	Yellow—bronze, ray-shaped petals; sturdy stem; 20-cm flower diam.
Izetka Köpenicker Barbarossa Goldkissen	German Democratic Republic, 1962 H. Jank, Inst. Zierpflanzenbau, Humboldt-Universität, Berlin	1—2.5 kR X-rays, 1958 [Barbarossa]	Bordeaux red with bright yellow centre, windflower-shaped petals, good stem, slightly susceptible to sprays

Izetka Köpenicker Barbarossa Roststern	German Democratic Republic, 1962 H. Jank, Inst. Zierpflanzenbau, Humboldt-Universität, Berlin	1—2.5 kR X-rays, 1958 [Barbarossa]	Dull red with yellow centre; other attributes as above
Dark Deep Tuneful	The Netherlands, 1969 S. de Bruyn Jr, Maasdijk *	1.5 krad X-rays, 1967 [Tuneful]	Red flower colour; other characters unchanged
Uncle Danny	The Netherlands, 1973 Riviera Plant Co., De Lier *	1.75 krad X-rays, 1971 [Beamsville Pink]	Yellow flower colour; growth and flower form same as in mother cultivar
Danny Boy	The Netherlands, 1973 Riviera Plant Co., De Lier *	1.75 krad X-rays, 1971 [Beamsville Pink]	Yellow flower colour; growth and flower form same as in mother cultivar
Danny's Cape	The Netherlands, 1973 Riviera Plant Co., De Lier *	1.75 krad X-rays, 1971 [Beamsville Pink]	Yellow flower colour; growth and flower form same as in mother cultivar
Danny's Pearl	The Netherlands, 1973 Riviera Plant Co., De Lier *	1.75 krad X-rays, 1971 [Beamsville Pink]	Yellow flower colour; growth and flower form same as in mother cultivar
Blue Winner	The Netherlands, 1975 Fides, De Lier *	1.75 krad X-rays, 1973 [Pink Winner]	Flower colour mutant; other characters unchanged
Bronze Winner	The Netherlands, 1975 Fides, De Lier *	1.75 krad X-rays, 1973 [Pink Winner]	Flower colour mutant; other characters unchanged
Coral Winner	The Netherlands, 1975 Fides, De Lier *	1.75 krad X-rays, 1973 [Pink Winner]	Flower colour mutant; other characters unchanged
White Winner	The Netherlands, 1975 Fides, De Lier *	1.75 krad X-rays, 1973 [Pink Winner]	Flower colour mutant; other characters unchanged
Yellow Winner	The Netherlands, 1975 Fides, De Lier *	1.75 krad X-rays, 1973 [Pink Winner]	Flower colour mutant; other characters unchanged
Gamma	Hungary, 1969 J. Simon	1.5 krad γ-rays, 1966 [Obuda]	?
Mikrop	The Netherlands, 1976 Chryveco, 's Gravenzande *	1.75 krad X-rays, 1975 [Pinkish sport of Horim]	Pink flower colour
Middelry	The Netherlands, 1976 Chryveco, 's Gravenzande *	1.75 krad X-rays, 1975 [Pinkish sport of Horim]	Bright yellow flower colour

TABLE 17 (continued)

Name of new cultivar	Place and date of release (or approval) and name of principal worker(s) and institute	Conditions and date of mutagenic treatment [Parent cultivar]	Main improved attributes of cultivar
Milava	The Netherlands, 1976 Chryveco, 's Gravenzande *	1.75 krad X-rays, 1975 [Pinkish sport of Horim]	Cream—yellow flower colour
Milonka	The Netherlands, 1976 Chryveco, 's Gravenzande *	1.75 krad X-rays, 1975 [Pinkish sport of Horim]	White flower colour
Dark Westland	The Netherlands, 1976 C.B.S.H., De Lier *	1.75 krad X-rays, 1975 [Westland]	Dark pink flower colour
Bright Westland	The Netherlands, 1976 C.B.S.H., De Lier *	1.75 krad X-rays, 1975 [Westland]	Pale pink flower colour
Bronze Westland	The Netherlands, 1976 C.B.S.H., De Lier *	1.75 krad X-rays, 1975 [Westland]	Bronze flower colour
Yellow Westland	The Netherlands, 1978 C.B.S.H., De Lier *	1.75 krad X-rays, 1975 [Westland]	Yellow flower colour
White Westland	The Netherlands, 1978 C.B.S.H., De Lier *	1.75 krad X-rays, 1975 [Westland]	White flower colour
Dark Charmette	The Netherlands, 1976 C.B.S.H., De Lier *	1.75 krad X-rays, 1975 [Charmette]	Dark pink flower colour
Bronze Charmette	The Netherlands, 1976 C.B.S.H., De Lier *	1.75 krad X-rays, 1975 [Charmette]	Bronze flower colour
Dark Oriette	The Netherlands, 1976 C.B.S.H., De Lier *	1.75 krad X-rays, 1975 [Oriette]	Dark pink flower colour
Blue Star	The Netherlands, 1977 Fides, De Lier *	1.75 krad X-rays, 1976 [Pink Star]	Darker pink flower colour
Bright Star	The Netherlands, 1977 Fides, De Lier *	1.75 krad X-rays, 1976 [Pink Star]	Bright pink flower colour
Bronze Star	The Netherlands, 1977 Fides, De Lier *	1.75 krad X-rays, 1976 [Pink Star]	Bronze flower colour

* In co-operation with the Association Euratom—ITAL, Wageningen.

C.B.S.H., Chrysanthemum Breeding and Selection Holland, an association of chrysanthemum breeding companies like Chryveco.

Sources: IAEA (1972, pp. 526–544) and *Mutation Breeding Newsletter*, 1 (1972) —9 (1977) and Association Euratom—ITAL,

The nature of the mutations is not completely clear. Many authors attribute it (partly) to changes in the number of chromosomes (Bowen et al., 1962; Bowen, 1965; Dowrick, 1953, 1958; Dowrick and El-Bayoumi, 1966a,b; Ichikawa et al., 1970; Walker, 1955). Some authors have tried to prove this by correlating the number of chromosomes in root tips with flower colour mutations. The main objection to this procedure is that the roots, from which the root tips are taken for the determination of the chromosome number, are known to develop from L-III tissue (Stangler, 1956), whereas flower colour mutants and sports are generally periclinal chimeras of which only L-I is genetically different from the original genotype (Bowen, 1965; Sampson et al., 1958; Stewart and Dermen, 1970b). Consequently, it does not make sense to correlate (radiation-induced) chromosome aberrations or differences in number with a flower colour change, unless one makes completely sure that the tissues compared are genetically identical. This also holds good when irregularities in the meiosis (L-II) are correlated with flower colour changes. Moreover, the numbers on which the conclusions are based are generally low. Sampson et al. (1958) tried to determine the number of chromosomes in developing petals to avoid the difficulty outlined above. Unfortunately they failed. In addition, they tried to demonstrate, with hot water treatment, the chimeric nature of their material. This did not work either and, as they stated, this does not prove that the plants were genetically homogeneous.

On the other hand, the positive correlation between chromosome number and flower size seems justified since the size of flowers might well be determined by L-III (and L-II?).

Other authors attach certain roles to gene mutations as well. Stewart and Dermen (1970b) attribute certain flower colour changes (e.g. pink to white) to true gene mutations, whereas the change from absence to presence of yellow chromoplasts might be caused by the loss (of part) of a chromosome which carries a dominant suppressor for yellow pigment formation. They also emphasized the significance of the rearrangement of chimeral layers as a cause of (phenotypic) variability.

A certain type of rearrangement, namely uncovering, is probably not always recognized as such and wrongly attributed to mutational events (Fujii and Mabuchi, 1961; Fujii, 1962; Shimotsuma and Sakurai, 1962; Yamakawa, 1970).

The high frequencies of identical "mutations", large sectors or "complete mutants", often indicate uncovering or related phenomena (Shapiro and Broertjes, 1961).

It seems justified to conclude that there is some confusion about the nature of (flower colour) mutations in *Chrysanthemum*; in some cases cell layer rearrangements may be involved (high frequencies, large sectors or "complete" mutants, especially after high doses) and/or true gene mutations, probably often more than one gene at a time (small chromosome aberrations;

low frequency, small sectors). Gross chromosome aberrations or even the gain or loss of complete chromosomes certainly plays a part as well (low frequency, small sectors). However, as long as one author tries to explain various phenomena on the basis of a tri-chimeral situation in 'Indianapolis Pink' (Weaver, 1963) while others claim that Indianapolis Pink is genetically homogeneous (Stewart and Dermen, 1970a,b), it remains difficult to decide what the main cause of radiation-induced variability is.

Whichever explanation is the right one, the mutation breeder has to reckon with the important fact that flower colour mutations follow a specific one-way system as the result of either true gene mutation, or smaller or larger chromosome aberrations (deletions), or the loss of complete chromosomes and, moreover, as if pink is based upon the maximum number of dominant genes. This may also be the case with other characters and it is consequently of importance to know the particular pathway for a particular character, either by experience or by collecting data about spontaneous mutations. Armed with this knowledge, the breeder can more easily decide on the strategy of his project.

The next question is what kind of plant material should be used for mutagenic treatment. Generally, rooted cuttings are irradiated, with or without shielding the roots and the base of the cutting. (In the case of shielding the base, one must be aware of the fact that unirradiated and so undamaged buds near the base may grow out at the expense of the fully irradiated buds.) Unrooted cuttings may also be irradiated but, since rooting is easily suppressed by radiation, comparatively low doses must be applied, unless the base is shielded.

The best mutagen seems to be radiation. Chemicals have occasionally been applied but without success. Bowen (1965), using ethyl methanesulphonate (EMS) and ethylene imine (EI), obtained very few mutations. This is explained by the relatively few chromosome aberrations that are generally induced by such a mutagenic chemical. However, the poor penetration of chemicals into vegetative plant parts in general certainly plays a part too. Therefore, notorious chromosome breakers, such as 8-ethoxy caffein, have also been tried (the results were not reported).

The optimum dose for rooted cuttings is reported to be around 1—2 krad X- or γ-rays (Bowen, 1965; Broertjes, 1966a; Chan, 1966; Crandell et al., 1966; Jank, 1957a,b). Some authors, however, report much higher doses, such as 25 krad γ-rays (e.g. Cawse, 1966), probably as the result of the application of a low dose rate. Every plant species has a certain critical dose rate at which the repair of the physiological damage is more or less in equilibrium with its induction by the radiation (Broertjes, 1972c), in the sense that much higher doses can be applied before the accumulating unrepairable damage ultimately kills the material. Yamakawa and Sekiguchi (1968) advised 10—20 kR (γ-rays; 1 kR/day) while Broertjes (1966a) reported that 8 krad γ-rays (125—150 rad/h) is not lethal by far. However, for practical reasons a high dose rate (short irradiation times) is also preferable since there are few indications that low dose rates are more advantageous than high ones.

Much less experience is available with neutrons. The optimum dose for thermal neutrons lies around $6-12 \times 10^{12}$ N_{th}/cm^2 (Bowen, 1965; Broertjes, 1966a), whereas for fast neutrons 300—500 rad is advised (Broertjes, 1966a; Love and Constantin, 1965). One would expect that neutrons, producing a dense ionization in the material and consequently causing relatively more chromosome aberrations, would be better mutagens than X- or γ-rays if aberrations played a greater role than gene mutations. Although there is a (slight) tendency for this to be the case (Broertjes, 1966a), there is not enough evidence to make a deliberate choice in favour of neutrons at present.

The radiosensitivity of different cultivars varies, though within certain limits; consequently, it is recommended to determine the optimum dose for each specific cultivar by applying a series of doses, such as, for example, 1—2.5 or 3 krad with steps of 250 rad using a comparatively high dose rate (> 50—100 rad/min).

The definition of the optimum dose is more complicated than one would expect. The advantages and disadvantages of low and high doses can be summarized as follows:

low dose

advantages:	(nearly) 100% survival;
	no problems in growing the plants and the subsequent cuttings.
disadvantages:	low mutation frequency;
	small sectors;
	few cases of rearrangement of a periclinal chimera, if any.

high dose

advantages:	high(er) mutation frequency;
	large(r) sectors and a higher percentage of complete periclinal chimeras;
	high percentage of rearrangements.
disadvantages:	low survival;
	bad growth;
	more mutations per cell, generally being disadvantageous.

Chromosome aberrations and aneuploidy also play an important part; there must be a correlation between the dose and the mutation spectrum, in the sense that a high dose would be preferred, for instance, to induce mutations based on loss or gain of chromosomes (size of flowers), whereas low doses would be chosen to induce certain flower colour mutations (Ichikawa et al., 1970). A similar discussion applies to the difference between neutrons, 8-ethoxy caffein etc., and X- and γ-rays, EMS etc., respectively.

After treatment, irradiated cuttings are potted (unrooted cuttings are rooted first) or planted, dependent on the cultivar. Then the question of how to handle the plants after they have regained growth arises. One way is to let the plants grow undisturbed. The result is that many small mutated sectors may be obtained on a relatively small area. Then one can try to recover the mutated sectors by cutting the stems, carrying the mutated sectors, into pieces and subsequently rooting the parts, thus stimulating the growth of axillary buds (long-day conditions just before flowering keeps axillary buds vegetative). It also seems possible (though unconfirmed) to culture parts of the mutated sectors (streaks in the single ray florets) in vitro and subsequently grow completely mutated plants. The disadvantage of the latter method, however, is that only very clearly visible mutated sectors are detected and that all minor changes in the flower colour, and changes in form and size etc., are easily overlooked.

The way practised most at present is to cut back the growing plant and its out-growing axillary buds several times and to plant the cuttings after rooting. The disadvantages are that small sectors may be lost and that this process takes up more room. The advantage, however, is that many mutated sectors, including those in the out-growing axillary buds, are given the opportunity to develop into complete periclinal chimeras which, by cutting, can be multiplied vegetatively. The result is that many more mutations, including those which influence much less obvious characters, can be detected. The repeated occurrence of several mutations guarantee their detection, a more critical selection and secure (and faster) propagation.

The exact number of times shoots should be cut back is hard to indicate since, to our knowledge, no one has ever compared the different possible treatments very precisely. From practical experience it appears that cutting back three or four times is the most efficient. It also seems reasonable to prefer a higher number of irradiated cuttings than an increase in the number of times fewer irradiated cuttings are cut back, given a certain growth area. A refinement that could be added is the selective use of basal cuttings, discarding the tip cuttings, if the observations of Bowen (1965) are correct. He claims that basal cuttings usually produce whole mutants and tip cuttings sectorial chimeras whereas axillary cuttings are intermediate (they may produce both).

A combination of in vivo and in vitro techniques is proposed by Matsubara et al. (1971). They used repeated cutting back to increase the sector size. Flower bud receptacles of mutated branches were then cut into pieces and cultivated in vitro until adventitious plantlets which grew out into complete (solid?) mutants were produced.

Another way to obtain complete mutants is to apply high doses and dose rates, thereby inactivating all initial cells but one. It results in an increased sector size and percentage of complete (periclinal) mutants. This is called "internal disbudding" by Yamakawa and Sekiguchi (1968) who studied sector size of the sporocyte tissue of tomato and *Antirrhinum majus.* They obtained one (?) entire shoot mutation in *Ch. morifolium* with a dose rate of γ-rays higher than 1 kR/day and a total dose of 10—20 kR. The significance of this method seems questionable, firstly since the dose rate applied (a few krads per day) for many species is not at all high and secondly because the sympodial growth of tomato and rice (automatic elimination of the chimeral construction) has not been taken into account.

Cutting back and all other measures to provide as many mutated cells as possible that grow out and take part in the formation of a shoot, would not be necessary if an adventitious bud technique was available by which solid mutants could be obtained. In vivo adventitious plantlets can be grown at the base of rooted leaves of some cultivars. Unfortunately, they develop on callus formed at the base of the petiole or on the upper part of the roots, in both cases obviously from more than one cell. For the majority of the plantlets carrying a mutation, the mutation has a chimeric structure like adventi-

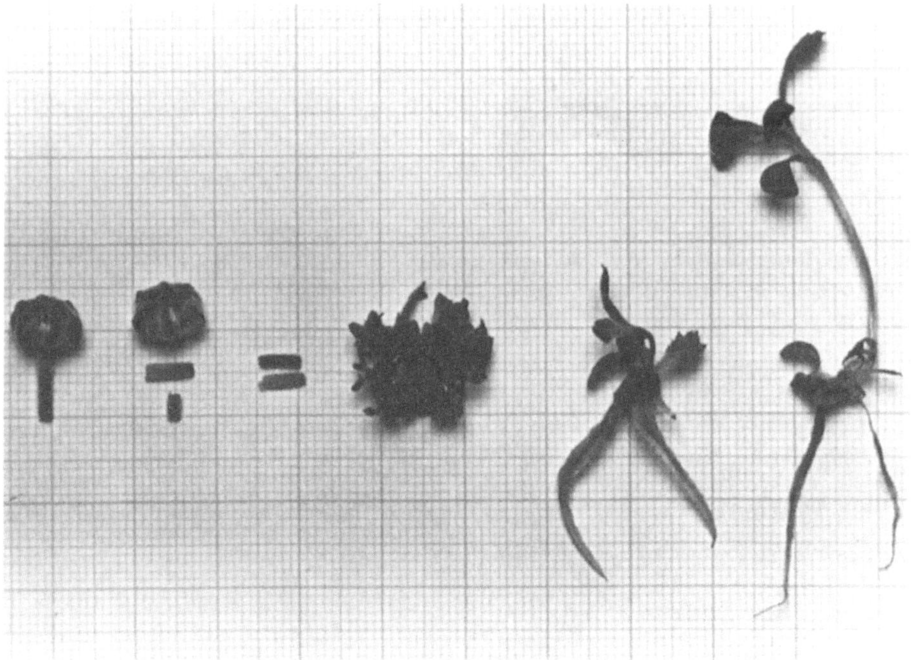

Fig. 20. Vegetative propagation of *Chrysanthemum morifolium* Ram. cv. Super Yellow in vitro. From left to right:
— pedicel with attached flower head;
— pedicel segment after detachment of the flower head and the basal stem portion;
— pedicel segment sectioned longitudinally into two pedicel explants;
— adventitious shoot formation on a pedicel explant, 4 weeks after incubation in vitro;
— plantlets obtained by root formation of two detached sub-cultured shoots, 2 weeks after shoot detachment.
(Roest, 1977, Fig. 1.) (Photograph, ITAL, Wageningen.)

tious shoots on de-budded plants (Stewart and Dermen, 1970a; Broertjes and Roest, 1976). In a second experiment, most mutants were apparently solid, however (C. Broertjes, unpublished data). Moreover, it takes too long (6–9 months) before the plantlets flower. In vitro, on the other hand, adventitious shoots can very rapidly be obtained using stem explants (Hill, 1968) or explants of small leaves, petioles, pedicels (Fig. 20), or decapitated young flower buds (Roest and Bokelmann, 1975) or other plant parts (Bush et al., 1974). In the case of pedicels, almost exclusively solid mutants were obtained. Consequently, it seems to be a more attractive method for mutant production (Broertjes and Roest, 1976; Roest, 1977) than the second part of the method described by Matsubara et al. (1971) in which receptacle pieces of irradiated flower buds could be used to produce apparently solid mutants. The only disadvantage of solid mutant production would be a restricted mutation spectrum, in the sense that phenotypes based on various arrangements of genotypically different groups of cell layers would be excluded. Which part of the mutation spectrum this might be with regard to the total one is unknown.

Summarizing, the following procedure could be used:

(1) careful definition of the problem and decision on strategy;

(2) in the case of mutation breeding: careful selection of the best genotype (pink or white for flower colour);

(3) the irradiation of rooted cuttings with the optimum dose (probably approximately 1 500–2 000 rad of X-rays);

(4) cutting back three to four times during the growth of the plants; planting of the rooted cuttings;

(5) selection of the desired genotype and subsequent stabilization by successive propagation and further testing.

(3) and (4) could be replaced by:

(3) irradiation of pedicels of young inflorescences with the optimum dose (approximately 800–900 rad X-rays);

(4) the production, in vitro, of adventitious shoots and subsequent production of plants in vivo by rooting the detached shoots.

Annual chrysanthemums

Some experience has been obtained with the diploid (2n = 18) seed propagated *Chrysanthemum carinatum* (Syn. *Ch. tricolour*) Schousb. (Rana, 1964a,b,c,d, 1965a,b; Jain et al., 1961). The disadvantage of this species is that it is self-incompatible and, therefore, very heterozygous. Consequently, irradiation experiments have to be carried out on a large scale and compared with sufficiently large control populations. Various flower form mutations have been induced, such as tubular and double type flowers as well as a single type flower with bisexual ray florets.

Literature Chrysanthemum

Anonymous, 1961. Chrysanthemums. Bull. Minist. Agric., Fish. Food, London, 92: 6.

Anderson, D., 1935. Sporting of the *Chrysanthemum*. Mo. Bot. Gard. Bull., 23: 161—163.

Bowen, H.J.M., 1965. Mutations in horticultural chrysanthemums. Radiat. Bot. 5 (Suppl.): 695—700.

Bowen, H.J.M., Cawse, P.A. and Dick, M.J., 1962. The induction of sports in chrysanthemums by gamma radiation. Radiat. Bot., 1: 297—303.

Broertjes, C., 1966a. Mutation breeding of *Chrysanthemum*. Euphytica, 15: 156—162.

Broertjes, C., 1966b. Mutatieveredeling bij *Chrysanthemum*. Meded. Dir. Tuinbouw, (Neth.), 29: 193—198.

Broertjes, C., 1972c. Use in Plant Breeding of Acute, Chronic or Fractionated Doses of X-rays or Fast Neutrons as Illustrated with Leaves of *Saintpaulia*. Thesis; Cent. Agric. Publ. Doc., Wageningen, Agric. Res. Rep. 776, 74 pp.

Broertjes, C., 1977a. Induced-mutant techniques in breeding asexually propagated plants. In: Manual on Mutation Breeding. IAEA, Vienna, 2nd edn, pp. 159—166.

Broertjes, C. and Roest, S., 1976. Mutation breeding of *Chrysanthemum morifolium* Ram. using in vivo and in vitro adventitious bud techniques. Euphytica, 25: 11—19.

Bush, S.R., Langhans, R.W. and Earle, E.D., 1974. In vitro propagation of the chimeral layers of *Chrysanthemum morifolium* Ramat 'Indianapolis'. HortScience, 9(3): 270.

Cawse, P.A., 1965. Production of *Chrysanthemum* sports by gamma radiation. Radioisot. Rev. Sheet, Wantage Res. Lab. (A.E.R.E.), Wantage.

Cawse, P.A., 1966. Using atomic radiation to produce colour sports in flowers. Commer. Grow., 3660: 381.

Chan, A.P., 1966. Chrysanthemum and rose mutations induced by X-rays. Proc. Am. Soc. Hortic. Sci., 88: 613—620.

Crandell, P.C., Clore, W.J. and Nilan, R.A., 1966. Dr. X and Sparkle. New *Chrysanthemums* for Washington. Wash. Agric. Exp. Stn, Vancouver, Wash., Circ. 467, 3 pp.

Das, P.K., Ghosh, P., Dube, S. and Dhua, S.P., 1974. Induction of somatic mutations in some vegetatively propagated ornamentals by gamma irradiation. Technology (Coimbatoire, India), 11(2,3): 185—188.

Dowrick, G.J., 1951. Sporting in chrysanthemums. Chrysanthemum, 1951: 152—155.

Dowrick, G.J., 1953. The chromosomes of *Chrysanthemum*. II. Garden varieties. Heredity, 7: 59—72.

Dowrick, G.J., 1958. Chromosome numbers and the origin and nature of sports in the garden chrysanthemum. Natl Chrysanthemum Soc. Yearb., 1958: 60—79.

Dowrick, G.J. and El-Bayoumi, A., 1966a. The origin of new forms of the garden *Chrysanthemum*. Euphytica, 15: 32—38.

Dowrick, G.J. and El-Bayoumi, A., 1966b. The induction of mutations in *Chrysanthemum* using X- and gamma radiation. Euphytica, 15: 204—210.

Fujii, T., 1962. Mutations induced by radiation in vegetatively propagated plants with special reference to flower color. Gamma Field Symp., 1: 51—59.

Fujii, T. and Mabuchi, T., 1961. Irradiation experiments with *Chrysanthemum*. Seiken Jihô, 12: 40—44.

Fujii, T. and Matsumura, S., 1967. Somatic mutations induced by chronic gamma irradiation. Seiken Jihô, 19: 43—46.

Gupta, M.N., 1966. Induction of somatic mutations in some ornamental plants. In: Proc. All India Symp. Hortic. pp. 107—114.

Gupta, M.N. and Shukla, R., 1971c. Mutation breeding of *Chrysanthemum*. I. Production of new cultivars by gamma ray induced somatic mutations in vM_1. In: Int. Symp., Use of Isotopes and Radiation in Agriculture and Animal Husbandry Research, New Delhi. pp. 164—174.

Hill, G.P., 1968. Shoot formation in tissue cultures of *Chrysanthemum* 'Bronze Pride'. Physiol. Plant., 21: 386—389.

IAEA, 1972. Induced Mutations and Plant Improvement. IAEA, Vienna, pp. 526—544.

Ichikawa, S., Yamakawa, K., Sekiguchi, F. and Tatsuno, T., 1970. Variation in somatic chromosome number found in radiation-induced mutants in *Chrysanthemum morifolium* Hemsl. cv. Yellow Delaware and Delaware. Radiat. Bot., 10: 557—562.

Jain, H.K., Bose, A.K., Sathpathy, D. and Sur, S.C., 1961. Mutation studies in annual chrysanthemum — I. Radiation-induced variation in flower form. Indian J. Genet. Plant Breed., 21: 68—74.

Jank, H., 1955. Zur Frage des "Sportens" beim *Chrysanthemum indicum*. Über den Umfang und die Auswirkungen der Farbsporte im *Chrysanthemum*-Sortiment. Dtsch. Gartenbau, 2: 265—268.

Jank, H., 1957a. Experimentelle Mutationsauslösung durch Röntgenstrahlen bei *Chrysanthemum indicum*. Züchter, 27: 223—231.

Jank, H., 1957b. Experimental Production of Mutations in *Chrysanthemum indicum* by X-rays. (Translation of Jank (1957a).) U.K. At. Energy Assoc., At. Energy Establ., Harwell, 36 pp.

Love, J.E. and Constantin, M.J., 1965. The response of some ornamental plants to fast neutrons. Tenn. Farm Home Sci., 56: 10—12.

Machin, B.J., 1971. The importance of mutation breeding in the year-round chrysanthemum industry. In: Eucarpia Meeting on Ornamentals. Inst. Hortic. Plant Breed., Wageningen, pp. 25—31.

Matsubara, H., Shigematsu, K., Suda, H. and Hashimoto, S., 1971. The isolation of the mutation plants from sectorial chimera induced by irradiation in begonia and chrysanthemum. In: Proc. 10th Jpn. Conf. Radioisot., Tokyo, pp. 374—376 (Abstr. Pap. 119,120).

Matsumura, S., Fujii, T., Kondo, S., Mabuchi, T. and Sakurai, N., 1961. Irradiation experiments with chrysanthemum. Jpn. J. Breed., 11: 240.

Nakajima, K. and Kawara, K., 1967. Induction of mutations in chrysanthemums by gamma-ray irradiation. Hoshasen Ikushujo Nenpo (Annu. Rep. Natl Inst. Radiat. Breed.), 1965: 46—48 (in Japanese).

Nybom, N., 1961. The use of induced mutations for the improvement of vegetatively propagated plants. Mutations and Plant Breeding. NAS—NRC, Publ. 891: 252—294.

Poole, R.T., 1962. Effect of gamma rays on *Chrysanthemum morifolium* 'Bluechip' grown at various nutritional levels. Proc. Fla. State Hortic. Soc., 75: 443—447.

Rana, R.S., 1964a. Multiple allelism in radiation-induced ray-floret mutants of annual *Chrysanthemum*. Curr. Sci., 33: 592—593.

Rana, R.S., 1964b. Radiation-induced chromosomal rearrangements in annual *Chrysanthemum*. Experientia, 20(11): 617.

Rana, R.S., 1964c. A radiation-induced chimera in annual *Chrysanthemum*. Naturwissenschaften, 51: 642—643.

Rana, R.S., 1964d. Phenotypic variability of an induced mutant of annual *Chrysanthemum*. Genetica (The Hague), 35: 236—240.

Rana, R.S., 1965a. Radiation-induced variation in ray-floret characteristics of annual *Chrysanthemum*. Euphytica, 14: 296—300.

Rana, R.S., 1965b. A radiation-induced useful mutant of annual *Chrysanthemum*. Curr. Sci., 34: 58.

Roest, S., 1977. Vegetative propagation in vitro and its significance for mutation breeding. Acta Hortic., 78: 349—359.

Roest, S. and Bokelmann, G.S., 1975. Vegetative propagation of *Chrysanthemum morifolium* Ram. in vitro. Sci. Hortic., 3: 317—330.

Ruprecht, H., 1961. Steigerung der Mutationsrate bei *Chrysanthemum* durch Röntgenstrahlen. Gartenwelt, 61: 219—220.

Sampson, D.R., Walker, G.W.R., Hunter, A.W.S. and Bragdø, M., 1958. Investigations in the sporting process in greenhouse chrysanthemums. Can. J. Plant Sci., 38: 346—356.
Satory, M., 1975. Chrysanthemenzüchtung mit Hilfe künstlicher Mutationsauslösung. Gartenwelt, 20: 433—435.
Shapiro, S. and Broertjes, C., 1961. Radiation-induced sports in floricultural plants. Brookhaven Natl Lab. Rep., BNL 5077. Upton, Long Isl., N.Y., pp. 1—2.
Shcherbakov, V., 1965. Chimères végétales. Tsvetovodstvo, 3: 18—20.
Sheenan, T.J. and Sagawa, Y., 1959. The effects of gamma radiation on *Chrysanthemum* and *Gladiolus*. Proc. Fla. State Hortic. Soc., 72: 388—391.
Shimotsuma, M. and Sakurai, N., 1962. Irradiation experiments with chrysanthemum. Seiken Jihô, 14: 106—110.
Simon, J., 1967. Elözetes adatok sugárzással indukált elváltozásokról disznövényeknél. Atomtech. Tájék., 10: 691—694.
Stangler, B.B., 1956. Origin and development of adventitious roots in stem cuttings of chrysanthemum, carnation and rose. Cornell Exp. Stn, Ithaca, N.Y., Mem. 342, pp. 3—24.
Stewart, R.N. and Dermen, H., 1970a. The origin of adventitious buds in *Chrysanthemum*. Am. J. Bot., 57: 734—735.
Stewart, R.N. and Dermen, H., 1970b. Somatic genetic analysis of the apical layers of chimeral sports in chrysanthemum by experimental production of adventitious shoots. Am. J. Bot., 57: 1061—1071.
Van Hoeck, F., 1962. Radio isotopes in agricultural research. Euratom Bull., 1: 15—18.
Walker, G.W.R., 1955. Chromosome numbers in *Chrysanthemum* sports. In: Dep. Agric. Hortic. Div., Progr. Rep. 1949—1953. pp. 69—70.
Wasscher, J., 1956. The importance of sports in some florist's flowers. Euphytica, 5: 163—170.
Weaver, G.M., 1963. The effect of Cesium 137 gamma radiation on plant growth and flower color of greenhouse chrysanthemum cultivars. Can. J. Genet. Cytol., 5: 73—82.
Yamakawa, K., 1970. Radiation-induced mutants of chrysanthemum and their somatic chromosome number. Inst. Radiat. Breed. Techn. Newsl., No. 6.
Yamakawa, K. and Sekiguchi, F., 1968. Radiation-induced internal disbudding as a tool for enlarging mutation sectors. Gamma Field Symp., 7: 19—39.
Yamaguchi, I. and Takato, S., 1970. The F_1 progenies of flower mutants in Chrysanthemums. Jpn. J. Breed., 20 (Suppl. 1): 109—110 (in Japanese).
Yoder Bros, Inc., 1967. Evolution of the Indianapolis family. Grow. Circ. Newsl., No. 51. Yoder Bros. Inc., Barberton, Ohio, pp. 1—4.

FREESIA

One of the most important cut flowers in Western Europe is the *Freesia*, family Iridaceae. Originally a spring crop, in recent years it has become possible to cut the flowers throughout most of the year. The genetic assortment consists of diploid and tetraploid seed propagated cultivars, and tetraploid vegetatively propagated ones, both groups having a variety of flower colours (Van Raalte, 1971).

The Euratom—ITAL Association at Wageningen in co-operation with several private plant breeders have irradiated thousands of tubers (C. Broertjes, unpublished data), so far with no success (i.e. mutation). Since spontaneous mutations are very rare as well, at least in the tetraploids, it is not known which genotype is most suitable for irradiation. The optimum dose seems to have a range of 2—5 krad X-rays, but depends on the cultivar and ploidy level.

In vitro propagation is being used to produce and maintain virus-free material by meristem culture. Vegetative propagation can also be carried out using callus and adventitious shoot formation in vitro (Bajaj and Pierik, 1974; Davies and Allen, 1973; Davies and Helsop, 1972; Pierik, 1976a; Pierik and Steegmans, 1975a,b, 1976b). Flower stalks were irradiated with a series of doses (50—2 000 rad X-rays) but no dose produced a mutant (H. Verboom, unpublished data).

The absence of (spontaneous) mutations in the relatively new tetraploid cultivars of *Freesia* could be attributed to buffering of the tetraploid genome. If this is so, it is possible that, in the course of time, the visible mutation frequency will increase as a result of continuous cross breeding and the accumulation of spontaneous and unnoticed mutations in genetic make-up. Consequently, the best chance of producing variability in a given excellent (sterile) cultivar would appear to exist when plant material is irradiated prior to adventitious bud formation in vitro.

Literature Freesia

Bajaj, Y.P.S. and Pierik, R.L.M., 1974. Vegetative propagation of freesia through callus cultures. Neth. J. Acric. Sci., 22: 153—159.

Davies, D.R. and Allen, Y., 1973. In vitro propagation of *Freesia*. John Innes Inst. Annu. Rep., 64: 67—68.

Davies, D.R. and Helsop, P., 1972. In vitro propagation of *Freesia*. John Innes Inst. Annu. Rep., 63: 64.

Pierik, R.L.M., 1976a. Nieuwe methoden voor vegetatieve vermeerdering van *Freesia* in kweekbuizen. Vakbl. Bloemisterij, 31(38): 61.

Pierik, R.L.M. and Steegmans, H.H.M., 1975a. Vegetatieve vermeerdering van *Freesia* in kweekbuizen. Vakbl. Bloemisterij, 30(24): 18—19; 21.

Pierik, R.L.M. and Steegmans, H.H.M., 1975b. Freesia plantlets from flower-buds cultivated in vitro. Neth. J. Agric. Sci., 23: 334—337.

Pierik, R.L.M. and Steegmans, H.H.M., 1976b. Vegetative propagation of *Freesia* through the isolation of shoots in vitro. Neth. J. Agric. Sci., 24: 274—277.

Van Raalte, D., 1971. Het Handboek voor de Bloemisterij. V. Snijbloemen. Born Periodieken N.V., Assen, Amsterdam, 616 pp.

GERBERA

Recently, flower production of *Gerbera* (family Compositae) in Western Europe has increased considerably and it has become one of the major cut flower plants (Van Raalte, 1971). Most cultivars are diploid (2n = 50) and hybrids of *Gerbera jamesonii*. Some tetraploids have been produced from diploids.

Gerbera has mainly been propagated by seed, probably because no fast vegetative propagation method was readily available (Garthwaite, 1960). Recently, the development of vegetative in vivo and in vitro propagation techniques has begun (Breutmann, 1976; Leffring and Hoogstrate, 1976; Murashige et al., 1974; Pierik and Segers, 1973; Pierik et al., 1973, 1974, 1975). These techniques, together with the increasing importance of clonal cultivars, have stimulated great interest in the improvement of selection procedures for high flower production (Leffring, 1973). In fact, investigations have already started in the form of co-operative programmes with the Association Euratom—ITAL, Wageningen, where plant irradiation (optimum dose 1.5 krad X-rays), has not yet met with success.

Spontaneous flower colour mutations are known to occur but generally they affect only a small part of the floret, they seldom result in complete periclinal chimeras. Irradiation of in vitro material followed by a series of sub-cultures might be more successful.

Literature Gerbera

Breutmann, B., 1976. Vegetative Vermehrung von *Gerbera jamesonii*. Gartenbaul. Versuchsber. Rheinland, 15: 216—219.

Garthwaite, J.M., 1960. *Gerbera* propagation. Exp. Hortic., 3: 65—69.

Leffring, L., 1973. Flower production in *Gerbera*. I. Correlations between shoot, leaf and flower formation in seedlings. Sci. Hortic., 1(3): 221—231.

Leffring, L. and Hoogstrate, J., 1976. Weefselkweekonderzoek *Gerbera* nog in volle gang. Vakbl. Bloemisterij, 31(47): 16—17.

Murashige, T., Serpa, M. and Jones, J.B., 1974. Clonal multiplication of gerbera through tissue culture. HortScience, 9(3): 175—180.

Pierik, R.L.M. and Segers, T.A., 1973. In vitro culture of midrib explants of *Gerbera*: adventitious root formation, callus induction. Z. Pflanzenphysiol., 69(3): 204—212.

Pierik, R.L.M., Steegmans, H.H.M. and Marelis, J.J., 1973. *Gerbera* plantlets from in vitro cultivated capitulum explants. Sci. Hortic., 1: 117—120.

Pierik, R.L.M., Jansen, J.L.M. and Maasdam, A., 1974. Vegetatieve vermeerdering van *Gerbera* in kweekbuizen. Vakbl. Bloemisterij, 29(39): 18—21.

Pierik, R.L.M., Jansen, J.L.M., Maasdam, A. and Binnendijk, C.M., 1975. Optimalization of *Gerbera* plantlet production from excised capitulum explants. Sci. Hortic., 3: 351—357.

Van Raalte, D., 1971. Het Handboek voor de Bloemisterij. V. Snijbloemen. Born Periodieken N.V., Assen, Amsterdam, 616 pp.

ORCHIDS

Many genera belong to the family of Orchidaceae, while several of them are widely grown as ornamental pot plants or for cut flower production. To our knowledge, no mutation breeding project is on record in the relevant literature, a fact which is not surprising in view of the large potentialities of cross breeding and the slow vegetative propagation of clones.

However, in several years, in vitro propagation was reported: at first in

Dendrobium and *Cymbidium* (Brown and Sagawa, 1965; Leffring, 1968; Marston, 1966—1967), but later also in *Lycaste*, *Miltonia*, *Odontoglossum* and *Phaius* (Marston, 1967) and several other genera (Churchill et al., 1971; Marston and Voraurai, 1967; Rao, 1977; Reisinger et al., 1976).

The described rapid in vitro propagation using protocorm or callus opens up possibilities for artificial mutation induction in valuable cultivars. In co-operation with the Association Euratom—ITAL a few projects with *Miltonia* and *Cymbidium* are in progress. So far, no results have been obtained since the plantlets formed from the irradiated material have not yet flowered (Broertjes, unpublished).

Literature orchids

Brown, A.J. and Sagawa, Y., 1965. Radiation effects on developing ovulary tissue in orchids. Radiat. Bot., 5(2): 107—108.

Churchill, M.E., Arditti, J. and Ball, E.A., 1971. Clonal propagation of orchids from leaf tips. Am. Orchid Soc. Bull., 40: 109—113.

Leffring, L., 1968. Vermeerdering van orchideeën d.m.v. meristeem cultuur. Meded. Dir. Tuinb. (Neth.), 31(10): 392—396.

Marston, M.E., 1966—1967. Vegetative propagation of plants using tissue culture techniques. Univ. Nottingham, School Agric. Rep., pp. 77—80.

Marston, M.E., 1967. Clonal multiplication of orchids by shoot meristem culture. Sci. Hortic., XIX: 80—86.

Marston, M.E. and Voraurai, P., 1967. Multiplication of orchid clones by shoot meristem culture. A review of the literature. Univ. Nottingham, Dep. Hortic., Misc. Publ. No. 17, 8 pp.

Rao, A.N., 1977. Tissue culture in the orchid industry. In: J. Reinert and Y.P.S. Bajaj (Editors), Plant Cell, Tissue, and Organ Culture. Springer Verlag, Berlin, pp. 44—69.

Reisinger, D.M., Ball. E.A. and Arditti, J., 1976. Clonal propagation of *Phalaenopsis* by means of flower-stalk node cultures. Orchid Rev., 84: 45—52.

ROSES

The genus *Rosa* (fam. Rosaceae) is a very large one. The exact number of species is still unknown and so is the history of our garden rose which has been developing from spontaneous and man-made hybridizations for centuries. Most roses are found in the temperate zones of the northern hemisphere. Asia seems to be the gene centre where the majority of species are found.

To classify the great number of species and hybrids the genus has been divided into several sections: *R. chinensis* (syn. *R. indica*), containing, amongst others, the miniature roses, *R. dilecta* to which many of the garden and greenhouse cultivars belong, *R. rehderiana* (synonym *R. polyantha*) with polyantha hybrids and floribunda roses, and many others.

The chromosome numbers vary from 2n = 14 to 2n = 56, but most species are diploid or tetraploid. The commercially grown rose cultivars are generally triploid or tetraploid.

The considerable interest of breeders in the application of mutation induction is not accidental. Spontaneous mutations occur frequently (Heslot, 1964) and have caused a valuable increase in genetic variation, especially in the tea hybrid group. Saakov (1960) studied the history of bud mutations in a number of cultivars and was able to draw some interesting conclusions concerning the time of mutation in relation to the age of the variety, the degree of mutations in relation to the genotype, as well as the tendency for the flower colour to mutate in relation to the flower colour of the initial cultivar. Red and pink flowers were generally found to be dominant over white, while the degree of filling of the flowers becomes less with mutations. Climbing types were found to result from mutation rather often.

Numerous authors have written about mutation breeding in roses. Chan (1966) irradiated five cultivars (Peace, Queen Elisabeth, Better Times, Baccara and Tropica) and found several mutations that affected flower colour and growth habit after X-ray doses of 7—8 krad or higher. Dommergues et al. (1967) compared the reactions of diploid cultivars (Gloria Mundi, Border King) with those of triploid (Orange Triumph) and tetraploid ones (Peace and several others) to a chemical mutagen (ethyl methanesulphonate (EMS)) and a physical mutagen (γ-rays). Both mutagens produced mutations in number and shape of the petals, flower colour and growth habit, two years after treatment of 1-year old plants with 8—9 krad X-rays or 0.5—0.8% EMS for 24 h at 20°C just before planting in the autumn. The survival, rearrangement of pre-existing periclinal chimeras and mutations in the cultivars were different, both between and within ploidy groups. Radiation seemed to be more effective in mutation induction, especially in polyploids.

Gupta (1966) and Gupta and Shukla (1970, 1971a,b) irradiated bud wood of cvs Montezuma and Super Star and several scented cultivars. Two years after irradiation (optimum dose 4—5 krad γ-rays) they observed mutations for flower colour and shape. Heslot (1966, 1968) treated plants of diploid, triploid and tetraploid cultivars with 4 or 8 krad γ-rays and with 8 ppm EMS. The roots were protected against radiation. High frequencies of "mutations" in cvs Orange Triumph and Gloria Mundi could be explained as consequences of pre-existing periclinal chimerism, whereas several mutants were obtained in other cultivars. Chromatographic analysis of pigments, which was carried out in co-operation with Dommergues (1976), demonstrated that mutants generally have either more or less of one or several of the pigments found in the control.

Kaicker and Swarup (1972) treated dormant buds of three cultivars (Christian Dior, Kiss of Fire and Queen Elisabeth) with γ-rays (5, 7.5 and 10 krad) or with one of several chemicals, including EMS, N-nitroso-N-methyl urethane (NMU) and ethylene imine (EI). Both groups seem to be equally effective in the production of flower colour mutations, but the numbers of mutations obtained were too low for definite conclusions to be drawn. Lata (1973) and Lata and Gupta (1971a,b, 1975) irradiated bud wood with

4 kR of γ-rays and stem cuttings with 2, 4, 6 and 8 kR of γ-rays. They studied the oil content of some scented cultivars of hybrid tea roses, survival in relation to dose and time of year of treatment, and cytological features of the cultivar Montezuma and two mutants as well as their breeding behaviour (Lata, 1975). Nakajima (1965, 1970, 1973) produced many sports from many cultivars (although special attention was paid to cv. Peace). He studied the effects of different dose rates, including an acute irradiation, with γ-rays and found that 8 kR (800 R/day) gave the highest mutation frequency. By the "one bud by one bud downward cutting-back" method he concluded that lateral buds develop from fewer cells than other buds and consequently are likely to produce apparently non-chimeric mutants. He obtained several promising flower colour mutants such as IRB 480-1 and IRB 480-2, a mutant of cv. Peace and a mutant of a mutant of cv. Peace, respectively, which are both on trial.

Streitberg (1964, 1966b,c,d, 1967a) carried out large-scale irradiation experiments with 42 cultivars. She found three sensitivity groups in which the optimum doses were, on the basis of an LD_{50}: 1.5—6, 1—3 and 0.5—1 krad (the relationship with ploidy, if any, was not mentioned). The type and number of mutants obtained depended on the cultivar, several mutants with different flower colour, flower form and size of the flower being produced. Usenbaev and Imankulova (1974) irradiated buds of four cultivars, grafted on *R. canina*, with 0.5—3 krad of γ-rays. After 5 years they obtained approximately 20 sports with changes in, for example, flower colour and petal number. A few are propagated for the State Variety Evaluation (GDR).

Only one publication was found in which adventitious shoots (on roots) were used (Zimmerman and Hitchcock, 1950—1952). Such shoots originate from the L-III layer and "sport back" to the original cultivar, such as a single flowering pink sport of cv. Briarcliff to the double flowering red cv. Briarcliff, cv. Better Times to cv. Briarcliff and cv. Souvenir (yellow) to cv. Talisman (red).

From the above literature review it is clear that roses are very suitable plants for mutation breeding, since many flower colour and other mutations can be produced without altering any other character of the original genotype. This was recently emphasized once again by Dommergues (1976). It is surprising that only three commercial mutants are known at present: namely, cv. Desi (Table 18), cv. Permoser, (references to this cultivar, if any, are unknown) and cv. Flamingo Queen, released in Canada (Harney, 1976), although several authors have described promising mutants. The problems seem to be that it takes a few years of cutting back before sizeable mutant sectors can be detected and that large-scale propagation is lengthy and expensive. Perhaps an in vitro propagation method is necessary to speed up mutation breeding as well as the propagation of interesting mutants, before a rapid and necessary increase in the variation of economically important cultivars can be obtained (Jacobs et al., 1969, 1970a,b).

Plate 7. *Chrysanthemum morifolium* cv. Westland (1) and the radiation-induced commercial mutants Yellow Westland (2), Dark Westland (3), Orange Westland (4) and White Westland (5).
(Slides from Fa. Fides, De Lier, who also sponsored this plate.)

TABLE 18

Commercial mutants of roses

Name of new cultivar	Place and date of release (or approval) and name of principal worker(s) and institute	Conditions and date of mutagenic treatment [Parent cultivar]	Main improved attributes of cultivar
Desi	German Democratic Republic, 1965 H. Rupprecht, Institut für Zierpflanzenbau, Humboldt-Universität, Berlin	3 kR X-rays, 1956 [Gloria Dei]	Intense colours, dark red stripes on yellow petals
Flamingo Queen	Canada, between 1966 and 1976 Agriculture Canada Research Station, Ottawa, Ont.	7—8 kR X-rays, 1964 [Queen Elizabeth]	Salmon pink flower colour
Permoser	German Democratic Republic, 1970 VEG Saatzucht Baumschuler, Dresden	Buds, 1500 R X-rays [Kordes Perfecta]	More intense colour of petal margin; otherwise like parent

Sources: IAEA (1972, pp. 526—544) and *Mutation Breeding Newsletter*, 1 (1972) —9 (1977).

In co-operative projects between the Association Euratom—ITAL at Wageningen, The Netherlands, and commercial plant breeders, several cultivars have been irradiated, so far without any marketable results. One project seems to have promise because a rapid multiplication system is used to speed up the production of solid (periclinal) sports (irradiated buds are grafted onto rootstocks in October, and the shoots are twice successively grafted onto rootstocks during the same winter). The project is too recent to predict on the results, however (C. Broertjes, unpublished data).

Literature roses

Chan, A.P., 1966. *Chrysanthemum* and rose mutations induced by X-rays. Proc. Am. Soc. Hortic. Sci., 88: 613—620.

Dommergues, P., 1976. Mutagénèse induite chez le rosier. Hortic. Fr., 74: 1—3.

Dommergues, P., Heslot, H., Gillot, J. and Martin, C., 1967. L'induction de mutations chez les rosiers. In: H. Stubbe (Editor), Induced Mutations and Their Utilization, Gatersleben, 1966. Akademie-Verlag, Berlin, pp. 319—349.

Gupta, M.N., 1966. Induction of somatic mutations in some ornamental plants. In: Proc. All India Symp. in Hortic. Calcutta, pp. 107—114.

Gupta, M.N. and Shukla, R., 1970. Mutation breeding of garden roses: studies on induction of somatic mutations after gamma irradiation of bud-wood. Prog. Hortic., 2(3): 69—83.

Gupta, M.N. and Shukla, R., 1971a. Mutation breeding of garden roses: recovery of somatic mutations after three years of gamma irradiation of bud-wood. Recent Adv. Hortic., Kampur (India), 1971: 81—86.

Gupta, M.N. and Shukla, R., 1971b. Mutation breeding of garden roses. Jpn. J. Breed., 21(3): 129—136.

Harney, P.M., 1976. Breeding of floricultural crops in Canada. In: K.C. Sink Jr (Editor), Int. Symp., Floriculture Plant Breeding and Genetics, East Lansing, Mich. Acta Hortic., 63: 19—26.

Heslot, H., 1964. L'induction expérimentale de mutations chez les plantes florales. C.R. Acad. Séances Agric. Fr., 1964 (16 December): 1281—1308.

Heslot, H., 1966. Induction de mutations par des agents mutagènes physiques et chimiques. In: Mutations in Plant Breeding. IAEA, Vienna, pp. 139—149.

Heslot, H., 1968. Mutation research done in 1967 on barley, roses and marigold. Mutations in Plant Breeding II. IAEA, Vienna, pp. 153—159.

IAEA, 1972. Induced Mutations and Plant Improvement. IAEA, Vienna, pp. 526—544.

Jacobs, G., Allan, P. and Bornman, C.H., 1969. Tissue culture studies on rose: use of shoot tip explants. I. Auxin: cytokinin effects. Agroplantae, 1: 179—188.

Jacobs, G., Allan, P. and Bornman, C.H., 1970a. Tissue culture studies on rose: use of shoot tip explants. II. Cytokinin: gibberellin effects. Agroplantae, 2: 25—28.

Jacobs, G., Allan, P. and Bornman, C.H., 1970b. Tissue culture studies on rose: use of shoot tip explants. III. Auxin: gibberellin effects. Agroplantae, 2: 45—50.

Kaicker, U.S. and Swarup, V., 1972. Induced mutations in roses. Indian J. Genet. Plant Breed., 32(2): 257—265.

Lata, P., 1973. Effects of ionizing radiation on roses with special reference to induction of mutations. Thesis, Kampur University, Kampur, 245 pp.

Lata, P., 1975. Hybridization in modern roses. IV. Hybrids between control and mutants of cultivar Montezuma. Sabrao J., 7(1): 103—108.

Lata, P. and Gupta, M.N., 1971a. Mutation breeding of garden roses. Effect of gamma rays on essential oil content in some scented roses. Parfüm. Kosmet., 52: 267—270.

Lata, P. and Gupta, M.N., 1971b. Effects of gamma rays on stem cuttings of essential oil bearing *Rosa* spp. Flavour Ind., 2(7): 421—425.

Lata, P. and Gupta, M.N., 1975. Effects of ionizing radiation on roses: meiotic studies on control and gamma ray induced mutants of rose cultivar Montezuma. Cytologia, 40: 213—219.

Nakajima, K., 1965. Induction of sports by gamma-ray irradiation. Gamma Field Symp., 4: 55—70.

Nakajima, K., 1970. Gamma ray induced sports from a rose variety (Peace). Tech. News, No. 4; Instit. of Radiat. Breed., 2 pp.

Nakajima, K., 1973. Induction of useful mutations of mulberry and roses by gamma rays. In: Induced Mutations in Vegetatively Propagated Plants. IAEA, Vienna, pp. 105—117.

Saakow, S.G., 1960. Die Sortenbildung bei Rosen durch Knospenmutationen. Arch. Gartenbau, VIII(8): 595—629.

Streitberg, H., 1964. Abschlussbericht zur Forschungsarbeit: Schaffung wirtschaftlich wertvoller Mutationen bei Zierpflanzen (*Rhododendron*- und *Rosa*-Arten) durch Behandlung mit Röntgenstrahlen. Teil b. Mutationen bei Rosen. Dtsch. Akad. Landwirtsch. Wiss. Berlin.

Streitberg, H., 1966b. Neue Rosen- und Azaleen-Sorten mit Hilfe der Röntgenbestrahlung. Dtsch. Gartenbau, 13(10): 267—268.

Streitberg, H., 1966c. Rosenzüchtung mit Hilfe der Röntgenbestrahlung. Arch. Gartenbau, 14(2/3): 81—88.

Streitberg, H., 1966d. Schaffung wirtschaftlich wertvoller Sprossvarianten bei Rosen durch Behandlung mit Röntgenstrahlen. Z. Pflanzenzücht., 55(2): 165—182.

Streitberg, H., 1967a. Production of economically valuable variation in roses and azaleas by means of X-irradiation. In: H. Stubbe (Editor), Induced Mutations and their Utilization, Gatersleben, 1966. Akademie-Verlag, Berlin, pp. 359—362.

Usenbaev, E.K. and Imankulova, K.I., 1974. Radiation mutants of roses. Proc. Int. Hortic. Congr., 19(1A): 109.

Zimmerman, P.W. and Hitchcock, A.E., 1950—1952. Rose "sports" from adventitious buds. Contrib. Boyce Thompson Inst., 16: 221—224.

Garden plants

CANNA

Canna, family Cannaceae, is an ornamental garden plant, widely grown in tropical countries. In addition, compact cultivars have been developed which are grown as pot plants. In total, over 1 000 cultivars were known at the beginning of the 20th century; however, several of them have disappeared.

They are propagated vegetatively by rhizomes since many cultivars are sterile, some of them being triploids. The rhizomes grow rapidly and the young shoots flower after only approximately 2 months growth.

The practice of mutation breeding is limited to a few breeders in India (Gupta, 1966; Khoshoo, 1968; Mukherjee and Khoshoo, 1970) and Thailand (Nakornthap, 1965). Mukherjee and Khoshoo (1970) compared the response to radiation of two diploid cultivars with that of two triploids as preliminary experiments before starting a mutation breeding programme which was designed to improved flower shape and colour and durability of blooms. Rhizomes were irradiated with 1, 2 and 3 kR γ-rays and the morphological and cytological properties of the irradiated clones were studied and compared with those of natural mutants.

Two observations are of interest to the mutation breeder: firstly, a dose of 3 kR γ-rays was (almost) lethal to the diploids but did not affect the triploids, and secondly, red was a better flower colour to start with if yellow-flowered mutants were required. Similar results were obtained by Gupta (1966) and by Nakornthap (1965). The latter irradiated 22 cultivars (three triploids and 18 diploids; 2n = 18) with 1.0, 1.5 and 2.7 krad ^{60}Co γ-rays and obtained a fairly large variability in flower colour, form and size.

No commercial mutant has so far been released, however, a fact that is not surprising since the experiments reported were probably on too small a scale. Rhizomes of dwarf cannas were irradiated with X-rays (C. Broertjes, unpublished data) to determine their radiosensitivity. The optimum dose for cultivars like Lucifer was much lower than that for cultivars used by other workers, namely 500—1 000 rad. No mutations have yet been found.

Literature Canna

Gupta, M.N., 1966. Induction of somatic mutations in some ornamental plants. In: Proc. All India Symp. Hortic., pp. 107—114.

Khoshoo, T.N., 1968. Genetic improvement of ornamentals in India. Indian J. Genet., 28A: 187—198.

Mukherjee, I. and Khoshoo, T.N., 1970. Genetic-evolutionary studies on cultivated cannas IV: parallelism between natural and induced somatic mutations. Radiat. Bot., 10(4): 351—364.

Nakornthap, A., 1965. Radiation-induced somatic mutations in the ornamental *Canna*. Radiat. Bot., 5 (Suppl.): 707—712.

TABLE 19

The commercial mutant of *Portulaca*

Name of new cultivar	Place and date of release (or approval) and name of principal worker(s) and institute	Conditions and date of mutagenic treatment [Parent cultivar]	Main improved attributes of cultivar
Karna phul	India, 1974 B.M. Desai, Bhabha Atomic Research Centre, Bombay	1 kR γ-rays [*P. grandiflora*]	*Gerbera*-type flower

Sources: IAEA (1972, pp. 526–544) and *Mutation Breeding Newsletter*, 1 (1972) –9 (1977).

PORTULACA

Perennial portulaca, *Portulaca grandiflora* Hook, family Portulacaceae, is native to eastern South America and is extensively cultivated in tropical and subtropical regions. Unlike the annual portulaca, the perennial type produces only double, deep carmine coloured flowers which do not seed (Lata and Gupta, 1971c). Therefore, several breeders such as Cotter (1963) and Ootani and Hagiwara (1969), have studied portulaca. Ootani and Hagiwara carried out an intensive investigation of the inheritance of flower colour and chymochromic pigments in the common portulaca.

In addition, several authors have reported the effects of radiation on, for example, flower colour and form. Banerjee (1967) irradiated cuttings and found that the optimum dose is in the range of 1—2.5 krad γ-rays. He found variation in flower types and a reversion to the annual plant. Gupta (1966, 1970) also irradiated cuttings with γ-rays and obtained semi-double, single, dissected and inward curved flower mutations. A dose of 4 krad was optimal. These different mutants were studied in more detail by Lata and Gupta (1971c). Finally, Desai (1974) has reported induction of several mutants which were mostly stabilized by subsequent vegetative propagation and include a *Gerbera*-type, a five petaled flower, and several flower colour mutants. To our best knowledge, only one mutant has been commercialized, the *Gerbera*-type (cv. Karna phul) (Table 19).

Literature Portulaca

Banerjee, S.K., 1967. Effect of gamma radiation on some perennial ornamentals. Indian J. Genet. Plant Breed., 27(3): 417—422.

Cotter, D.J., 1963. Effects of radiation on the photoperiodism of *Portulaca smallii*. Radiat. Bot., 3(3): 265—269.

Desai, B.M., 1974. New cultivars of perennial *Portulaca* through gamma radiation. Indian Hortic., 19(2): 19, 23.

Gupta, M.N., 1966. Induction of somatic mutations in some ornamentals. In: Proc. All India Symp. Hortic., pp. 107—114.

Gupta, M.N., 1970. Use of gamma irradiation in the production of new varieties of perennial *Portulaca*. In: Proc. Symp., Radiation and Radiomimetic Substances in Mutation Breeding. Food Agric. Comm. Dep. At. Energy, Bombay, pp. 206—214.

IAEA, 1972. Induced Mutations and Plant Improvement. IAEA, Vienna, pp. 526—544.

Lata, P. and Gupta, M.N., 1971c. Morphological and cytological studies on gamma ray induced mutants of perennial *Portulaca grandiflora*. Jpn. J. Breed., 21(3): 121—128.

Ootani, S. and Hagiwara, T., 1969. Inheritance of flower colors and related chymochromic pigments in F_1 hybrids of common portulaca, *Portulaca grandiflora*. Jpn. J. Genet., 44(2): 65—79.

Woody plants

GENERAL

Woody ornamentals, arboreous plants and conifers, as well as woody perennial and forest trees, broad leaved forest trees and coniferous forest trees have several points in common. A great many of them are (close to) real botanical species and are, therefore, not very heterozygous even if they are propagated vegetatively. This means that mutation breeding does not have very great potential in these crops. Moreover, the size of the mature plants and the generally long vegetative period discourage large-scale mutation breeding programmes. Thus, it is not surprising that the number of reports about mutation breeding of these plants is rather limited.

As will be commented on in the introductory remarks to Chapter 6, many woody plants are propagated by seeds, although there is an increasing interest in vegetative propagation for various reasons (Pierik, 1975b). This may ultimately lead to consideration of the use of mutation breeding.

It is a pity that, in general, Russian work is relatively inaccessible because of the language barrier. We don't know much about their work, for example, at the Institute of Cytology and Genetics in Novosibirsk, where G.F. Privalov studied several problems on trees and shrubs, such as the effects of ionizing radiation versus those of chemical mutagens, the modification of the effects of chemical mutagens by plant hormones, and other subjects (see IAEA, 1972). He mainly uses *Acer negundo* (Privalov, 1967).

Other Russians work with chemicals: Samoshkin (1974) and Shchepot (1974) experiment with *Fraxinus* and *Quercus*, respectively. Conifers are also subjected to mutagenic treatments (Maskin et al., 1974). There must be many more publications to which we have no access.

TREES AND SHRUBS

Azalea and *Rhododendron*, which strictly speaking belong to the group of woody ornamentals, have already been discussed in the section on ornamental pot plants.

The few other literature references that were found concerning arboreous ornamentals were about *Abelia*, *Forsythia*, *Potentilla* and *Weigelia*. In addition, a few reports of preliminary investigations in a number of other species exist.

One commercial mutant of *Abelia grandiflora* Rehd., an ornamental shrub, has been reported by Suda and Matsubara (Anonymous, 1977) (Table 20).

Rooted cuttings of *Forsythia* × *intermedia* were irradiated with various doses of X-rays. The optimum dose was found to be 5—6 kR. Although several mutations which altered, for example, leaf size and form and flower

TABLE 20

The commercial mutant of *Abelia grandiflora*

Name of new cultivar	Place and date of release (or approval) and name of principal worker(s) and institute	Conditions and date of mutagenic treatment [Parent cultivar]	Main improved attributes of cultivar
Mei-Fu-Hana-Tsukubane-Utsugi	Japan, 1976 H. Suda and H. Matsubara, Tokyo Metropolitan Isotope Research Center, Tokyo	Rooted young stem cuttings, 3 kR ^{60}Co γ-rays, 1972 [Hanazono-Tsukubane-Utsugi]	More fixed variegated plants of deep green having brighter yellow colour on leaf margins than other variegated varieties (original variety, green); very dwarf, bright variegated leaves throughout the year; very strong

Sources: IAEA (1972, pp. 526—544) and *Mutation Breeding Newsletter*, 1 (1972) —9 (1977).

colour (paler, darker yellow) were found, none of them was an improvement. Several of them, however, were considered to be valuable material for further breeding work (Knuth, 1974; Knuth and Kaufmann, 1974).

Buds from a dozen other cultivars were irradiated with 3—5 and 7 krad γ-rays by Cadic and Decourtye (1974). They report that the tetraploid cultivars Beatrix Farrand, Karl Sax and Tetragold yielded very few mutants but that a much higher frequency of aberrant shoots were produced from the diploid species and cultivars.

The effects of radiation upon (apomictic) species of *Potentilla* were studied by Asker (1966a,b). Their radioresistance observed was extremely high and an LD_{50} of 60—80 krad or (much) higher was normal, varying, however, with the ploidy level of the species. Several aberrant types were found, generally the result of a changed chromosome number.

In *Weigelia*, after treating seeds with ethyl methanesulphonate (EMS), the mutated sector size in leaves was studied as well as alterations in the ratio between the length and width of leaves etc. So far, no valuable mutation has been obtained (Poszwinska, 1974).

In vitro propagation and its use for the production of tetraploids or mutants by the application of respectively colchicine or gamma rays is reported by Duron (1975) and Duron and Decourtye (1976).

Our own experience with woody ornamentals, mostly in co-operation with the Research Station for Arboriculture at Boskoop, The Netherlands, is also rather limited. We have irradiated rooted cuttings or plants of *Buddleya* cv. Royal Red (optimum dose 2—3 krad X-rays), *Clematis* cv. Jackmanii (0.2—0.3 krad X-rays), *Hydrangea* cv. Grandiflora (2 krad), *Laburnum watereri* cv. Vossii (3 krad), *Potentilla fruticosa* cv. Tangerine (on a large scale; 7.5 krad) and *Syringa vulgaris* cv. Primrose (3 krad) without any success. Irradiation of a "sport" of *Forsythia intermedia spectabilis* with an unusual flower colour resulted in variations in flower colour, the mutant flowers generally being paler in colour.

CONIFERS

The number of publications about mutation induction in ornamental conifers is very limited and in fact only one author, Pohlheim, could be traced (Pohlheim, 1971a,b,c,d,e, 1972a,b, 1973). Most of his studies, however, concern the behaviour of pre-existing variegated periclinal chimeras of various Cupressaceae, such as *Chamaecyparis*, *Cupressus*, *Thuja* and *Juniperus*. They often have completely white, chlorophyll-deficient shoots, that are either spontaneously or X-ray induced, from which conclusions about the histology of the two-layered apex and the behaviour of individual cells can be drawn. The haploid *Thuja gigantea gracilis* is particularly useful for this purpose, since (chlorophyll-deficient) recessive mutations can be seen in the irradiated material itself (Pohlheim, 1968).

Mutation breeding for practical purposes has not yet been reported. This is somewhat surprising since the spontaneous variability in the colour of the needles as well as the form and size of trees is quite large, while vegetative propagation, by one method or another, is generally uncomplicated and should therefore yield results within a reasonable period of time. To test this hypothesis, rooted cuttings of *Chamaecyparis lawsoniana*, *Cham. obtusa* (several cultivars), *Juniperus chinensis*, *J. communis*, *J. horizontalis* and *J. squamata* (several cultivars) were irradiated with a series of doses of X-rays to determine their radiosensitivities (generally very low to low doses are used since these plants have very limited radioresistances). This is to be followed by large-scale irradiation of each cultivar with the optimum dose and a search for mutant shoots during a number of years (Miss A. Heyting, Research Station for Arboriculture, Boskoop, unpublished data).

Literature woody plants

Anonymous, 1977. List of mutant varieties. Mutat. Breed. Newsl., 9(1): 14—17.

Asker, S., 1966a. Effects of mutagen treatment on some apomictic *Potentilla* species. Hereditas, 55(2-3): 249—265.

Asker, S., 1966b. The occurrence of aberrants in some apomictic *Potentilla argentea* — biotypes. Hereditas, 56(1): 54—70.

Cadic, A. and Decourtye, L., 1974. Recherche de formes nouvelles dans le genre *Forsythia* par mutagénèse provoquée. Proc. Int. Hortic. Congr., 19(1A): 99.

Duron, M., 1975. La culture in vitro de méristèmes de quelques cultivars de *Weigelia* (Thunb.) hybrides. C.R. Acad. Sci. Sér. D, 281: 985—987.

Duron, M. and Decourtye, L., 1976. Mutagenesis with plantlets of *Weigelia* sp. grown in vitro. In: Proc. Workshop Eur. Comm. — Israel, The Use of Ionizing Radiation, Wageningen, March 22—24. Assoc. Euratom—ITAL, Wageningen, pp. 479—484.

IAEA, 1972. Induced Mutations and Plant Improvement. IAEA, Vienna, pp. 526—544.

IAEA, 1975. Study Tour Report of the IAEA: Isotopes and Radiation in Agricultural Research in the Soviet Union. IAEA, Vienna.

Knuth, M., 1974. Züchterische Ergebnisse durch Röntgenbehandlung an *Forsythia* × *intermedia* Zab. Arch. Gartenbau, 22(5/6): 397—406.

Knuth, M. and Kaufmann, H.G., 1974. Breeding results after X-ray treatment of *Forsythia* × *intermedia* Zab. Proc. Int. Hortic. Congr., 19(1A): 101.

Mashkin, S.I., Guliaeva, E.M. and Grabor, G.A., 1974. Experimental mutagenesis in conifers. In: I.A. Rapoport (Editor), Uspekhi Khimicheskogo Mutageneza v Selektsii. pp. 300—302 (in Russian).

Pierik, R.L.M., 1975b. Vegetative propagation of horticultural crops in vitro with special attention to shrubs and trees. Acta Hortic., 54: 71—82.

Pohlheim, F., 1968. *Thuja gigantea gracilis* Beissn. — ein Haplont unter den Gymnospermen. Biol. Rundsch., 6(2): 84—86.

Pohlheim, F., 1971a. Untersuchungen zur Sprossvariation der Cupressaceae, 1. Nachweis immerspaltender Periklinalchimären. Flora (Jena) Abt. B, 160: 264—293.

Pohlheim, F., 1971b. Untersuchungen zur Sprossvariation der Cupressaceae, 2. Ploidiechimären an der haploiden *Thuja gigantea gracilis* nach spontaner Diploidisierung. Flora (Jena) Abt. B, 160: 294—316.

Pohlheim, F., 1971c. Untersuchungen zur Sprossvariation der Cupressaceae, 3. Quantitative Auswertung des Scheckungsmusters immerspaltender Periklinalchimären. Flora (Jena) Abt. B., 160: 360—372.

Pohlheim, F., 1971d. Untersuchungen zur Sprossvariation der Cupressaceae, 5. Das Scheckungsverhalten zweier unterschiedlicher Klone von *Juniperus chinensis variegata*. Wiss. Z. Pädagog. Hochsch. Potsdam, 15(1): 67—64.

Pohlheim, F., 1971e. Untersuchungen zur Sprossvariation der Cupressaceae, 6. Zum Wechsel von Nadel- und Schuppenform des Blattes. Wiss. Z. Pädagog. Hochsch. Potsdam, 15(1): 65—75.

Pohlheim, F., 1972a. Untersuchungen zur Sprossvariation der Cupressaceae, 4. Zur Auslese von Mutations-Chimären und Mutanten an der Haploiden *Thuja gigantea gracilis*. Arch. Züchtungsforsch., 2(3): 223—235.

Pohlheim, F., 1972b. Überlebungsrate und Sprossvariation durch Mutation nach Röntgenbestrahlung haploider und diploider Pflanzen von *Thuja plicata*. Biol. Rundsch., 10(3): 200—201.

Pohlheim, F., 1973. Untersuchungen zur Sprossvariation der Cupressaceae, 7. Auslösung von Perforationen durch Röntgenbestrahlung an der immerspaltenden Periklinalchimäre *Juniperus sabina 'Variegata'*. Wiss. Z. Pädadog. Hochsch. Potsdam, 17(1): 79—86.

Poszwinska, J.L., 1974. Effects of EMS on two vegetative generations of *Weigelia* cultivars. Proc. Int. Hort. Congr., 19.

Privalov, G.F., 1967. Experimental mutations in woody plants. In: H. Stubbe (Editor), Induced Mutations and Their Utilization. Erwin Baur Gedächtnisvorlesungen IV. Akademie-Verlag, Berlin, pp. 383—386.

Samoshkin, E.N., 1974. Effect of chemical mutagens on the growth of red ash (*Fraxinus pennsylvanica*). In: I.A. Rapoport (Editor), Uspekhi Khimicheskogo Mutageneza v Selektsii. pp. 303—305 (in Russian).

Shchepot, F.L., 1974. Chemical mutagenesis in oak (*Quercus*). In: I.A. Rapoport (Editor), Uspekhi Khimicheskogo Mutageneza v Selektsii. pp. 298—300 (in Russian).

Chapter 6

WOODY PERENNIAL AND FOREST TREES

GENERAL

Though several forest trees, arboreous plants or conifers, can be propagated asexually, most basically belong to the group of plants that are propagated by seed; consequently, mutation breeding can only be used if a second generation can be raised to obtain the double recessive mutants. This obviously takes a lot of time and space and is, therefore, far from attractive. Moreover, in contrast to ornamentals, the objectives of (mutation) breeding of forest trees are mainly concerned with characters that are not directly visible, like yield and resistance, which are much more difficult to select than characters of ornamental value. Despite the foreseen difficulties mutation breeding programmes for such trees may sometimes be inevitable if the necessary variability for resistance to pests and diseases is to be obtained or increased (Rudolph, 1972).

Many publications concerning woody perennials deal with studies on the radiosensitivity of these plants (Nishida et al., 1967; Ohba, 1964; Heaslip, 1967; Sparrow and Sparrow, 1965; Sparrow et al., 1968; Witherspoon, 1969). General aspects of mutation induction were discussed by Gustafsson (1960) and Ohba (1971a) who both refer to a large number of publications. Ohba reports that mutations are frequent in species like *Chamaecyparis* and *Cryptomeria*, whereas the mutation frequency in *Pinus* is apparently very restricted. No mutations have been found in *Alnus*, *Larix* and *Quercus*, in contrast to *Betula* and *Populus* in which some somatic variegation has been observed.

Seeds of several species have been subjected to mutagenic treatment, as reviewed in an article by Lynn (1967), who discussed publications concerning several trees, such as *Pinus*, *Cryptomeria*, *Chamaecyparis*, *Acer negundo*, *Betula*, *Quercus*, *Populus* and *Ulmus*. No beneficial mutant has been reported by any of the various authors, except for one of *Ulmus* which is resistant to Dutch elm disease (Quellet and Pomerleau, 1965).

Privalov (1963, 1968b) has also reported irradiation of seeds of many arboreous plants but, because some publications lack English summaries, it is not clear what the details of procedure and objectives were.

Further details are discussed on the following pages where tree crops have been grouped into two parts: (1) broad leaved trees and (2) coniferous trees.

Literature woody perennial and forest trees — general

Gustafsson, Å., 1960. Polyploidy and mutagenesis in forest-tree breeding. In: Proc. 5th World For. Congr., II. Genetics and Tree Improvement. pp. 793—806.

Heaslip, M.B., 1967. Radiosensitivity of deciduous tree seed to different ratios of fast neutron and gamma radiation. Radiat. Bot., 7(5): 415—428.

Lynn, M., 1967. Ionizing radiations in forests and forestry (excluding the use of radioactive tracers). For. Abstr., 28(1): 1—18.

Nishida, T., Nakajima, K., Ohba, K. and Takato, S., 1967. Radiosensitivity and induction of somatic mutations in woody perennials under chronic gamma ray irradiation. Gamma Field Symp., 6: 20—42.

Ohba, K., 1964. Studies on radiosensitivity and induction of somatic mutations in forest trees. Gamma Field Symp., 3: 111—139.

Ohba, K., 1971a. Studies on the Radiation Breeding of Forest Trees. Bull. Inst. Radiat. Breed., 2, Ohmiya, Ibaraki, 102 pp.

Privalov, G.F., 1963. Experimental mutations of the vegetative organs of trees. Dokl. Akad. Nauk SSSR, 150: 661—664 (in Russian).

Privalov, G.F., 1968b. Investigations of experimental mutagenesis in arboreous plants. Genetika, 4(6): 144—158 (in Russian with English summary).

Quellet, C.E. and Pomerleau, R., 1965. Recherches sur la résistance de l'orme d'amérique au *Ceratocystis ulmi*. Can. J. Bot., 43: 85—96.

Rudolph, T.D., 1972. Mutation breeding in forest trees. In: M.J. Constantin (Editor), Mutation Breeding Workshop, Knoxville, Tenn. University of Tennessee, Knoxville, Tenn.

Sparrow, R.C. and Sparrow, A.H., 1965. Relative radiosensitivities of woody and herbaceous spermatophytes. Science, 147 (3604): 1449—1451.

Sparrow, A.H., Rogers, A.F. and Schwemmer, S.S., 1968. Radiosensitivity studies with woody plants. I. Acute gamma irradiation survival data for 28 species and predictions for 190 species. Radiat. Bot., 8(2): 149—187.

Witherspoon, J.P., 1969. Radiosensitivity of forest tree species to acute fast neutron radiation. Nucl. Sci. Abstr., 23(10): No. 18128.

There are only two known publications about *Acer negundo* L., a seed-propagated species, that deal with the use of mutation techniques (Privalov, 1965, 1968a). The procedures and objectives of the work are obscure because the greater part is written in Russian. Further work was carried out by Agafonova et al. (1974), who studied esterase isoenzymes in various tissues of mutants of *Acer negundo* by electrophoretic analysis.

Of *Alnus*, another seed-propagated plant, one report was found concerning the reaction of plants to chronic irradiation in a γ-field (Ohba and Murai, 1966).

Fig. 21. Adventitious plantlets growing from callus on split branches (A), or on wounded, rooted cuttings (B) of *Populus*. (Photographs, ITAL, Wageningen.)

Betula plants were also subjected to chronic γ-irradiation (Ohba and Murai, 1966), while seeds of *Betula verrucosa* and *B. pubescens* were treated with X-rays (Scholz, 1957), resulting in chlorophyll mutants in M_1. Because of the high sterility of the M_1 plants, a second generation could not be raised.

The only literature of importance for *Populus* concerns in vivo or in vitro propagation techniques which might ultimately be used for mutation breeding purposes as well (Brand and Venverloo, 1973; Venverloo, 1973; Winton, 1970, 1971; Winton and Mathes, 1973). Ohba and Murai (1966) placed *Populus* in a γ-field to determine its radioresistance. They found somatic mutations, such as variegated leaves and narrow light green leaves after 50—70 R/day, the optimum dose rate for *Populus* and other broad leaved trees.

Since *Populus* can easily be propagated by splitting a branch into two parts or by making a long cut through a branch and subsequently allowing development of adventitious shoots on the callus formed (Fig. 21), the method might well be suitable for producing (non-chimeral?) mutants.

Hardly any literature can be found about irradiation experiments with *Quercus*, which almost exclusively is propagated by seed. The LD_{50} for seeds of *Quercus alba* and *Q. rubra* is around 6 kR (Stairs, 1963).

Ulmus could be one of the examples in which mutation breeding might provide necessary variability, in this instance to select for resistance to *Ceratocystis ulmi* (Buis) C. Moreau, the Dutch elm disease, though this resistance does not exist in nature. Radiation-induced resistance has been reported (Quellet and Pomerleau, 1965) and mutants could be used either directly, or indirectly by crossing with existing species (Holmes, 1967).

Other, similar programmes are in sight since it seems possible to propagate *Ulmus* from callus in vivo (Kramer, 1967; Tchernoff, 1963) and in vitro, as was reported in the fifties by Jacquiot (1949, 1951, 1955, 1966), and very recently by Durzan and Campbell (1974a,b) and Durzan and Lopushanski (1975). One wonders whether the in vitro techniques could be combined with sophisticated selection methods to screen very early for such characters as disease resistance.

Considering broad leaved trees as a group, it is worthy of note that there is a tendency to propagate several species vegetatively, clonally. This will ultimately shift the breeding methods for these plants towards those for other vegetatively propagated species, which means that the use of mutation breeding to create or increase variability in certain characters will be within reach. Mutation breeding will also be made more feasible by the fast, in vitro propagation techniques which are being or soon will be developed for a considerable number of woody species (Bonga, 1977; Durzan and Campbell, 1974a).

Literature broad leaved trees

Agafonova, O.V., Korochkin, L.I. and Privalov, G.F., 1974. Study of the esterase isoenzymes in various tissues of mutants of *Acer negundo* L. (in Russian). Plant Breed. Abstr., 47: No. 1692.

Bonga, J.M., 1977. Application of tissue culture in forestry. In: J. Reinert and Y.P.S. Bajaj (Editors), Plant Cell, Tissue, and Organ Culture. Springer Verlag, Berlin, pp. 93—108.

Brand, K. and Venverloo, C.J., 1973. The formation of adventitious organs, II. The origin of buds formed on young adventitious roots of *Populus nigra* L. 'Italica'. Acta Bot. Neerl., 22(4): 399—407.

Durzan, D.J. and Campbell, R.A., 1974a. Prospects for the mass production of improved stock of forest trees by cell and tissue culture. Can. J. For. Res., 4(2): 151—174.

Durzan, D.J. and Campbell, R.A., 1974b. Prospects for the introduction of traits in forest trees by cell and tissue culture. N. Z. J. For. Sci., 4(2): 261—266.

Durzan, D.J. and Lopushanski, S.M., 1975. Propagation of American elm via cell suspension cultures. Can. J. For. Res., 5(2): 273—277.

Holmes, F.W., 1967. Radiation Mutation Program of the Biology Department, Brookhaven National Laboratory. Biol. Dep. Brookhaven Natl Lab., Upton, Long Isl., N.Y., (July), pp. 114—115.

Jacquiot, C., 1949. Observations sur la néoformation de bourgeons chez le tissu cambial d'*Ulmus campestris* cultivé in vitro. C. R. Acad. Sci., 229: 529—530.

Jacquiot, C., 1951. Action du méso-inositol et de l'adenine sur la formation de bourgeons par le tissu cambial d'*Ulmus campestris* cultivé in vitro. C. R. Acad. Sci., 233(15): 815—817.

Jacquiot, C., 1955. Formation d'organes par le tissu cambial d'*Ulmus campestris* L. et de *Betula verrucosa* Gaertn. cultivés in vitro. C. R. Acad. Sci., 240: 557—558.

Jacquiot, C., 1966. Plant tissues and excised organs cultures and their significance in forest research. J. Inst. Wood Sci., 16: 22—34.

Kramer, A., 1967. Het vermeerderen van iepen door stekken afkomstig van wortelcallus. Beplant. Boomkwekerij, 2: 34—35.

Ohba, K. and Murai, M., 1966. Studies on radiosensitivity of forest trees 2. Growth and somatic mutations of *Alnus*, *Populus* and *Betula* species under chronic gamma irradiation in a gamma field. J. Jpn. For. Soc., 48(1): 12—19 (in Japanese with English summary).

Privalov, G.F., 1965. Somatical radiomutations by *Acer negundo* L. Bjcell. Mosk. Obšč. Isp. Prir., Otd. Biol., 70(1): 153—160.

Privalov, G.F., 1968a. Mutability in box-elder (*Acer negundo* L.) induced by gamma- and X-rays. Genetika, 4(5): 39—42.

Quellet, C.E. and Pomerleau, R., 1965. Recherches sur la résistance de l'orme d'amérique au *Ceratocystis ulmi*. Can. J. Bot., 43: 85—96.

Scholz, E., 1957. Röntgenmutationen bei der Birke. Züchter, 27(1): 54—60.

Stairs, G.R., 1963. Acute gamma irradiation of *Quercus* seed. Its effect on germination and seedling growth. Silvae Genet., 12(6): 192—200.

Tchernoff, V., 1963. Vegetative propagation of elms by means of shoots cut from callused roots. Acta Bot. Neerl., 12: 40—50.

Venverloo, C.J., 1973. The formation of adventitious organs, I. Cytokinin-induced formation of leaves and shoots in callus cultures of *Populus nigra* L. 'Italica'. Acta Bot. Neerl., 22(4): 390—398.

Winton, L.L., 1970. Shoot and tree production from aspen tissue cultures. Am. J. Bot., 57(8): 904—909.

Winton, L.L., 1971. Tissue culture propagation of European aspen. For. Sci., 17(3): 348—350.

Winton, L.L. and Mathes, M.C., 1973. Aspen callus. In: Tissue Culture Methods and Applications. Academic Press, New York, N.Y., pp. 161—165.

CONIFEROUS TREES

The remarks made in discussing the possibilities of mutation breeding in woody perennials and forest trees in general, apply especially to coniferous forest trees; most species are seed propagated and have a long vegetative period so there is a long period of time between mutation and the raising of a second generation for selection of mutants.

Most publications concerning the effects of radiation on coniferous trees, therefore, only deal with the stage in which the radiosensitivity is determined either of seeds to acute radiation or of growing plants to chronic γ-irradiation (Capella and Conger, 1967; Davis, 1962; Donini, 1967; Mergen and Thielges, 1966; Miksche et al., 1961, 1962; Murai and Ohba, 1966; Sparrow, 1963; Sparrow et al., 1963b, 1965), while few articles describe effects in later generations (Rudolph, 1967).

It does not seem to be important to apply mutation breeding to most coniferous trees and they hardly belong to the group of vegetatively propagated plants. However, since some conifers may sooner or later be propagated in in vitro cultures (Campbell and Durzan, 1975; Durzan and Campbell, 1974a; Sommer et al., 1975; Winton, 1972; Winton et al., 1974), as is already done with a few broad leaved trees, the most important coniferous forest trees will be discussed, though shortly and superficially. More details can be found in a publication by Mergen (1963) and especially one by Ohba (1971a), who conclude that, so far, seed irradiation has not yielded commercial mutants, whereas some variability has been obtained by chronic irradiation of growing trees. Most mutants, either radiation-induced or spontaneous, are of no importance for silviculture but are, at best, useful as ornamentals (dwarfs, chlorophyll variation, waxless).

Since conifers have large chromosomes their radiosensitivity is very high. The optimum dose is a few hundred rads of X-rays for acute irradiation of plants and 10 R/day if growing plants are irradiated with γ-rays. The growing point appears to be less complicated than that in angiosperms which undoubtedly has an influence upon the recovery of mutants from mutated sectors.

A concluding remark made by Ohba (1971a) was that mutation breeding needs effective screening techniques for economic traits in trees that can be used at earlier stages than is now possible, very urgently.

We know a lot, relatively speaking, about *Cryptomeria* (Sugi) because of Japanese work (Kukimura et al., 1975, 1976; Ohba, 1971a, 1971b; Ohba and Maeta, 1973). The Japanese achieved segregation of spontaneous mutants from Sugi elite clones. Somatic mutation frequency was optimum at approximately 8 R/day for chlorophyll mutations. For waxless and morphological changes a somewhat lower dose rate is preferable.

Very little is known about *Cupressus*, *Juniperus* (see section on ornamental conifers, Chapter 5), *Larix* (Eriksson et al., 1966; Murai and Ohba, 1966) and *Picea*.

Most publications concerning *Pinus* deal with its radiosensitivity (Davis, 1962; Donini, 1967; Mergen and Thielges, 1966; Miksche et al., 1961; Rudolph, 1967; Sparrow et al., 1963b). In *Pinus*, chlorophyll-deficient and dwarf mutants were recovered in M_1 after seed irradiation but were mostly of limited value (Mergen, 1963).

In *Taxus* again only few publications of minor importance for mutation breeding are known (Miksche et al., 1961, 1962) while in *Thuja* some work has been done by Pohlheim (see section on ornamental conifers, Chapter 5).

Literature coniferous trees

Campbell, A. and Durzan, D.J., 1975. Induction of multiple buds and needles in tissue cultures of *Picea glauca*. Can. J. Bot., 53(16): 1652—1657.

Capella, J.A. and Conger, A.D., 1967. Radiosensitivity and interphase chromosome volume in the gymnosperms. Radiat. Bot., 7(2): 137—151.

Davis, T.S., 1962. Effect of cobalt-60 gamma radiation on pine seed and one-year-old seedings. For. Sci., 8(4): 411—412.

Donini, B., 1967. Effects of chronic gamma irradiation on *Pinus pinea* and *Pinus halepensis*. Radiat. Bot., 7(3): 183—193.

Durzan, D.J. and Campbell, R.A., 1974a. Prospects for the mass production of improved stock of forest trees by cell and tissue culture. Can. J. For. Res., 4(2): 151—174.

Eriksson, G., Ekberg, J., Ehrenberg, L. and Bevilacqua, B., 1966. Genetic changes induced by semi-acute γ-irradiation of pollen mother cells in *Larix leptolepis* (Sieb et Zucc) Gord. Hereditas, 55(2-3): 213—226.

Kukimura, H., Ikeda, F., Fujita, H., Maeta, T., Nakajima, K., Katagiri, K., Nakahira, K. and Somegou, M., 1975. Genetical, cytological and physiological studies on the induced mutants with special regard to effective methods for obtaining useful mutants in perennial woody plants. In: Improvement of Vegetatively Propagated Plants through Induced Mutations, Tokai, 1974. IAEA, Vienna, pp. 83—104.

Kukimura, H., Ikeda, F., Fujita, H., Maeta, T., Nakajima, K., Katagiri, K., Nakahira, K. and Somegou, M., 1976. Genetical, cytological and physiological studies on the induced mutants with special regard to effective methods for obtaining useful mutants in perennial woody plants. — II. In: Improvement of Vegetatively Propagated Plants and Tree Crops through Induced Mutations, Wageningen, 1976. IAEA, Vienna, pp. 93—137.

Mergen, F., 1963. Evaluation of Spontaneous, Chemical and Radiation-induced Mutations in the Pinaceae. BNL-6896, Brookhaven Natl Lab., Upton, Long Isl., N.Y., 27 pp.

Mergen, F. and Thielges, B.A., 1966. Effects of chronic exposures to Co^{60} radiation on *Pinus rigida* seedlings. Radiat. Bot., 6(3): 203—211.

Miksche, J.P., Sparrow, A.H. and Rogers, A.P., 1961. Effects of chronic gamma irradiation on the apical meristems of *Pinus strobus* and *Taxus media*. Am. J. Bot., 48: 529.

Miksche, J.P., Sparrow, A.H. and Rogers, A.F., 1962. The effects of chronic gamma irradiation on the apical meristem and bud formation of *Taxus media*. Radiat. Bot., 2(2): 125—131.

Murai, M. and Ohba, K., 1966. Studies of radiosensitivity of forest trees III. Effects of chronic gamma irradiation on growth and foliar cell division of Japanese larch, *Larix*

leptolepis Gord. in a gamma field. J. Jpn. For., 48: 62—68 (in Japanese).
Ohba, K., 1971a. Studies on the Radiation Breeding of Forest Trees. Bull. Inst. Radiat. Breed. Ohmiya, Ibaraki, 2, 102 pp.
Ohba, K., 1971b. Segregation of spontaneous mutants from Sugi elite clones *Cryptomeria japonica* D. Don. Tech. News, No. 9; Inst. Radiat. Breed., Ohmiya, Ibaraki.
Ohba, K. and Maeta, T., 1973. Induction of somatic mutations and cross fertility of the mutants in Sugi, *Cryptomeria japonica* D. Don. Gamma Field Symp., 12: 19—36.
Rudolph, T.D., 1967. Effects of X-irradiation of seed on X_1 and X_2 generations in *Pinus banksiana* Lambert. Radiat. Bot., 7(4): 303—312.
Sommer, H.E., Brown, C.L. and Kormanik, P.P., 1975. Differentiation of plantlets in longleaf pine (*Pinus palustris* Mill.) tissue cultured in vitro. Bot. Gaz. (Chicago), 136(2): 196—200.
Sparrow, A.H., 1963. A survey of radiosensitivities of plant species (*Juniperus*, *Picea*, *Pinus* and *Taxus*) and a method of predicting radiosensitivity. Health Phys., 9(8): 876.
Sparrow, A.H., Schairer, L.A., Sparrow, R.C. and Campbell, W.F., 1963b. The radiosensitivity of gymnosperms I. The effect of dormancy on the response of *Pinus strobus* seedlings to acute gamma irradiation. Radiat. Bot., 3: 169—173.
Sparrow, A.H., Schairer, L.A. and Woodwell, G.M., 1965. Tolerance of *Pinus rigida* trees to a ten-year exposure to chronic gamma irradiation from Cobalt-60. Radiat. Bot., 5: 7—22.
Winton, L.L., 1972. Callus and cell cultures of Douglas-fir. For. Sci., 18(2): 151—154.
Winton, L.L., Parham, R.A., Johnson, M.A. and Einspahr, D.W., 1974. Tree improvement by callus, cell, and protoplast culture. Tappi, 57(10): 151—152.

TREES USED IN SERICULTURE (*MORUS*)

Sericulture, the production of natural silk by caterpillars which feed on the leaves of certain trees, is a very old industry. It probably originates from China and India and was introduced to Japan and other countries at the beginning of the 20th century. Because it is a labour-intensive industry, it has disappeared in several countries and is declining in Italy and Japan, whereas the silk production in China, India and Korea is rapidly increasing (De Vries, 1977) as it is particularly attractive to small farmers and families in such developing countries. Yearly production is estimated to be 20 000—30 000 metric tons with a value of approximately US \$ 500×10^6 in 1972.

There are various types of silk worm which feed on a variety of trees, but the most important is *Bombyx mori*, chiefly feeding on the leaves of white and black mulberry (*Morus alba*, *M. latifolia* and *M. bombysis*). Many cultivars of these species are diploid (2n = 28); several excellent ones are triploid (2n = 42), however. Autotetraploids have been produced by colchicine treatment and irradiation, often as cyto-chimeras (Hazama, 1968a; Katagiri, 1976a,b,c; Kukimura et al., 1976).

The main objectives of mulberry breeders are a higher leaf yield, better leaf quality and resistance to diseases or climatic hazards, while ability to adapt to different climatic regions, the rooting ability of cuttings, the time of sprouting (in order to escape late frost damage), dry matter per unit leaf area and other characters are of importance as well (Hazama, 1967, 1968a,b; Nakajima, 1973; Kukimura et al., 1975, 1976).

Selection for some of the mentioned characters is relatively simple: leaf yield, for example, depends on visible characters, like the number of branches, their length, speed of shoot growth, leaf size, leaf thickness and the number of leaves per branch (correlated with internode length). Leaf quality and disease resistance are, on the other hand, much harder to select for (Fujita, 1974; Hazama, 1968a,b; Katagiri and Wada, 1971; Nakajima, 1973).

Mulberry seems to be fairly heterozygous which, together with the high number of (sterile) triploid cultivars, makes cross breeding unattractive. Therefore, interest in the possibilities of mutation induction is considerable in a country like Japan where most of the mutation work has been carried out since 1957 when irradiation facilities became available.

As in other woody plants (ornamental trees and shrubs; fruit crops), the main problem is to determine which method is most efficient in recovering the highest mutation frequency and the widest mutation spectrum. Consequently, many of the publications known to us are concerned with the development of suitable methods. For instance, Nakajima (1973) used roses as test plants to study the best time for irradiation and the best dose and dose rate while paying special attention to cutting-back methods to increase the percentage of "wholly" mutated mutants.

Using mulberry, he found that a dose of 7.5 kR γ-rays (0.05—5 kR/day) administered to grafted plants, followed by cutting shoots back three times in succession, resulted in 73% of mutants being "wholly" mutated (whether so-called wholly mutated mutants are really wholly mutated mutants or in fact periclinal chimeras has been investigated very rarely).

Similar work on methodology has also been reported by others, such as Fujita and Nakajima (1973), Hazama (1968b) and Katagiri (1970, 1973, 1976a). In general, a total dose of 5—10 kR of γ-rays for grafted plants or rooted cuttings seems to be optimum, while repeated cutting back for 3 successive years provides optimum chances for mutated cells in lateral and sub-lateral buds to form wholly mutated shoots (or complete periclinal chimeras). The mutation frequency in secondary and tertiary shoots does not only depend on the dose but also on the dose rate and the part of the shoot which develops from an irradiated main bud (Katagiri, 1973).

Re-irradiation of mutants is also practised to increase the mutation frequency and widen the mutation spectrum (Nakajima, 1973). After 10 kR γ-rays (5—10 kR/day) and cutting shoots back twice, a higher mutation frequency was reported but no different spectrum was observed. Other authors also refer to wider mutation spectra. Again, it is a little surprising that it is very rarely clearly mentioned that most mutants are expected to be periclinal chimeras and that re-irradiation may not only induce new mutations but also uncover existing ones by the re-arrangement of genetically different tissues within the chimera. This is demonstrated, for example, by the high figures for "reverses" in Katagiri's Table 6 (1973). The fact that both radiation and colchicine almost exclusively produce cyto-chimeras (mainly 2-4-4, 4-2-2 and 2-4-2) also indicates that most mutants must be periclinal chimeras (Katagiri, 1976a,b,c).

Various attempts have been made to induce resistance to diseases such as the Dagare or die-back disease (*Diaporthe nomurai* Hara). No resistance was found by Kukimura et al. (1975), whereas various degrees of susceptibility were reported by Nakajima (1973).

Another problem seems to be that mutants, being promising from a certain point of view, often carry additional unfavourable traits as well. Kukimura et al. (1975, 1976) report mutants with altered time of sprouting and rooting ability. Later as well as earlier sprouting mutants were found but the change was always in combination with changes in leaf shape, which was unacceptable.

Many promising mutants, mainly of the leading cultivars Ichinose and Kairyo-nezumigaeshi, have been reported and described (Hazama, 1968b; Nakajima, 1973). Most of them possess thicker leaves, shorter internodes, higher leaf yield (in kilograms per hectare) and sometimes have an improved ability to tolerate climatically different regions. None, however, has been used on a commercial scale, probably because of accompanying unfavourable traits, which explains the interest in re-irradiation of mutants.

*Literature trees used in sericulture (*Morus*)*

De Vries, A.C., 1977. Sericulture, bee-keeping and mushroom growing. Koninklijk Instituut voor de Tropen, Amsterdam, Dep. Agric. Res., Bull., 299: 9—23 (116 references!).

Fujita, H., 1974. Induction of die-back disease resistant bud mutations by gamma irradiation in mulberry. Mutat. Breed. Newsl. 4: 6.

Fujita, H. and Nakajima, K., 1973. Re-treatment of induced mulberry mutants with γ-rays. Gamma Field Symp., 12: 49—59.

Hazama, K., 1967. On the useful mutants of mulberry induced by radiations. In: Proc. 8th Jpn Conf. Radioisot., pp. 415—417.

Hazama, K., 1968a. Breeding of mulberry tree. Jpn Agric Res. Q., 3: 15—19.

Hazama, K., 1968b. Adaptability of mutant in mulberry tree. Gamma Field Symp., 7: 79—85.

Katagiri, K., 1970. Varietal differences of mutation rate and mutation spectrum after acute gamma ray irradiation in mulberry. J. Sericult. Sci. Jpn, 39(3): 194—200.

Katagiri, K., 1973. Radiation damage in winter buds and relation of shoot cutting-back to mutation frequencies and spectra in acutely gamma-irradiated mulberry. Gamma Field Symp., 12: 63—79.

Katagiri, K., 1976a. Radiation damage and induced tetraploidy in mulberry (*Morus alba* L.). Environ. Exp. Bot., 16(2/3): 119—130.

Katagiri, K., 1976b. Polyploidy induction in mulberry by gamma irradiation. Mutat. Breed. Newsl., 8: 11—12.

Katagiri, K., 1976c. Radiosensitivity, mutation and tetraploid induction in the gamma-ray-irradiated growing shoots in mulberry, *Morus alba* L. J. Sericult. Sci. Jpn, 45 (6): 461—468.

Katagiri, K. and Wada, M., 1971. Induction of bud mutation resistant to stem blight disease in mulberry by chronic gamma-irradiation. Jpn J. Breed., 21 (Suppl. 2): 156.
Kukimura, H., Ikeda, F., Fujita, H., Maeta, T., Nakajima, K., Katagiri, K., Nakahira, K. and Somegou, M., 1975. Genetic, cytological and physiological studies on the induced mutants with special regard to effective methods for obtaining useful mutants in perennial woody plants. In: Improvement of Vegetatively Propagated Plants through Induced Mutations, Tokai, 1974. IAEA, Vienna, pp. 83—104.
Kukimura, H., Ikeda, F., Fujita, H., Maeta, T., Nakajima, K., Katagiri, K., Nakahira, K. and Somegow, M., 1976. Genetical, cytological and physiological studies on the induced mutants with special regard to effective methods for obtaining useful mutants in perennial woody plants. II. In: Improvement of Vegetatively Propagated Plants and Tree Crops through Induced Mutations, Wageningen, 1976. IAEA, Vienna, pp. 93—137.
Nakajima, K., 1973. Induction of useful mutations of mulberry and roses by gamma rays. In: Induced Mutations in Vegetatively Propagated Plants. IAEA, Vienna, pp. 105—117.

Chapter 7

FRUIT CROPS

Temperate Fruit Crops

Tree fruits

GENERAL

For various reasons the possibilities of mutation breeding in woody perennials like fruit crops appear to be promising. The long time lapse between one generation and the next, the large space requirements and the high degree of heterozygosity in these plants make it very difficult to produce successful cultivars using conventional breeding methods. This is demonstrated by the unusually high ratio of commercial cultivars that have originated from spontaneous bud mutations (Doorenbos, 1977) and chance seedlings, to cultivars produced by deliberate cross breeding (Faedi and Rosati, 1975; Donini, 1974, 1975, 1976b,c; Donini and Veglio, 1976; Nishida, 1973; Sadamori, 1965; Thompson, 1976). Consequently, it is not surprising that a great deal of interest in the application of mutation induction in fruit crops has existed for many years, as illustrated by extensive articles, for instance, those by Nybom (1961) and Nybom and Koch (1965).

The important point with regard to mutation breeding in such crops is to decide on the best possible mutagenic technique. To restrict space and to save time it is essential to produce the highest possible mutation frequency and, at the same time, large mutated sectors (Broertjes, 1977c).

As Donini (1976c, 139 references!) and Donini and Veglio (1976) discussed in very extensive review articles, the best mutagenic treatments seem to be those administering radiation with X-rays, γ-rays or fast neutrons, acutely or semi-acutely (Nishida, 1973). The optimum dose depends on the radiosensitivity of the cultivar involved as well as on the developmental stage of the material. A high dose, though supposed to induce relatively more chromosome aberrations than gene mutations, is attractive because some of the initial cells in the apex are inactivated. Thus, fewer apical initial cells participate in the formation of the shoot, increasing the share a mutated cell has in it. This is later on expressed by a larger mutated "sector" than that resulting when lower doses are used.

It is important to know which part of the shoot that has developed from an irradiated bud should be used for further vegetative propagation and testing. As has been pointed out by several authors (Gupta and Samata, 1967;

Katagiri, 1973) the respective developmental stages of lateral primordia of an original bud have apices of increasing size and it is obvious that a vital mutated cell has optimum chances to form a large mutated sector in a (very) small apex. On the other hand, the absolute number of mutated cells in a small apex is lower than that in a large apex. Somewhere on a vM_1 shoot between the parts where the smallest and the largest apices are situated, is the optimum part of the shoot that gives an acceptable mutation frequency and, above all, a sector size which is large enough to ensure detection and recovery after repeated cutting back and vegetative propagation.

The position of this optimum part of the shoot depends on the dose. A higher dose — and a higher dose rate — not only causes a higher mutation frequency, but also more damage. This shifts the optimum zone to a lower part of the shoot, viz., to older and larger primordia, indicating that the radiation-induced damage is dominant over the increased mutation frequency. In most crops studies about this subject are fairly new and it will take some time before the best treatment for every important crop has been determined, since different crops seem to have a different size and number of primordia in the buds.

The major practical breeding objectives in tree fruit crops are the induction of "spur" types, disease resistance, self-fertility, improvement of fruit quality and many other commercially important characteristics. While considering objectives, we refer to the very comprehensive handbook of Janick and Moore (1975) in which very detailed information can be found concerning the breeding of a great number of fruit crops. A second useful source of information for aspects concerning mutation breeding, radiosensitivity etc. is written by Donini (1976b).

Despite the great number of reports concerning promising mutants obtained from a considerable number of projects in several parts of the world, very few commercial mutants have, as yet, been released. Perhaps more advanced techniques should be developed, such as irradiation of very early ontogenetic stages with moderate doses (Lapins et al., 1969), or irradiation of material immersed in water (this provides a more even dose distribution) as proposed by Lacey (1976), who also recommended at least three successive generations of vegetative propagation to ensure genetic stabilization, or the use of in vitro techniques (Abbott and Whiteley, 1976; Broertjes, 1977c; Pech and Fallot, 1974).

Even with all these precautions, mutation breeding will not always be successful. The necessity of keeping the physical extent of the plant material in a breeding programme within bounds, and consequently, the relatively small number of mutants that can be produced, limit the possibilities unless mutation breeding is restricted to more easily recognized or more specific characters, such as fruit colour or disease resistance (Lapins, 1973).

Literature tree fruit — general

Abbott, A.J. and Whiteley, E., 1976. Culture of *Malus* tissues in vitro. 1. Multiplication of apple plants from isolated shoot apices. Sci. Hortic., 4(2): 183—189.

Broertjes, C., 1977c. Artificially induced mutations in fruit trees. Acta Hortic., 75: 19—26.

Donini, B., 1974. Induction and isolation of somatic mutations in vegetatively propagated plants. Proc. Int. Hortic. Congr., 19: 189—200.

Donini, B., 1975. Induction and isolation of somatic mutations in vegetatively propagated plants. In: Improvement of Vegetatively Propagated Plants through Induced Mutations, Tokai, 1974. IAEA, Vienna, pp. 35--51.

Donini, B., 1976b. The use of radiations to induce useful mutations in fruit trees. In: Improvement of Vegetatively Propagated Plants and Tree Crops through Induced Mutations, Wageningen, 1976. IAEA, Vienna, pp. 55—67.

Donini, B., 1976c. Breeding methods and applied mutagenesis in fruit plants. In: Proc. Workshop Europ. Comm. — Israel, The Use of Ionizing Radiation, Wageningen, 22—24 March, 1976. Assoc. Euratom—ITAL, Wageningen, pp. 445—478 (139 references!).

Donini, B. and Veglio, P., 1976. Impiego della mutagenesi per le miglioramento della piante da frutto. In: Tecniche Nucleari per il Miglioramento e la Difesa dei Fruttiferi. Vol. 1: Impiego della Mutagenesi per il Miglioramento delle Piante a Moltiplicazione Vegetativa (fruttiferi e piante da fiori). Casaccia, 1974. Com. Naz. Energ. Nucl., Cent. Stud. Nucl., pp. 13—56.

Doorenbos, J., 1977. Spontaneous mutations as a source of clonal variation in deciduous fruits. Acta Hortic., 75: 13—18.

Faedi, W. and Rosati, P., 1975. Indagini sul melo. 7 Rome Beauty, suoi mutanti e suoi incroci. Frutticoltura, 37(5): 25—34.

Gupta, M.N. and Samata, Y., 1967. The relationship between developmental stages of flower-buds and somatic mutations induced by acute X- and chronic gamma-irradiation in *Cosmos bipinnatus.* Radiat. Bot., 7: 225—240.

Janick, J. and Moore, J.N., 1975. Advances in Fruit Breeding. Purdue University Press, West Lafayette, Ind., 623 pp.

Katagiri, K., 1973. Radiation damage in winter buds and relation of shoot cutting-back to mutation frequencies and spectra in acutely gamma-irradiated mulberry. Gamma Field Symp., 12: 63—81.

Lacey, C.N.D., 1976. Propagation and irradiation techniques used to produce and isolate mutants in fruit trees. In: Proc. Workshop Europ. Comm. — Israel, Use of Ionizing Radiation, Wageningen, 22—24 March, 1976, pp. 485—495.

Lapins, K.O., 1973. Induced mutations in fruit trees. In: Induced Mutations in Vegetatively Propagated Plants. IAEA, Vienna, pp. 1—19.

Lapins, K.O., Bailey, C.H. and Hough, L.F., 1969. Effects of gamma rays on apple and peach leaf buds at different stages of development. I. Survival, growth and mutation frequencies. Radiat. Bot., 9: 379—389.

Nishida, T., 1973. Induction of somatic mutations in deciduous fruit trees by gamma irradiation. Gamma Field Symp., 12: 1—15.

Nybom, N., 1961. The use of induced mutations for the improvement of vegetatively propagated plants. In: Symp. Mutations and Plant Breeding, Cornell University, Ithaca, N.Y. NAS—NCR, Publ. 891: 252—294.

Nybom, N. and Koch, A., 1965. Induced mutations and breeding methods in vegetatively propagated plants. Radiat. Bot., 5 (Suppl.): 661—678.

Pech, J.C. and Fallot, J., 1974. Intérêt des cellules de pommes cultivées in vitro. Obtention et production. Fruits, 29(11): 771—776.

Sadamori, S., 1965. Utilization of the spontaneous mutation in deciduous fruit trees. Gamma Field Symp., 4: 1—23 (125 references).

Thompson, M.M., 1976. Use of ionizing radiation in fruit breeding. Fruit Var. J., 30(1): 20.

APPLE

According to Brown (1975b), the apple, *Malus pumila* Mill., family Rosaceae, sub-family Pomoideae, is "undoubtedly the most ubiquitous of all fruits". It is a fruit tree mainly grown in temperate regions, which has been cultivated in Europe and Asia since ancient times and has been spread by man to almost every part of the world. The centre of its diversity, according to Vavilov (1931), is to be found in Southwest Asia. Annual production is estimated to be between 20×10^6 and 30×10^6 metric tons.

There is a considerable amount of variation, which may be explained for the most part by the fact that intercrossing is easy between different species of the genus *Malus*, which, according to most authors, has 25—30 distinguishable species and many subspecies. Self-incompatibility is common.

According to Brown (1975b) the spontaneous mutation rate is low. Bud sports displaying several important characters have been described (Shamel and Pomeroy, 1936): for example, the red sports of the most widely grown cultivar, Golden Delicious. In addition to the bud sports which have resulted from genetic changes in one or a few genes only and which led to different characteristics of the tree and the fruit, ploidy chimeras are known to occur. The latter have led to "giant" sports (Einset et al., 1946).

The majority of cultivated apples are diploids ($2n = 2x = 34$), but triploid cultivars ($2n = 3x = 51$), which have arisen from unreduced gametes, are also known. Triploids are often more vigorous and produce larger fruits but, from a breeding point of view, they are difficult to handle. The occurrence of spontaneous tetraploid sports has been known for a long time (Einset, 1952; Einset and Imhofe, 1947). In practice, vegetative propagation is used; the majority of the bud sports are periclinal chimeras.

At present, the most important breeding objective is to achieve high production of good quality apples at low costs. People are interested in firm, juicy fruits with a good flavour and attractive colours. All-over russet types are appreciated but small traces of russetting are not favoured. Large fruits are usually preferred. Storage quality should be good. There is an increasing interest in breeding new cultivars which are resistant to most important pests and diseases on account of a growing uneasiness about spraying. From an economic point of view, shortening the juvenile phase is important. Much interest is also being taken in "spur" types. These are mutants with compact growth, reduced internode length and limited side branching on shoots. Such types, which are often, though not completely correctly, named compact types or dwarfs, have short juvenile phases and relatively high yields in early years (Brown, 1975b).

Classical breeding consists of collecting pollen from favoured trees, emasculation of the female parent and subsequent cross-pollination in order to combine favourable characters of both parents. Afterwards, the tedious job of selecting the most promising seedlings begins. It is clear that the evaluation of several characters, like fruit quality, cannot take place before the juvenile phase has terminated. Most important genetic characters inherit polygenically.

Efforts to induce genetic changes in apple with mutation techniques have been made for a fairly long time. Bishop (1954a) even refers to some early, unpublished, trials in the period 1930—1932 which were, however, unsuccessful. In 1944, X-irradiation experiments (and somewhat later experiments with other mutagenic agents as well) were started in Sweden by Gustafsson and Nilsson (Granhall, 1949, 1951, 1953; Granhall et al., 1949; Ehrenberg and Granhall, 1952). In Canada, Bishop and Aalders started using X-rays in 1948 and, in 1953, they began with neutron irradiation (Bishop, 1954a,b, 1956, 1957, 1959a,b; Bishop and Aalders, 1955).

Granhall (1953) reported the production of three fruit colour sports in apples of the (triploid) cultivar Gravenstein after irradiation of dormant scions with 5 kR of X-rays. At this dose, about 50% of the irradiated diploid cultivars survived after grafting, as compared to about 80% of the control plants. Triploid and especially tetraploid cultivars show a higher degree of resistance to radiation.

The primary radiation effects observed in apple include scion death, delayed growth and bifurcation, sometimes occurring after a number of years had passed since radiation.

As part of this early work, Bishop X-irradiated dormant scions with 3.5—4 kR and framework-grafted them to grown apple trees with up to 200 scions per tree. Four different cultivars were used. The advantage of using the framework method is, of course, that large numbers of scions can be handled and that some fruits are obtained in the second season of growth, while in the third or fourth year almost a full crop is produced.

A red "over colour" was considered as an especially desirable character to induce. According to Bishop, such changes happen when mutations towards dominant occur, which makes it very doubtful that such mutations can be induced. In later work, Bishop (1957) mentions another possibility, that is, the mutation of a suppressor which allows the dominant character for "over colour" to manifest itself. In the second year after irradiation, the experiment yielded a number of giant, irregularly shaped fruits and a number of colour changes with more, as well as less, pigment.

It was realized that, in most cases, single-gene mutations are confined to (parts of) single histogenic layers. All colour changes and most other mutations were reported to be of the "sectorial" (i.e. mericlinal) type. Granhall (1953), on the other hand, reported the occurrence of complete mutants.

In one of his earliest publications, Bishop (1954b) referred to the occurrence of short, relatively thick shoots in the first year after grafting.

In an experiment with X-rays and thermal neutrons, it was found that neutron treatment produced twice as many bifurcations as X-rays if dosages giving an equal amount of shoot growth inhibition were compared (Bishop, 1956, 1957; Bishop and Aalders, 1955). The authors postulated that shoot bifurcations were caused mainly by the effect of radiation on the chromosomes, whereas shoot growth inhibition was an index of total damage. If this were true, neutrons would be more effective in the induction of gene mutations than X-rays. In this experiment dormant cuttings, 7—9 inches in length and aged 1 year, were irradiated and subsequently cut into two scions for grafting. X-ray dosages varied from 3.5 to 5 kR and neutrons (flux 5.4×10^8 neutrons per cm^2 per sec) were administered over 2—8 h. This time, complete (i.e. visually chimera-free) dark red sports of cv. Cortland were obtained from the neutron treatment.

X-ray induced mutations for other characters, for example, less russet in cv. Golden Russet and the opposite in cv. Sandow, were also observed, but the authors do not exclude the possibility of a natural cause for these phenomena. Between 1948 and 1958, at least 13 000 dormant scions were irradiated in Canada.

Another mutation in cv. Cortland, very probably induced by (an unknown dosage of) thermal neutrons administered to dormant scions, was reported by Einset and Pratt (1959). Irradiation took place in 1953 at Brookhaven National Laboratory and, after grafting that same year onto a 12-year old tree of cv. McIntosh, one graft was observed to have irregularly shaped fruits in 1958.

Irradiation of dormant scions was not the only method used. In 1952, Zwintzscher (1955, 1959, 1962) started experiments to develop suitable procedures to handle the starting material. He claimed that the low mutation frequencies observed in previous experiments could be mainly explained by negative selection (diplontic selection) of induced mutations. Zwintzscher used young trees from nursery stocks and only the lower parts of the main branch and two other shoots, each carrying about 3—4 buds. The trees were irradiated with 4 kR of X-rays. In the next season, five buds from the lower parts of the shoots which had developed after the irradiation treatment were occulated on 1-year old trees. The process of cutting back was repeated every year for the next years. The method takes a lot of time but, according to Zwintzscher, the advantages are considerable limitation of chimerism and an increased chance of obtaining mutants.

Dormant buds of cv. Ontario were irradiated with 5 kR in Würzburg, Germany, in 1952, by Breider (1959). In the first year, a variant with smaller leaves was observed, but further reports are not available.

Gröber (1959) reported work in Gatersleben (German Democratic Republic), which started in 1955. About 2 800 scions of three different cultivars, including cvs Ontario and Gravensteiner, were X-irradiated with doses of 4—7.5 kR. In addition, about 2 000 buds were treated with slightly lower doses during the summer. Gröber used a system of localized irradiation, that is scions of about 15 cm, carrying an average of 4—5 buds, were partly protected by a lead shield in such a way that only the upper two buds were exposed to X-rays. Thus, physiological damage to the lower part of the scions remained low and this increased the frequency of successful grafts. Later, the shoots which had developed from the lower, protected buds after grafting, were removed. It was calculated that during three seasons a mutation rate of 4.8% was obtained for red fruit colour in cv. Weisser Klarapfel (or White Transparent), which normally has green fruits. Other changes observed included some in ripening time, fruit shape and fruit size. It was proved that those changes were caused by real mutations.

In a later publication Gröber (1967) discussed some possible explanations for the occurrence of the (supposedly) dominant mutation for red fruit colour, such as chimerism, suppressor genes, duplications, etc. It was claimed at that time that more than 50 000 buds or scions had been treated with X-rays, ethylene methanesulphonate (EMS) or radioactive phosphorus in Gatersleben. No later reports from this institute are known to us.

Different types of radiation damage in apple shoot apices were described by Pratt et al. (1959). Histogenic effects leading to a change in ploidy level of certain layers after X-irradiation were described in 1960 (Pratt, 1960). Techniques to treat seeds, seedlings, cuttings and embryos with radioactive isotopes were decribed by Stankovic (1962).

Starting in 1959, X-, γ- and neutron irradiation experiments were performed at the Summerland Station in Canada by Lapins and colleagues, mainly with the intention of inducing "spur" mutants. Doses of 4—5 kR of X-rays and 4—7 $\times$ 10^{12} N_{th}/cm^2 were applied to dormant scions (Fisher and Lapins, 1966; Lapins, 1963b). Each of the lowest ten buds of a shoot produced by an irradiated bud were budded to mature trees. In this way, up to 10% of buds obtained were mutated. Several compacts were found in cultivars in which such sports had not yet been found in nature (N.B. Lapins makes no clear distinction between "spurs" and "compacts"). During selection for compactness, trees with undesirable fruit shape etc. were discarded.

In the ensuing years, a considerable number of publications were devoted to apple mutation breeding work (Lapins, 1963b, 1965a, 1969, 1971a, 1972a; Lapins et al., 1969; McIntosh and Lapins, 1966). A schematic drawing of irradiated scions presented by Lapins et al. (1969) is shown in Fig. 22. The essentials were summarized in a review (Lapins, 1973), presented during an IAEA meeting in Vienna in 1972.

When choosing the starting material for their experiments, Lapins and many others have worked according to the ideas outlined by Kaplan (1953), which imply that the lower buds on a shoot originating from an irradiated bud produce more mutations than buds situated higher on the shoot. To explain

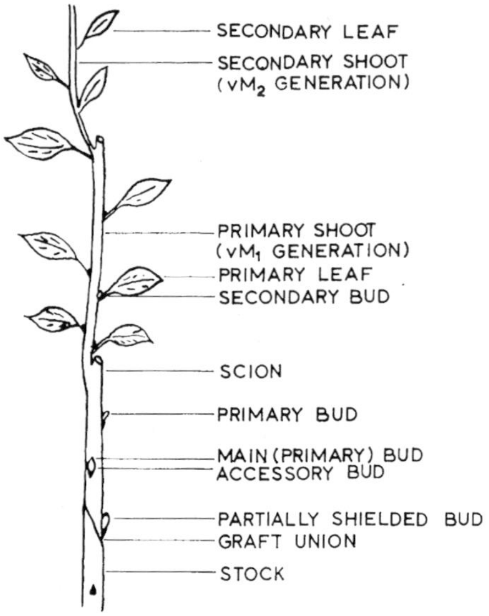

Fig. 22. Schematic presentation of irradiated scions and primary and secondary buds, shoots and leaves in apple. (Lapins et al., 1969, Fig. 1.)

this, one must remember the fact that those lower buds are already present as primordia in the irradiated buds. On the other hand, mutant sectors are larger in the higher secondary buds which are still undifferentiated at the time of irradiation.

Coming back to the highly desired tree types with reduced growth, it was found that the character for compact growth could not in general be transferred to sexual progeny, probably because the mutated character was confined to only one histogenic layer. Lapins (1973) mentions that the use of adventitious bud techniques may help to solve this problem. In a number of cases adventitious buds (of L-III origin) were obtained in compact types.

Other interesting induced mutations concerning differences in susceptibility to powdery mildew (*Podosphaera leucotricha*) were recorded among selected mutants with a favoured growth type from irradiated material of cv. McIntosh (McIntosh and Lapins, 1966). Mutations for partial gametic sterility occur at high frequencies (Lapins, 1973). They can be useful, as certain cultivars tend to have heavy fruit sets, in which case a small reduction in fertility is desirable. Up to now most mutations for fruit characteristics have not been useful (Lapins, 1972a). In cv. McIntosh, some mutants with a more intense red colour were obtained. Reduced russetting in the fruit stem cavity was also observed (Lapins, 1971a).

Lapins (1973) advises future investigators of apple mutations to concentrate on one or a few characters which can be mutated and detected rela-

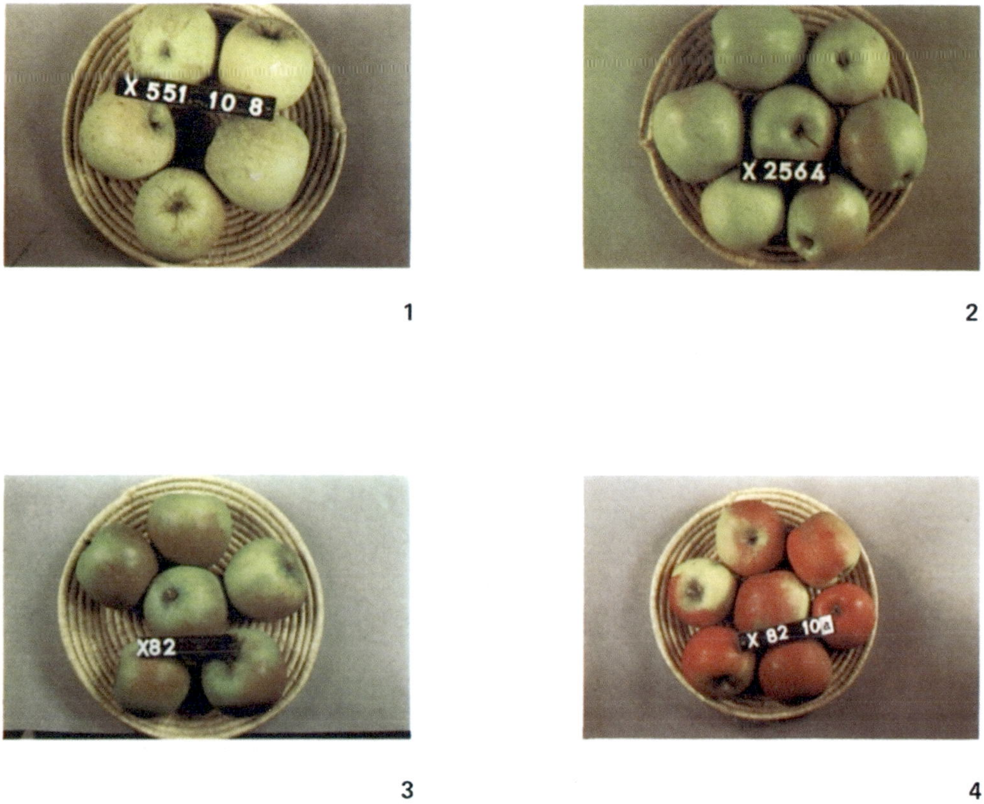

Plate 8. The apple cultivar Golden Delicious (1), a commercial mutant of it, cv. Lysgolden (2), which lacks russeting; the cultivar Reine des reinettes (3), and its commercial mutant, cv. Belrène (4), which has a much redder fruit skin and ripens earlier.
(Slides from L. Decourtye, Station de Recherches d'Arboriculture Fruitière, Angers.)

tively easily, such as, for example, compact growth. Improved fruit characteristics are less likely to be obtained in this way.

Starting around 1960, Decourtye and colleagues studied mutation in apple in Angers (France) for practical breeding purposes as well as from a more fundamental point of view, for example, to investigate suspected chimeric structures from observations of the effects of irradiation (Decourtye, 1963, 1964, 1967a,b, 1970; Decourtye and Lantin, 1969, 1971). Dormant scions with two or three buds were the usual material they treated, with 3—8 krads of γ-rays. Observations dealt with vM_2, using not the oldest, basal buds on the shoots which had developed from irradiated buds, but the ones situated higher on the shoots and which had developed after irradiation. In order to obtain the maximum number of mutants per progeny, a maximum of three scions with many descendants were used. The work resulted in the induction of mutations for fruit colour, russetting, sterile gametes and compact growth.

Irradiation of dormant scions of spurs of cvs Golden Delicious and Starking Delicious did not confirm the hypothesis of a periclinal structure for such mutants (Decourtye and Lantin, 1969). In an article about the methodology of mutation induction in fruit trees, Decourtye and Lantin (1971) reported that repeated irradiation administered to dormant vM_1 shoots induced greater genetic variability in vM_2 than that in material which had only been irradiated once. In most of the experiments, 15% of the progeny were mutants (Decourtye, 1970). No correlation was found between primary damage in vM_1, and genetic changes in later generations. Three mutants have been released and are grown commercially in France, namely cvs Lysgolden (Goldenir), Belrène and Blackjon BA 2 520 (Table 21) (Le Lezec, 1973; L. Decourtye, personal communication, 1977).

Mutation work on apple in Wageningen (The Netherlands) started in 1965. Dormant scions were X-irradiated, 3 krad being found to be the most efficient dose (Visser, 1973; Visser et al., 1969, 1971; De Vries et al., 1970). Much attention was paid to developing a system of pre-selection, or selection at the nursery stage, in order to determine the most effective dose of X-rays. Irradiated scions were kept under water in glass jars in a greenhouse during early spring. It appeared that there was a direct relationship between the survival rate of the buds on those scions after 3—4 weeks and the percentage of surviving grafts after two seasons in the field. Selection was mainly directed towards producing compact types (which the authors clearly distinguish from spurs). Selection was carried out during three successive seasons with shoots being cut back after the first and second season. About 70% of the surviving trees of cvs Golden Delicious, Belle de Boskoop, Tydeman's Early and Cox's Orange Pippin produced compact shoots (Visser et al., 1971). Per treatment, 100—200 dormant scions from each cultivar investigated were irradiated yearly. Compact shoots yielded at least four times as many distinct compact clones as visually normal shoots. Another important observation was that four of every five selected clones were stable which is a

TABLE 21

Commercial mutants of apple

Name of new cultivar	Place and date of release (or approval) and name of principal worker(s) and institute	Conditions and date of mutagenic treatment [Parent cultivar]	Main improved attributes of cultivar
McIntosh 8F-2-32	Canada, 1970 K.O. Lapins, Canadian Department of Agriculture, Research Station, Summerland, B.C.	Shoots, γ-rays [McIntosh]	Improved skin colour; resistance to *Podosphaera leucotricha* and *Venturia inaequalis*
Lysgolden (Goldenir)	France, 1970 L. Decourtye and B. Lantin, I.N.R.A., Angers	Dormant trees, 5 krad γ-rays, 1963 [Golden Delicious]	Fruit free of russetting; yield somewhat reduced
Belrène	France, 1970 L. Decourtye and B. Lantin, I.N.R.A., Angers	Growing shoots, 1% EMS, 1961 [Reine des reinettes]	Earlier maturing, more coloured and bigger fruit; yield somewhat reduced
Blackjon BA 2 520	France, 1970 L. Decourtye and B. Lantin, I.N.R.A., Angers	Dormant trees, 5 krad γ-rays, 1963 [Jonathan Blackjon]	Improved and more regular red coloured fruit

I.N.R.A., Institut National de Recherches Agronomiques.
Sources: IAEA (1972, pp. 526—544) and *Mutation Breeding Newsletter*, 1 (1972) —9 (1977) and Association Euratom—ITAL, Wageningen.

very high proportion if compared, for example, with the results of Lapins (1965a). Visser prefers the use of selected rootstocks for frameworking for reasons of uniformity. In contrast to the method advocated by most researchers, Visser starts selection as early as vM_1. According to his calculations (Visser, 1973), the efficiency of this method is seven times larger than that of selection in vM_2, assuming that all irradiated clones are planted in the orchard. The work of Visser and colleagues was reviewed recently by Visser and Verhaegh (1977). Irradiation work ceased in 1971 because of lack of funds, but selection of the material continued. As regards the production of a stable compact apple tree, it was found that such types very rarely combine with acceptable fruit and production properties. Of 7 300 irradiated scions propagated on rootstocks in the nursery (vM_0), only 41 acceptable vM_2 selections could be obtained, of which only one spur of cv. Boskoop shows sufficient promise to be retested. The authors calculate that at least 20 years are necessary before a final decision can be made as to the practical use of a mutant, and from an economic point of view this period of time is, of course, too long. They conclude that mutation breeding in apple should be used only reluctantly, but that is does deserve lasting attention in case other methods are not available or more laborious for certain objectives.

Since 1967, irradiation work on apple has been in progress at the Long Ashton Research Station, United Kingdom. Only virus-free material is used in experiments intended to improve single undesirable traits. Campbell and Lacey (1973) irradiated 50 dormant 1-year old shoots of cv. Bramley's Seedling with 5—10 krad of γ-rays. During radiation, the basal parts of the 20-cm long shoots were covered with a lead shield and the shoots were protected by 0.5—1 cm of water. The lead shield protection had no positive effect on survival. In the under water treatments it was found that the survival rate was higher and the mutation rate better than when shoots were irradiated in the air (Campbell and Lacey, 1974a,b). Several promising compact types resulted from these treatments, some of them having commercial prospects (Campbell, 1976; Campbell and Lacey, 1975; Lacey and Campbell, 1977). Campbell (1976) presented a multiplication scheme that is used at Long Ashton Research Station (Fig. 23). Experiments to induce compact types of Cox and other cultivars were also in progress. Another goal at this research station has been to produce a Cox with more regular cropping features.

Recently, Lacey (1977) studied the mutation spectrum of Cox's Orange Pippin, irradiated with 7 krad Cobalt-60 γ-rays (approximately 1.5 krad/h). All the buds on 747 vM_1-shoots were grafted resulting in over 13 000 plants, of which 272 showed mutations for dwarf type or other growth characters. In 60 cases, nearly all the mutants produced from one vM_1 bud stick appeared to be the same, whereas 24 vM_1 bud sticks produced two different mutations each, the two sorts of mutant originating from well-separated parts of the stick. Fruit changes were not found in plants which were selected for changes other than vegetative ones, whereas in trees selected on vegetative grounds the fruit was different from the control in nearly every case. Lacey's main conclusions were that dwarf types are easily produced by γ-rays, that the majority do not produce enough good quality fruit to become commercial but that acceptable compact forms with an acceptable fruit quality can be found when sufficient mutants are produced. Moreover, some pre-selection in the vM_1, a small vM_2, obtained by using only a few buds from each vM_1 shoot, as well as a more rigorous selection at the vM_2 stage is advised.

Fideghelli et al. (1977) report a practical project in Italy concerning the improvement of the Annurca apple variety. This popular but local variety has several shortcomings, such as bare wood, tremendous vigour and very short peduncles. Several mutants with improved characters were produced by irradiation, and sometimes mutants had several favourable new characteristics.

In addition to the work we have already mentioned, mutation studies con-

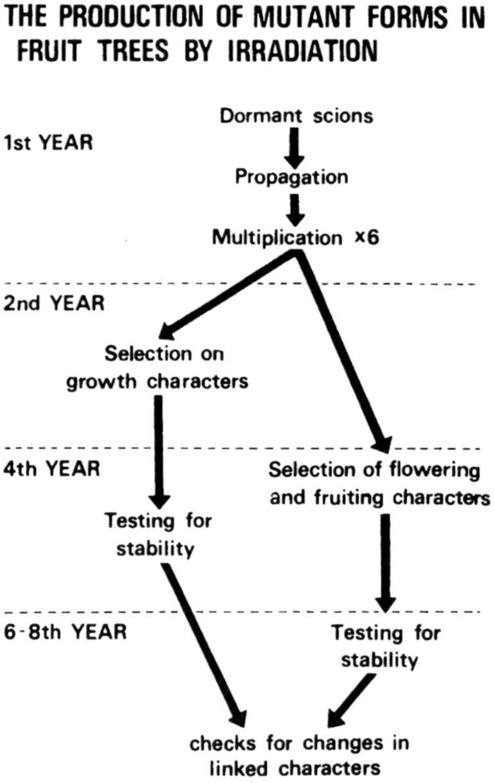

Fig. 23. The multiplication system and the time scale of the mutation breeding programmes for apples at Long Ashton Research Station, Bristol. (Campbell, 1976, Fig. 1.)

cerning apples have been performed by several others, notably workers in Eastern Europe, India and Japan. The Russian work in particular is rather inaccessible. The primary effects of γ-irradiation have been described by Vondráček and Ullmann (1963) and by Mišić (1963) in Yugoslavia and by Thakur and Verma (1967) in India. Apple seeds and cuttings have been irradiated since 1957 by Dryagina et al. (1967) using different sources. In certain instances chemical mutagens have also been applied. The experiments of Zagaja and Przybyła in Poland started in 1968 (Zagaja, 1975; Przybyła and Zagaja, 1974; Zagaja and Przybyła, 1973, 1976c). Most work dealt with the induction of compactness. Of the approximately 3 000 shoots studied, 44 mutants with significantly shorter internode lengths were found. Six of them produced compact trees. The procedures described by Zwintzscher (1962) were followed. The authors found that at least three propagations were necessary to obtain stable compact mutants in each case. Recently, investigations have been initiated to produce adventitious shoots from apple roots in vivo.

In the U.S.A., Pratt et al. (1972) irradiated several sports to induce changes in skin colour in order to study the relationship between skin colour changes and the genetic structure of the apical histogenic layers.

Recently, Ikeda (1974) reported the induction of some improved skin colour mutants in the well-known Japanese apple cultivar Fuji. Later, it was found that several of the mutants obtained were incompatible for grafting on the rootstock which is most suitable for shoots of normal cv. Fuji (Ikeda, 1976a; Kukimura et al., 1976). Poll (1974) described the induction of compact types in different cultivars using γ-irradiation in Norway.

In Russia, Kolontaev (1973a, 1974b) described chromosomal aberrations after X-irradiation. In addition to irradiation, a number of chemicals like diethyl sulphate (DES) and n-nitroso-methylcarbamide were used (Kolontaev, 1973b, 1974a,c). Mutations were classified according to different characteristics (e.g. growth rate, degree of branching) and a selection was made for hardiness. In some cases, chemicals and X-rays were given in combination and found to be more effective than separate treatments (Kolontaev, 1974d).

Recently, Blažek (1976a,b), in Czechoslovakia, described the usefulness of irradiating summer buds with 4 krad of γ-rays, especially in cultivar McIntosh.

In summary it seems that several methods of treatment and selection have led to fairly high mutation frequencies (up to 10%) and that most technical problems have been solved. Despite this and the many thousands of buds or scions that have been irradiated over more than 30 years there have, so far as is known, only been four cultivars that are the result of mutation treatment, commercially released up to now (see Table 21 and Plate 8). However, it is difficult to say at present whether mutation methods or other breeding methods are better for specific goals in apple breeding. An important breakthrough in favour of mutation techniques may take place when the use of adventitious bud techniques becomes more common. One should not forget that the release of a new apple cultivar, produced using the classical breeding methods, may also take 20 years or more, and that systematic work on induced mutations in apple can be concluded in less than two decades.

Literature apple

Bishop, C.J., 1954a. Mutations in apples induced by X-irradiation. J. Hered., 45: 99—104.

Bishop, C.J., 1954b. X-ray induced bud sports in apples. Fruit Var. Hortic. Dig., 9(2): 25—28.

Bishop, C.J., 1956. The production of budsports in apples through the use of artificial radiation. Proc. Int. Congr. Hortic., 14(1): 740—745.

Bishop, C.J., 1957. Genetic changes in apples induced by thermal neutrons. Can. J. Plant Sci., 37: 55—58.

Bishop, C.J., 1959a. Radiation induced fruit color mutations in apples. Can. J. Genet. Cytol., 1: 118—123.

Bishop, C.J., 1959b. Color sports of apples induced by radiation. Fruit Var. Hortic. Dig., 14(2): 37—39.

Bishop, C.J. and Aalders, L.E., 1955. A comparison of the morphological effects of thermal neutron and X-irradiation of apple scions. Am. J. Bot., 42: 618—623.

Blažek, J., 1976a. McIntosh and Cox's Orange Pippin apple mutants induced by gamma radiation of summer buds. Fruit Sci. Rep., 3(3): 9—15.

Blažek, J., 1976b. Gamma-radiation induced growth mutations in the cherry cultivar Napoleon. Vedecke Prace Ovonarske, 5: 31—43 (in Czechoslovakian with English summary); Hortic. Abstr., 47: No. 7200.

Breider, H., 1959. Röntgeninduzierte Mutationen bei *Vitis*, *Malus* und *Cyclamen*. Bayer. Landwirtsch. Jahrb., 36: 396—401.

Brown, A.G., 1975b. Apples. In: J. Janick and J.N. Moore (Editors), Advances in Fruit Breeding. Purdue University Press, West Lafayette, Ind., pp. 3—37.

Campbell, A.I., 1976. Compact apple trees produced by irradiation for use in the meadow orchard. Compact Fruit Tree, 9: 43—48.

Campbell, A.I. and Lacey, C.N.D., 1973. Compact mutants of Bramley's Seedling apple induced by gamma irradiation. J. Hortic. Sci., 48: 397—402.

Campbell, A.I. and Lacey, C.N.D., 1974a. Radiation mutants in apple can produce made-to-measure trees. Grower, December 21/28: 1172; 1174.

Campbell, A.I. and Lacey, C.N.D., 1974b. Mutation induction in fruit tree cultivars. In: Rep. Eucarpia Meet. Fruit Tree Breeding, Canterbury, 1974. pp. 40—46.

Campbell, A.I. and Lacey, C.N.D., 1975. Induction and selection of mutant forms of fruit plants. Long Ashton Rep., 1975: 22—24.

Decourtye, L., 1963. Action des radiations isonisantes sur une chimère de pommier. Ann. Amélior. Plant., 13(2): 133—140.

Decourtye, L., 1964. The use of radiations in apple and pear breeding. Proc. Balsgård Fruit Breed. Symp. Balsgård Fruit Breed. Inst., Fjälkestad, pp. 40—43.

Decourtye, L., 1967a. Action des rayons gamma sur des variétés de poirier et de pommier en chimère. Bull. Soc. Bot. France, Coll. Morphol. Exp., 1966: 48—54.

Decourtye, L., 1967b. Russet-free sectors on fruits of 'Golden Delicious' apple trees after gamma irradiation. Proc. Am. Soc. Hortic. Sci., 91: 73—77.

Decourtye, L., 1970. Methodology in induced mutagenesis and results. In: Proc. Angers Fruit Breed. Symp. Eucarpia Fruit Sect., Editions S.E.I., C.N.R.A., Versailles, pp. 161—174.

Decourtye, L. and Lantin, B., 1969. Contribution à la connaissance des mutants spur de pommier; hérédité du caractère. Ann. Amélior. Plant., 19(3): 227—238.
Decourtye, L. and Lantin, B., 1971. Considérations méthodologiques sur l'isolement de mutants provoqués chez le pommier et le poirier. Ann. Amélior. Plant., 21(1): 29—44.
De Vries, D.P., Verhaegh, J.J. and Visser, T., 1970. Nursery selection for spur types in X-ray treated apple and pear varieties. Proc. Angers Fruit Breed. Symp. Eucarpia Fruit Sect., Editions S.E.I., C.N.R.A., Versailles, pp. 147—160.
Dryagina, I.V., Kazarinov, G.E. and Kalinichenko, I.M., 1967. Mutations in fruit and ornamental plants induced by the action of neutrons and gamma-rays Co^{60}. In: Z. Landa (Editor), G. Mendel Memorial Symp. Academia Publishing House, Prague, pp. 131—134.
Ehrenberg, L. and Granhall, I., 1952. Effects of beta-irradiating isotopes in fruit trees. Hereditas, 38: 385—419.
Einset, J., 1952. Spontaneous polyploidy in cultivated apples. Proc. Am. Soc. Hortic. Sci., 59: 291—302.
Einset, J. and Imhofe, B., 1947. Chromosome numbers of apple varieties and sports. Proc. Am. Soc. Hortic. Sci., 50: 45—50.
Einset, J. and Pratt, C., 1959. Spontaneous and induced apple sports with misshapen fruit. Proc. Am. Soc. Hortic. Sci., 73: 1—8.
Einset, J., Blaser, H.W. and Imhofe, B., 1946. A chromosomal chimera of the Northern Spy apple. J. Hered., 37: 265—266.
Fideghelli, C., Monastra, F., Della Strada, G., Quarta, R. and Donini, B., 1977. Induced mutations in Annurca apple variety. Acta Hortic., 75: 43—50.
Fisher, D.V. and Lapins, K.O., 1966. Spur-type fruit varieties resulting from natural and induced mutation. In: P.M. Synge (Editor), Fruit — Present and Future. Meet. R. Hortic. Soc., pp. 72—77.
Granhall, I., 1949. Mutationsforskningens tillämpning på frukträden. Sver. Pomol. Fören. Årsskr., 51: 105—126.
Granhall, I., 1951. Mutationsforädling av frukträd. Nord. Jordbrugsforsk., 31-32: 401—411.
Granhall, I., 1953. X-ray mutations in apples and pears. Hereditas, 39: 149—155.
Granhall, I., Gustafsson, Å., Nilsson, F. and Olden, E.J., 1949. X-ray effects in fruit trees. Hereditas, 35: 269—279.
Gröber, K., 1959. Mutationsversuche an Kulturpflanzen. XI. Die Erzeugung von somatischen Mutationen mittels Röntgenstrahlen beim Obst. Kulturpflanze, 7: 37—54.
Gröber, K., 1967. Some results of mutation experiments in apple and black currant. In: H. Stubbe (Editor). Induzierte Mutationen und ihre Nützung. Erwin-Baur-Gedächtnisvorlesungen IV, 1966. Akademie Verlag, Berlin, pp. 377—382.
IAEA, 1972. Induced Mutations and Plant Improvement. IAEA, Vienna, pp. 526—544.
Ikeda, F., 1974. Radiation-induced fruit color mutation in the apple var. Fuji. Inst. Radiat. Breed., Ohmiya, Ibaraki, Tech. News. No. 15, 2 pp.
Ikeda, F., 1976a. Graft incompatibility with induced mutants in apple. Mutat. Breed. Newsl. (IAEA, Vienna), 8: 6.
Kaplan, R.W., 1953. Über Möglichkeiten der Mutationsauslösung in der Pflanzenzüchtung. Z. Pflanzenzücht., 32: 121—131.
Kolontaev, V.M., 1973a. Experimental mutagenesis of apple-tree. I. The character of the effect of X-irradiation on the development and genotype of *Malus domestica*. Genetics (USSR), 9(4): 12—21 (in Russian with English summary).
Kolontaev, V.M., 1973b. Experimentally induced mutations of apple tree. Genetics (USSR), 9(8): 158—160 (in Russian with English summary).
Kolontaev, V.M., 1974a. Specificity of the growth of cuttings of experiment mutations. Skh. Biol., 9(1): 50—60 (in Russian with English summary).

Kolontaev, V.M., 1974b. Experimental mutagenesis in apple trees. II. Characteristics of development and cytogenetic changes of *Malus domestica* induced by dimethyl sulphate. Soviet Genet., 10(3): 17—27.

Kolontaev, V.M., 1974c. Experimental mutagenesis in apple. III. Morphological and genetics effects in *Malus domestica*, induced by N-nitroso-N-methylurea. Soviet Genet., 10(4): 427—433.

Kolontaev, V.P., 1974d. Experimental mutagenesis in apple. IV. Phenotypic and genotypic changes in *Malus domestica*, induced by low doses of X-rays and nitrosomethylurea. Soviet Genet., 10(5): 578—583.

Kukimura, H., Ikeda, F., Fujita, H., Maeta, T., Nakajima, K., Katagiri, K., Nakahira, K. and Somegow, M., 1976. Genetical, cytological and physiological studies on the induced mutants with special regards to effective methods for obtaining useful mutants in perennial woody plants, II. In: Improvement of Vegetatively Propagated Plants and Tree Crops through Induced Mutations, Wageningen, 1976. IAEA, Vienna, pp. 93—137.

Lacey, C.N.D., 1977. The mutation spectrum of Cox's Orange Pippin resulting from gamma irradiation. In: Eucarpia Meet., Tree Fruit Breeding, Wageningen, 1976, pp. 46—52.

Lacey, C.N.D. and Campbell, A.I., 1977. Character changes in induced mutants of Bramley's seedling apple. Acta Hortic., 75: 51—62.

Lapins, K.O., 1963b. Spur types of apple and cherry produced by ionizing radiation. Proc. Wash. State Hortic. Assoc., 59: 93—95.

Lapins, K.O., 1965a. Compact mutants of apple induced by ionizing radiation. Can. J. Plant Sci., 45: 117—124.

Lapins, K.O., 1969. Tree growth habits in radiation-induced mutants of McIntosh apple. Can. J. Plant Sci., 49: 483—487.

Lapins, K.O., 1971a. Mutants of Golden Delicious apple by ionizing radiation. Can. J. Plant. Sci., 51: 123—131.

Lapins, K.O., 1972a. Mutations induced in McIntosh apple by ionizing radiation. Can. J. Plant Sci., 52: 209—214.

Lapins, K.O., 1973. Induced mutations in fruit trees. In: Induced Mutations in Vegetatively Propagated Plants. IAEA, Vienna, pp. 1—19.

Lapins, K.O., Bailey, C.H. and Hough, L.F., 1969. Effects of gamma rays on apple and peach leaf buds at different stages of development. I. Survival, growth and mutation frequencies. Radiat. Bot., 9: 379—389.

Le Lezec, M., 1973. Quelques variétés de pommier pour un renouvellement du verger Français. Pomol. Fr., 15(5): 111—125.

McIntosh, D.L. and Lapins, K., 1966. Differences in susceptibility to apple powdery mildew observed in McIntosh clones after exposure to ionizing radiation. Can. J. Plant Sci., 46: 619—623.

Mišic, P.D., 1963. A contribution to the effect of acute gamma radiation of Jonathan apple scions. In: Genetics Today. Proc. XI Int. Congr. Genet. Vol. 1. p. 217.

Poll, L., 1974. The production and growth of young apple compact mutants induced by ionizing radiation. Euphytica, 23: 521—533.

Pratt, C., 1960. Changes in structure of a periclinal chromosomal chimera of apple following X-irradiation. Nature (London), 186: 255—256.

Pratt, C., Einset, J. and Zahur, M., 1959. Radiation damage in apple shoot apices. Am. J. Bot., 46: 537—544.

Pratt, C., Way, R.D. and Ourecky, D.K., 1972. Irradiation of color sports of 'Delicious' and 'Rome' apples. J. Am. Soc. Hortic. Sci., 97: 268—272.

Przybyła, A. and Zagaja, S.W., 1974. Reaction of apple and cherry root cuttings to mutagen treatments. Fruit Sci. Rep., 1(1): 60—66.

Shamel, A.D. and Pomeroy, C.S., 1936. Bud mutations in horticultural plants. J. Hered., 27: 487—494.

Stankovic, D., 1962; Application of ionizing irradiation in the breeding of fruit trees. Proc. Int. Hortic. Congr., 16(3): 117–120.
Thakur, D.R. and Verma, B.R., 1967. Some observations on the effect of chronic gamma radiation on young apple trees. Sci. Cult., 33: 76–77.
Vavilov, N.I., 1931. Wild progenitors of the fruit trees of Turkistan and the Caucasus and the problems of the origin of fruit trees. Proc. Int. Hort. Congr., 9: 271–286.
Visser, T., 1973. Methods and results of mutation breeding in deciduous fruits, with special reference to the induction of compact and fruit mutations in apple. In: Induced Mutations in Vegetatively Propagated Plants. IAEA, Vienna, pp. 21–33.
Visser, T. and Verhaegh, J.J., 1977. Review on tree fruit breeding carried out at the Institute for Horticultural Plant Breeding at Wageningen from 1951–1976. In: Eucarpia Meet., Tree Fruit Breeding, Wageningen, 1976, pp. 113–133.
Visser, T., De Vries, D.P. and Verhaegh, J.J., 1969. Pre-determination of survival rate of apple and pear scions after X-ray treatment. Euphytica, 18: 352–354.
Visser, T., Verhaegh, J.J. and De Vries, D.P., 1971. Pre-selection of compact mutants induced by X-ray treatment in apple and pear. Euphytica, 20: 195–207.
Vondráček, I.J. and Ullman, J., 1963. Primary effects of gamma rays on apples. Ved. Pr. Ovocn. Holovousch., 2: 197–211 (with German summary).
Zagaja, S.W., 1975. Induced mutations in apple and sour cherry cultivars. In: Improvement of Vegetatively Propagated Plants through Induced Mutations, Tokai, 1974. IAEA, Vienna, pp. 129–130.
Zagaja, S.W. and Przybyła, A., 1973. Gamma-ray mutants in apples. Induced Mutations in Vegetatively Propagated Plants. IAEA, Vienna, pp. 35–40.
Zagaja, S.W. and Przybyła, A., 1976c. Compact type mutants in apples and sour cherries. In: Improvement of Vegetatively Propagated Plants and Tree Crops through Induced Mutations, Wageningen, 1976. IAEA, Vienna, pp. 171–184.
Zwintzscher, M., 1955. Die Auslösung von Mutationen als Methode der Obstzüchtung. I. Die Isolierung von Mutanten in Anlehnung an primäre Veränderungen. Züchter, 25: 290–302.
Zwintzscher, M., 1959. Die Auslösung von Mutationen als Methode der Obstzüchtung. In: Proc. II Congr. Eucarpia, pp. 202–211.
Zwintzscher, M., 1962. Methoden zur Isolierung induzierter Mutanten der Baumobstgehölze. Mitt. Obst. Garten, Ser. B., 1962: 125–134.

PEAR

The pear, *Pyrus*, is a member of the Rosaceae, sub-family Pomoideae. There are over 20 species, all indigenous to Europe and Asia. In the Northern hemisphere *Pyrus communis* (diploid; 2n = 34) is the main species used commercially, but other species and hybrids are grown as well. In other parts of the world, a variety of different species and hybrids are cultivated. World production, inclusive of fresh, cooked, dried and preserved fruits, is second to apples among the deciduous tree fruits (Layne, 1975).

The breeding objectives are numerous, but the more important ones are improved climatic adaptation, disease resistance, for example to fire blight, *Erwinia amylovora*, the induction of spur types, precocity, self-compatibility, higher productivity, and improvement of various fruit characteristics.

Spontaneous mutations are reported, for instance, by Gardner et al. (1933) who studied a striped bud sport of the Bartlett pear and by Reimer

(1951) who described bud mutations of the Bartlett pear and of the cultivar Beurré Hardy. The latter cultivar produced a mutant with red fruits, i.e. 'Royal Red Hardy'. Similar fruit colour mutants were also found in other cultivars, such as 'Starkrimson', 'Max Red Bartlett' and 'Red Anjou', whereas russetted fruit mutants were also found and commercialized, for instance, 'Russet Bartlett' (see Donini, 1976c).

Radiation was used in pear, as a tool to induce plant tissue rearrangements of existing periclinal chimeras (sports) by Decourtye (1964, 1967a). He tried to obtain genetically homogeneous shoots as well as to reveal mutations, hidden in deeper lying cell layers, by the irradiation of dormant buds with 3—5 krad ^{60}Co γ-rays. Earlier, Dommergues (1961) carried out similar work using cv. Max Red Bartlett which is heterozygous for red colour in the epidermis. About 30% of the shoots growing from dormant buds irradiated with 6 kR ^{60}Co γ-rays showed green sectors, proving that cv. Max Red Bartlett is a periclinal chimera and that radiation is a useful tool to uncover or rearrange a periclinal chimera.

The use of mutagenic agents to induce mutations in pears has been reported by several authors, including M.R. Celestre (who worked with A. Pirovano in Bologna, Italy) who, as early as 1946, published the results of a study of the effect of electricity or radiation on the genotype of this species (Celestre, 1946). Most publications date from later years (Granhall, 1953; Granhall et al., 1949) and many of them are very recent (Decourtye, 1970; Decourtye and Lantin, 1971; De Vries et al., 1970; Lacey, 1975; Morettini, 1963; Pratt, 1967; Roby, 1972a,b; Visser et al., 1971). Several kinds of mutations have been found, for example, those concerning blooming time, blossom colour, ripening time and fruit colour (Decourtye, 1970; Roby, 1972a,b) as well as compact types (Lacey, 1975; Visser et al., 1971). The latter emphatically distinguish between compact types (shoots thicker than normal for their length or, in other words, shoots with relatively short internodes but similar diameters to normal ones) and spur types (which have a large number of fruiting spurs in addition) (De Vries et al., 1970).

Spontaneous spurs are rarely found in pears and it seems to be important to induce them, either for direct use or for indirect use as a parent in a crossbreeding programme (provided the spur mutation is present in the L-II). After irradiation, spur types were found, but in much smaller percentages than in apple. Visser et al. (1971) reported 0.5% in the pear cultivars Beurré Hardy and Doyenné du Comice and 7% in apples (various cultivars).

Compact types seem difficult to select, one of the causes being that the direct irradiation damage appears to last longer in pears than in apples (Lacey, 1975). Since Gardner et al. (1933) had already noticed that continued selection is necessary to isolate and maintain mutant types, it is not surprising that much emphasis is given to the application of the optimum method of mutant induction and its recovery by post-irradiation treatment.

The train of thought about the questions concerning the methodology largely runs parallel with that discussed earlier in the section Tree Fruits — General and Apple. Since a bud contains approximately ten bud primordia (Pratt, 1967) a decision has to be made on which shoots produced by these buds are to be used for selection. Buds which are differentiated after having had mutagenic treatment give fewer mutants but larger sectors, whereas axillary buds at the base of a vM_1 shoot which are differentiated during treatment give more mutations but smaller mutated sectors. Therefore, the latter group must be propagated vegetatively 1 year more than the former. Consequently, the most efficient way seems to be to grow only three trees per "family", from the buds situated high on a shoot, and to use as many "families" as possible (Decourtye and Lantin, 1971). The most effective way to obtain stabilization and uniformity within a "family" is to drastically cut the vM_2 and vM_3 shoots back (De Vries et al., 1970).

Most authors irradiate dormant buds, either with X- or γ-rays and generally with 5—7 krad, the optimum dose depending on the cultivar. Experiments on the application of chemical mutagens to this species are not known to us.

Although some authors have mentioned potentially promising mutants, none of them has been put on the market, to our knowledge.

Literature pear

Celestre, M.R., 1946. Genetica del pero. Interventi elettrici e radianti. Ital. Agric., 83: 455—460.

Decourtye, L., 1964. The use of radiation in apple and pear breeding. In: Proc. Balsgård Fruit Breed. Symp. Balsgård Fruit Breed. Inst., Fjälkestad, pp. 40—43.

Decourtye, L., 1967a. Action des rayons gamma sur des variétés de poirier et de pommier en chimère. In: Coll. Morphol. Exp. Bull. Soc. Bot. France, 1966: 50—54.

Decourtye, L., 1970. Methodology in induced mutagenesis and results. In: Proc. Eucarpia Fruit Breed. Symp., Angers, 1970. Eucarpia Fruit Sect., Editions S.E.I., C.N.R.A., Versailles, pp. 161—174.

Decourtye, L. and Lantin, B., 1971. Considérations méthodologiques sur l'isolement de mutants provoqués chez le pommier et le poirier. Ann. Amélior. Plant., 21(1): 29—44 (with English summary).

De Vries, D.P., Verhaegh, J.J. and Visser, T., 1970. Nursery selection for 'spur' types in X-ray treated apple and pear varieties. In: Proc. Eucarpia Fruit Breed. Symp., Angers, 1970. Eucarpia Fruit Sect., Editions S.E.I., C.N.R.A., Versailles, pp. 147—160.

Dommergues, P., 1961. Action des rayons gamma sur les bourgeons de la variété de poirier Max Red Bartlett. Ann. Amélior. Plant., 11(3): 349—356 (with English and German summaries).

Donini, B., 1976c. Breeding methods and applied mutagenesis in fruit plants. In: Proc. Workshop Eur. Comm. — Israel, The Use of Ionizing Radiation, Wageningen, 1976. Assoc. Euratom—ITAL, Wageningen, pp. 445—478.

Gardner, V.R., Crist, J.W. and Gibson, R.E., 1933. Somatic segregation in a sectorial chimera of the Bartlett pear. J. Agric. Res., 46(11): 1047—1057.

Granhall, I., 1953. X-ray mutations in apples and pears. Hereditas, 39: 149—155.

Granhall, I., Gustafsson, Å, Nilsson, F. and Oldén, E.J., 1949. X-ray effects in fruit trees. Hereditas, 35: 269—279.

TABLE 22

The commercial mutant of apricot

Name of new cultivar	Place and date of release (or approval) and name of principal worker(s) and institute	Conditions and date of mutagenic treatment [Parent cultivar]	Main improved attributes of cultivar
Early Blenheim	Canada, 1970 K.O. Lapins, Canadian Department of Agriculture, Research Station, Summerland, B.C.	Shoots treated with thermal neutrons, 1955 [Blenheim]	Matures one week earlier than parent; annual yield; pollen self-incompatibility; large fruit

Sources: IAEA (1972, pp. 526—544) and *Mutation Breeding Newsletter*, 1 (1972) —9 (1977).

Lacey, C.N.D., 1975. Induction and selection of mutant forms of fruit plants. Long Ashton Rep., 1975: 22—24.
Layne, R.E.C., 1975. Pears. In: J. Janick and J.N. Moore (Editors), Advances in Fruit Breeding. Purdue University Press, West Lafayette, Ind., pp. 38—70.
Morettini, A., 1963. L'impiego di agenti mutageni per la produzione di mutazioni gemmarie nelle colture arboree. Riv. Ortoflorofrutticolt. Ital., 47(1): 5—11 (with English summary).
Pratt, C., 1967. Axillary buds in normal and irradiated apple and pear. Radiat. Bot., 7(2): 113—123.
Reimer, F.C., 1951. A genetic bud mutation in the pear. J. Hered., 42(2): 93—94.
Roby, F., 1972a. Doce mutaciones en el peral Williams obtenidas por injertos de ramitas irradiadas. Rev. Invest. Agropec., Ser. 2, 9(3): 55—64 (with English summary).
Roby, F., 1972b. Mutaciones inducidas por irradiacion en el peral Packham's Triumph. In: Induced Mutations and Plant Improvement, Buenos Aires, 1970. IAEA, Vienna, pp. 475—483 (with English summary).
Visser, T., Verhaegh, J.J. and De Vries, D.P., 1971. Pre-selection of compact mutants induced by X-ray treatment in apple and pear. Euphytica, 20: 195—207.

APRICOT

Apricot, *Prunus armeniaca* L., family Rosaceae, sub-family Prunoideae, is grown in the temperate regions of central Asia and China as well as in Eastern Europe, where annual production is almost as high as in the rest of the world. Most cultivated apricots belong to *Prunus armeniaca* L. (diploid; 2n = 16) but several other species are grown as well (Bailey and Hough, 1975).

The breeding objectives in this species are to improve climatic tolerance and hardiness, resistance to disease and to improve the pomological fruit characteristics, such as size, appearance and shape, depending on the use of the fruits (fresh, canned or dried) and the needs of the region where they are grown.

Spontaneous mutants apparently occur very seldom since no record of such events could be found in the literature.

Only two artificially produced mutants have been reported (Donini, 1976c) both of which were produced by Lapins (1972b, 1973) and one is being marketed, namely, cv. Early Blenheim (Table 22). Later, potentially useful mutations altering the ripening time of the fruits were obtained following colchicine treatment of three diploid apricot cultivars (Lapins, 1975).

As has been discussed elsewhere (see Temperate Tree Fruit Crops — General and the sections on other tree fruits: apple, pear and cherry), it is important to use the most efficient method of mutation induction and screening (e.g. irradiating buds at the most appropriate stage of development) to ensure large mutated sectors and easy recovery of stable periclinal chimeras through successive vegetative propagation.

As far as we know, no application of in vitro propagation techniques have been reported.

Literature apricot

Bailey, C.H. and Hough, L.F., 1975. Apricots. In: J. Janick and J.N. Moore (Editors), Advances in Fruit Breeding. Purdue University Press, West Lafayette, Ind., pp. 367—383.
Donini, B., 1976c. Breeding methods and applied mutagenesis in fruit plants. In: Proc. Workshop Eur. Comm. — Israel, The Use of Ionizing Radiation, Wageningen, 1976. Assoc. Euratom—ITAL, Wageningen, pp. 445—477.
IAEA, 1972. Induced Mutations and Plant Improvement. IAEA, Vienna, pp. 526—544.
Lapins, K.O., 1972b. New Fruits from Summerland, B.C., 1956—1970. Can. Dept. Agric., Publ. 1471, Summerland, B.C.
Lapins, K.O., 1973. Induced mutations in fruit trees. In: Induced Mutations in Vegetatively Propagated Plants. IAEA, Vienna, pp. 1—19.
Lapins, K.O., 1975. Polyploidy and mutations induced in apricot by colchicine treatment. Can. J. Genet. Cytol., 17(4): 591—599.

CHERRY

The two most important commercially grown cherry species are sweet cherry, *Prunus avium* L., usually diploid (2n = 16) (although triploid and tetraploid forms are found as well), and sour cherry, *Prunus cerasus* L. which is tetraploid (2n = 32). They are widely grown in Europe and North America. Numerous other species of cherry exist and have occasionally been utilized in breeding programmes (Fogle, 1975).

The modern breeding objectives are to obtain genotypes with better adaptation to climatic conditions, resistance to diseases, for example, bacterial canker, viruses and mildew, resistance of the fruit to cracking after rain, improved fruit size, firmness, colour and/or quality, and higher productivity. The induction of self-compatibility and spur-type growth is often the objective of a mutation breeder. Reduced growth is important since, in contrast to apple for instance, only a restricted number of dwarfing rootstocks or interstocks is available. It can be obtained either by inducing spur-type mutants or by producing dwarfing rootstocks.

Spontaneous mutations, affecting all kinds of characters, including chlorophyll content and ripening time (Baev, 1968), stem length, fruit size and shape and ripening time (Gardner, 1949) have frequently occurred. A crinkled sweet cherry was studied by Pratt et al. (1968), who concluded from the high frequency of shoots with normal foliage, obtained after irradiation with 2—8 kR γ-rays, that the crinkled form was a periclinal chimera, as is expected of most "sports". Kursakov and Kursakova (1974) studied 105 cultivars of sour cherry and hybrids between sour cherries and sweet cherries and found that several showed bud mutations. Several characteristics of the mutant shoots differed from those of the initial material.

Induced mutations have been reported by several authors (see later), either after exposure to radiation or, very rarely, following chemical treatment (Mashkin and Fufaeva, 1974; Przybyła and Zagaja, 1974). Most authors have used X- or γ-rays at doses between 2.5 and 5 krad and, more specifically, those close to the optimum dose of 4 krad (Donini, 1976b; Fideghelli et al., 1976; Roselli and Donini, 1976; Kolesnikova, 1970, 1974; Lapins, 1963a, 1974b; and others). Using such doses fairly high frequencies of mutations are induced, depending on the type of mutation scored and the cultivar treated. Donini (1976b), for instance, lists the percentage of morphological mutations induced by exposure of various cultivars to 3.5—4.5 kR X- or γ-rays (Table 23). Various other types of mutations, apart from self-compatibility and compact growth (see later), such as higher fruit set (Georgiev, 1975), weeping types (Lacey, 1975) and cracking resistance (Milenkov, 1975b) have also been obtained.

Self-compatibility was obtained by applying irradiated pollen to incompatible female parents (Lewis and Crowe, 1954; Crowe, 1956; Matthews and Lapins, 1967; Oldén, 1959). Some self-fertile plants obtained were used in further breeding programmes (Lapins, 1974a) which resulted in the commercial self-fertile sweet cherry cultivar Stella.

The main objective of most mutation breeding programmes in cherry, however, is the induction of spur type growth, either for direct use or for further breeding, as was reported by Bavtuto (1973), Blažek (1976b), Donini (1975, 1976b), Georgiev (1975), Kolesnikova (1970, 1974), Lacey (1975), Matthews and Dow (1975), Milenkov (1975a), Roselli and Donini (1976),

TABLE 23

Morphological mutations induced in different cherry varieties after X- or γ-irradiations (exposure rates 471—1 110 R/h; total exposures 3.5—4.5 kR) (Donini, 1976b, Table 1)

Cultivar	Number of treated buds	Number of vM_2 shoots analysed	Morphological mutations induced	
			Number	Percent
Bigarreau Moreau	2 710	2 510	40	1.60
Bigarreau Napoleon	2 507	8 223	11.5	1.40
Bigarreau Burlat	757	830	50	6.02
Durona di Vignola I	406	548	39	7.11
Durona di Vignola II	1 056	1 955	136	6.95
Durona di Vignola III	549	94	6	6.38
Mora di Vignola	939	1 273	93	7.30
Saccoccia	120	47	3	6.38
Ravenna precoce	150	26	1	3.84
Ravenna nana	60	26	0	—
Ravenna	210	32	1	3.12
Merton Bounty	290	576	3	0.52
Merton Heart	178	288	1	0.35
Mora di Cazzano	2 297	4 040	37	0.91
F12/1	958	1 867	88	4.71

Zagaja (1975) and Zagaja and Przybyła (1976a,b).

Lapins successfully treated dormant scions with X-rays or thermal neutrons and grafted onto trees. He obtained several spur types from both the Lambert cherry and the self-fertile cultivar Stella. One from each cultivar was made commercially available, namely 'Compact Lambert' and 'Compact Stella', respectively (Lapins, 1963a,b, 1965b, 1973, 1974b).

As in all tree fruits, as has been pointed out in the general introduction to Tree Fruits, much attention has been paid to the improvement of efficiency in mutagen treatment and post treatment of the material in order to restrict the disadvantageous consequences of chimera formation. Several authors have tried to determine which part of the vM_1 shoot is the best part to expose to radiation, in the sense that, relatively speaking, the highest mutation frequency and the largest possible sector size are obtained (Donini, 1975, 1976b,c; Donini et al., 1973; Fideghelli et al., 1976; Lapins, 1971b, 1973; Zwintzscher, 1967) (see also Fig. 24). Since a cherry bud appears to have several bud primordia (Donini, 1976b; Pratt, 1968), a choice must be

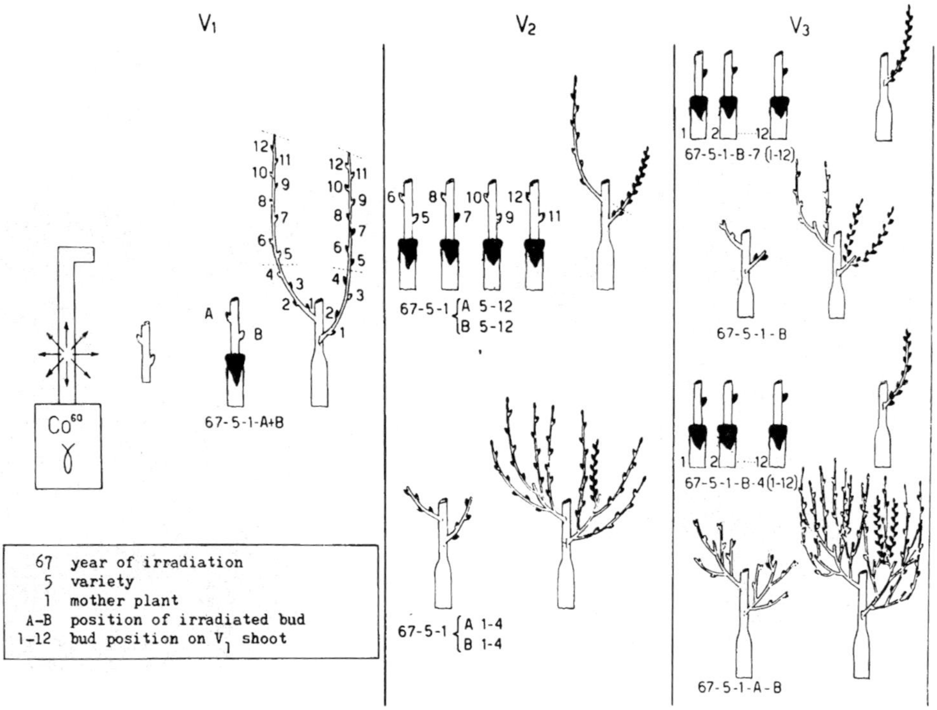

Fig. 24. Methodology scheme used for isolation of mutations in cherries. (Donini, 1975, Fig. 5.)

made between vM_1 buds 1—5, 6—10 and 11—15. Most authors prefer buds 5—10 and, more particularly, 5 and 6 (Table 24; from Donini, 1976b) (Donini, 1976b; Donini et al., 1973; Lapins, 1971b, 1973). Accessory buds can be forced to produce shoots by the removal of the main bud and such production of shoots can be promoted by the application of 6-benzylamino purine (BAP) with adenine or adenine and indolyl acetic acid (IAA). These accessory buds have the potential to produce larger sectors (Lapins, 1971b; Katagiri and Lapins, 1974).

It is important to induce as high percentages of mutations and to obtain as large mericlinal sectors as possible, while the post-irradiation treatment should be designed to attain homogeneous and stable periclinal clones. As has been pointed out by Gardner (1949) and Baev (1968) careful and continued selection is necessary to fix the original type and to obtain a relatively uniform cultivar, a procedure that clearly applies to (radiation-induced) mutants.

Another way to restrict the disadvantageous complications involved in chimera formation, which automatically follows treatment of multicellular apices, is the irradiation of buds at the earliest possible ontogenetic stage. Root cuttings of rootstocks or of cultivars which are not grafted, are mentioned in this context and used as experimental material by Przybyła and Zagaja (1974) and Zagaja (1975). They concluded from their preliminary work that bud development was more advanced than expected. This was demonstrated by albinic sectors of leaves, although entire shoots with deviating chlorophyll content were obtained as well.

So far, relatively few mutants are of commercial interest (Table 25), partly due to the fact that most breeding programmes have only recently started. The high number of mutants, mainly spur types, which are at present under trial, allows us to cherish the hope that new cherry cultivars will soon be released for commercial use.

TABLE 24

Frequency of vM_2 mutated shoots with reference to their origin from the buds along the vM_1 shoots (Donini, 1976b, Table 2)

Origin of vM_2 shoot from the buds along the vM_1 shoot	Number of vM_2 shoots analysed	Morphological mutations induced	
		Number	Percent
1—2	1 590	81	5.09
3—4	1 368	90	6.57
5—6	832	70	8.41
7—8	1 038	77	7.41
9—10	799	62	7.75
11—12	257	15	5.83

TABLE 25

Commercial mutants of cherry

Name of new cultivar	Place and date of release (or approval) and name of principal worker(s) and institute	Conditions and date of mutagenic treatment [Parent cultivar]	Main improved attributes of cultivar
Compact Lambert	Canada, 1964 K.O. Lapins, Canadian Department of Agriculture, Research Station, Summerland, B.C.	Scions, 4 kR X-rays, 1958 [Lambert]	Desirable compact and dwarf growth habit combined with very early and very heavy cropping; fruit size slightly reduced
Stella	Canada, 1968 K.O. Lapins, Canadian Department of Agriculture, Research Station, Summerland, B.C.	Lambert × John Innes Seedling 2420 (self-fertile mutant by X-ray treatment of pollen) 1956	First good quality self-fertile sweet cherry
Compact Stella	Canada, 1972 K.O. Lapins, Canadian Department of Agriculture, Research Station, Summerland, B.C.	Scions, 3 kR X-rays, 1964 [Stella]	Semi-dwarf; heavily spurred growth; self-fertile

Sources: IAEA (1972, pp. 526—544) and *Mutation Breeding Newsletter,* 1 (1972) —9 (1977).

Literature cherry

Baev, H., 1968. Knospenmutation bei den Sauerkirschensorten spät-Amarelle und Topolischka. Gradinar. Lozar. Nauka, 5(7): 17—26 (in Russian with German summary).

Bavtuto, G.A., 1970. X-ray mutants of hybrids between sweet and sour cherry (in Russian). Plant Breed. Abstr., 46; No. 10472.

Blažek, J., 1976b. Gamma-radiation induced growth mutations in the cherry cultivar Napoleon. Vedecke Prace Ovonarske, 5: 31—43 (in Czechoslovakian with English summary); Hortic. Abstr., 47: No. 7200.

Crowe, L.K., 1956. X-rays make self-fertile fruit trees. Grower, 45(24): 1494—1495.

Donini, B., 1975. Induction and isolation of somatic mutations in vegetatively propagated plants. In: Improvement of Vegetatively Propagated Plants through Induced Mutations, Tokai, 1974. IAEA, Vienna, pp. 35—51.

Donini, B., 1976b. The use of radiation to induce useful mutations in fruit trees. In: Improvement of Vegetatively Propagated Plants and Tree Crops through Induced Mutations, Wageningen. IAEA, Vienna, pp. 55—67.

Donini, B., 1976c. Breeding methods and applied mutagenesis in fruit plants. In: Proc. Workshop Eur. Comm. — Israel, Use of Ionizing Radiation, Wageningen, 1976. Assoc. Euratom — ITAL, Wageningen, pp. 445—478.

Donini, B., Fideghelli, C. and Rosati, P., 1973. Mutani compatti indotti con radiazioni in varietà di ciliegio. In: Atti del 2. Convegno del Ciliegio. pp. 1—9 (with English summary).

Fideghelli, C., Rosati, P. and Donini, B., 1976. Mutanti compatti indotti con radiazioni ionizanti in varietà di ciliegio. Tecniche Nucleari per il Miglioramento e la Difesa dei Fruttiferi. I. Com. Naz. Energ. Nucl., Cent. Stud. Nucl. Casaccia, 1974, pp. 57—70.

Fogle, H.W., 1975. Cherries. In: J. Janick and J.N. Moore (Editors), Advances in Fruit Breeding. Purdue University Press, West Lafayette, Ind., pp. 348—366.

Gardner, V.R., 1949. Type and variability in bud mutations of the cherry and apple. Q. Bull. Mich. State Coll. Agric. Exp. Stn, 32: 28—41.

Georgiev, V., 1975. Gamma irradiation of softwood scions of cherry cultivars. Preliminary communication. Gradinar. Lozar. Nauka, 12(6): 3—12 (in Bulgarian with English summary); Hortic. Abstr., 46: No. 9038; Plant Breed. Abstr., 46: No. 3686.

IAEA, 1972. Induced Mutations and Plant Improvement. IAEA, Vienna, pp. 526—544.

Katagiri, K. and Lapins, K.O., 1974. Development of gamma-irradiated accessory buds of sweet cherry, *Prunus avium* L. Radiat. Bot., 14(3): 173—178.

Kolesnikova, A.F., 1970. Induced mutations in sour cherry. Skh. Biol., 5(4): 554—559 (in Russian with English summary).

Kolesnikova, A.F., 1974. Ultraviolet and γ-rays in the clonal breeding of sour cherry (in Russian) Plant Breed. Abstr., 46: No. 1775.

Kursakov, G.A. and Kursakova, L.E., 1974. Methodology and results of a study of spontaneous bud mutations in sour cherry (in Russian). Plant Breed. Abstr., 47: No. 1590.

Lacey, C.N.D., 1975. Induction and selection of mutant forms of fruit trees. Long Ashton Rep., 1975: 22—24.

Lapins, K.O., 1963a. Note on compact mutants of Lambert cherry produced by ionizing radiation. Can. J. Plant Sci., 43(3): 424—425.

Lapins, K.O., 1963b. Spur types of apple and cherry produced by ionizing radiation. Proc. Wash. State Hortic. Assoc., 59: 93—95.

Lapins, K.O., 1965b. The Lambert Compact cherry. Fruit Var. Hortic. Dig., 19: 23.

Lapins, K.O., 1971b. Mutation frequencies in vegetative shoots derived from two zones of irradiated buds of sweet cherry, *Prunus avium* L. Radiat. Bot., 11(3): 197—200.

Lapins, K.O., 1973. Induced mutations in fruit trees. In: Induced Mutations in Vegetatively Propagated Plants. IAEA, Vienna, pp. 1—19 (37 references).

Lapins, K.O., 1974a. Progress in breeding self-compatible sweet cherries. Mutat. Breed. Newsl., 3: 14.

Lapins, K.O., 1974b. Compact Stella sweet cherry introduced. Mutat. Breed. Newsl., 4: 18.

Lewis, D. and Crowe, L.K., 1954. The induction of self-fertility in tree fruits. J. Hortic. Sci., 29: 220—225.

Mashkin, S.I. and Fufaeva, L.D., 1974. The experimental induction of somatic changes in sweet and sour cherry by treating cuttings with chemical mutagens. Plant Breed. Abstr., 46: No. 10471.

Matthews, P. and Dow, P., 1975. Induction of small tree spur mutants in bacterial canker resistant selections. John Innes Inst. Annu. Rep., 63: 16—18.

Matthews, P. and Lapins, K.O., 1967. Self-fertile sweet cherries. Fruit Var. Hortic. Dig., 21: 36—37.

Milenkov, M., 1975a. The effect of gamma rays on cherries. Gradinar. Lozar. Nauka, 12(4): 19—26 (in Bulgarian with English summary); Hortic. Abstr., 46: No. 4290.

Milenkov, M.M., 1975b. Possibilities of producing polyploid forms of cherry. Gradinar. Lozar. Nauka, 12(8): 3—9 (in Russian with English summary); Hortic. Abstr., 47: No. 1201.

Oldén, E.J., 1959. Cherry breeding at Balsgård. Comm. Balsgård Fruit Breeding Institute, Balsgård, No. 48, 14 pp. (in Swedish with English summary).

Pratt, C., 1968. Radiation damage in shoots of sweet cherry (*Prunus avium* L.) Radiat. Bot., 8(4): 297—306.

Pratt, C., Gilmer, R.M. and Way, R.D., 1968. Occurrence of normal foliage in irradiated sweet cherries with crinkle. Plant Dis. Rep., 52(4): 268—271.

Przybyła, A. and Zagaja, S.W., 1974. Reaction of apple and cherry root cuttings to mutagen treatment. Fruit Sci. Rep., 1(1): 60—66.

Roselli, G. and Donini, B., 1976. Mutazioni radioindotte in ciliegio, pesco e olivo. In: Techniche Nucleari per il Miglioramento e la Difesa dei Fruttifera, I. Com. Naz. Energ. Nucl., Cent. Stud. Nucl. Casaccia, pp. 71—83.

Zagaja, S.W., 1975. Induced mutations in apple and sour cherry cultivars. In: Improvement of Vegetatively Propagated Plants through Induced Mutations, Tokai, 1974. IAEA, Vienna, pp. 129—130.

Zagaja, S.W. and Przybyła, A., 1976a. Compact type mutants in apple and sour cherries. Mutat. Breed. Newsl., 8: 8—9.

Zagaja, S.W. and Przybyła, A., 1976b. Compact type mutants in apple and sour cherries. In: Improvement of Vegetatively Propagated Plants and Tree Crops through Induced Mutations, Wageningen, 1976. IAEA, Vienna, pp. 171—184.

Zwintzscher, M., 1967. On the variability of isolates of the sour cherry variety 'Schattenmorelle' from a mutation trial. In: Induzierte Mutationen und ihre Nützung. Erwin Baur Ged. Vorlesungen, IV, 1966. Akademie Verlag, Berlin, pp. 363—367.

OLIVE

Olive, *Olea europea*, family Oleaceae, is widely grown in the Mediterranean countries for its fruit, an important source of oil and also eaten as a pickle or relish. Its hard, yellow wood is used in wood turning. Different cultivars are known, cultivars for both oil and pickled olives generally being diploid (2n = 46).

Spontaneous mutants, that were shown to be chimeral, were reported by Roselli (1972).

The main breeding objective in olive is to produce spur-type growth habit mutants of outstanding cultivars to be used in dense plantations and to enable easy fruit picking. Also, it seems to be possible to influence the growth habit by using certain rootstocks or interstocks (Hartmann and Whisler, 1970). Another objective is the induction of self-fertility.

Present-day mutation breeding of olive is almost exclusively an Italian affair, carried out by Donini and his associates, sometimes in co-operation with a few breeders from other institutes (Donini, 1975, 1976a,b; Donini and Roselli, 1972; Donini et al., 1975; Petruccioli et al., 1976; Roselli and Donini, 1976). They have irradiated young trees in a γ-field, and dormant and rooted scions of several cultivars were acutely irradiated with 1—6 kR ^{60}Co γ-rays, 3—4 kR being the optimum dose. The mutations were scored in vM_2 shoots after being repeatedly cut back. Because olive has a strong apical dominance, scions from each of buds 5—12 were rooted in addition.

The mutation frequency was different in the different cultivars used and this was explained by differences in the degree of heterozygosity. The highest mutation frequencies were observed in shoots growing from primordia which had been present in the buds at the time of irradiation. However, nothing was said about the size of mutated sectors.

Various types of mutations were found, including shortened internodes,

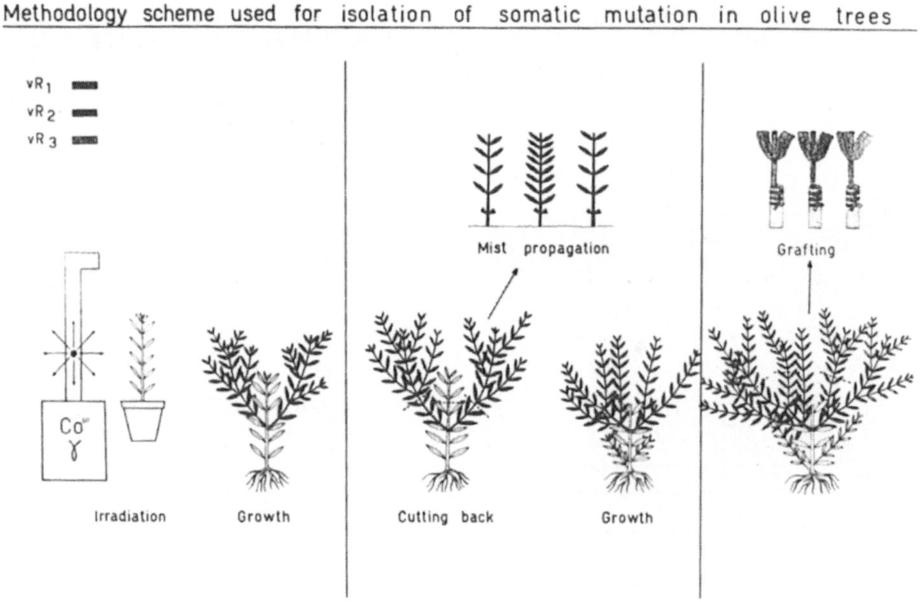

Fig. 25. Methodology scheme used for isolation of mutations in olives. (Donini, 1975, Fig. 4.)

loss of apical dominance with lateral shoot growth, and changes in shoot growth habit and leaf morphology (Donini, 1976a).

On the basis of their experience, schemes for the isolation of somatic mutants in olive were developed, for example, irradiation of plants from which lateral shoots are obtained when cutting back and either rooted or used for grafting (Fig. 25). The procedure ensures a rapid increase in mutated sector size so that complete periclinal chimeras are obtained in as short a time as possible.

At present there are no known commercial mutants.

Literature olive

Donini, B., 1975. Induction and isolation of somatic mutations in vegetatively propagated plants. In: Improvement of Vegetatively Propagated Plants through Induced Mutations, Tokai, 1974. IAEA, Vienna, pp. 35—51.

Donini, B., 1976a. Use of radiations to induce useful mutations in fruit trees. Mutat. Breed. Newsl., 8: 7—8.

Donini, B., 1976b. The use of radiations to induce useful mutations in fruit trees. In: Improvement of Vegetatively Propagated Plants and Tree Crops through Induced Mutations, Wageningen, 1976. IAEA, Vienna, pp. 55—67.

Donini, B. and Roselli, G., 1972. Mutazioni indotte nell'olivo per irraggiamento di talee autoradicate. Genet. Agrar., 26(1-2): 149—160 (with English summary).

Donini, B., Petruccioli, G. and Roselli, G., 1975. The use of radiations for inducing useful mutations in olive trees. In: 2° Seminario Oleicolo Internazionale Cordoba. Madrid Servicio de Publicationes Agrarias, Madrid, 7 pp.

Hartmann, H.T. and Whisler, J.E., 1970. Some rootstock and interstock influences in the olive (*Olea europaea* L.) cv. Sevillano. J. Am. Soc. Hortic. Sci., 95(5): 562—565.

Petruccioli, G., Filippucci, B. and Donini, B., 1976. Impiego della mutagenesi per l'ottenimento di forme nanizatte nell'olivo. In: Tecniche Nucleari per il Miglioramento e la Difesa dei Fruttiferi, I. Com. Naz. Energ. Nucl., Cent. Stud. Nucl., Casaccia, 1974, pp. 95—100 (with English summary).

Roselli, G., 1972. Mutazioni spontanee a piante chimeriche nell'olivo. Genet. Agrar., 26(1-2): 62—74.

Roselli, G. and Donini, B., 1976. Mutazioni radioindotte in ciliegio, pesco e olivo. In: Tecniche Nucleari per il Miglioramento e la Difesa dei Fruttifera, I. Com. Naz. Energ. Nucl., Cent. Stud. Nucl. Casaccia, 1974, pp. 71—83 (with English summary).

PEACH

The peach and its smooth-skinned mutant, the nectarine, *Prunus persica* (L.) Batsch. (diploid; 2n = 16), belong to the family Rosaceae, sub-family Prunoideae, genus *Prunus* L. and subgenus *Amygdalus*. All commercial cultivars, belonging to the species *Pr. persica*, are widely grown in southern Europe, the Near East, southern North America, Latin America and several other countries (Hesse, 1975).

The most important breeding objectives are improvement of climatic tolerance, earlier maturity, better processing characteristics for the canning industry, improved tree structure and disease resistance.

The interest of the peach breeder in artificial induction of mutations as a means to enlarge the variability seems to be increasing, partly because of the results obtained in other tree fruit crops but also because spontaneous mutations are frequently found in peach. Several authors have studied spontaneous sports. Work published in this field includes that of Dermen (1956; colour and fuzzless sport), Monet and Salesses (1975; compact growth), Oberle and Nicholson (1953; nectarine associated with reduced size of fruit), Pejkic (1972/1973; a more vigorous, more frost resistant and more productive later ripening sport), Thompson et al. (1953; an unstable nectarine-type) and Yeager and Meader (1956; a flesh-coloured mutant).

Donini (1976c) lists more than ten cultivars which originate from spontaneous bud mutations which affected ripening time of the fruit and other fruit characteristics (Table 26). By comparing these mutants with the parental varieties he concluded that (a) spontaneous mutations occurred in varieties originally produced by both self- and cross-pollination, (b) only a few characters were changed, (c) the changes involved monofactorial as well as polyfactorial characters, and (d) the changes were from dominant to recessive.

Several of the afore-mentioned authors reported that sports are sometimes unstable. From the work of Dermen, who studied the histogenetics and the

TABLE 26

Spontaneous bud mutations isolated in peaches (Donini, 1976c, Table 4)

Cultivar	Bud mutation or new variety	New character
Mayflower	Madame Kira Evreinoff	Late fruit ripening (18 days), changes in fruit texture
	Madame Pujet	Early fruit ripening (24 days)
Morettini 5/14	Pietro	Late fruit ripening (1 week)
Elberta	Red Elberta	Early fruit ripening (10 days), yellow—rose fruit skin
	Sullivan Red Elberta	Early fruit ripening (1 week)
July Elberta	Starking Delicious	Changes in fruit texture
Coronet	Early Coronet	Early fruit ripening (1 week)
Red Haven	Stark Earliglo	Early fruit ripening (1 week)
	Early Red Haven	Early fruit ripening (1 week)
	Sunshine	Increased fruit size
Vesuvio	Mazzocchella	Nectarine
Valiant	Fisher	Early fruit ripening (3 weeks)
Vedette	Two mutations	Changes in fruit texture, early fruit ripening (2 weeks)

TABLE 27

The commercial mutant of peach

Name of new cultivar	Place and date of release (or approval) and name of principal worker(s) and institute	Conditions and date of mutagenic treatment [Parent cultivar]	Main improved attributes of cultivar
Magnif 135	Argentina, 1968 L.B.C. de Terraciano, Instituto de Fitotecnia, Inst. Nac. Technol. Agropecu., Castelar	Chronic irradiation in γ-field, 1962—1963 [Magnif 43]	Bigger fruit with deeper red skin colour; ripening 7 days earlier

Sources: IAEA (1972, pp. 526—544) and *Mutation Breeding Newsletter*, 1 (1972)—9 (1977).

origin of tissues of either sports or colchicine-induced cyto-chimeras of a large variety of plants, including peach and other fruit crops, we know that the cell layers in peach have three groups (Dermen, 1947c, 1953, 1954a, 1956). Consequently, spontaneous as well as artificially induced mutations lead to mericlinal chimeras which are unstable and explains the fact that repeated vegetative multiplications are necessary to obtain a stable and periclinal mutant clone.

This also explains the interest in the development of efficient methods to induce and recover stable mutants, as explained in the sections on other tree fruits (Apple, Pear, Cherry). Hough et al. (1965), Lapins and Hough (1970) and Lapins et al. (1969) report the irradiation of very young seedlings from embryo culture and summer buds with γ-rays at several doses and dose rates. Mutations in morphology, and deficiencies of anthocyanins and chlorophyll in the leaf were studied in the primary and secondary shoots which grew from the irradiated buds. The highest mutation frequencies were obtained in secondary shoots arising between leaf 6 and 10 of the primary shoot. In seedlings, increasing mutation frequencies were obtained with increasing dose in the lower zones.

Artificially induced mutants have been studied by several authors. Hough and Weaver (1959) irradiated five cultivars with ^{60}Co γ-rays in a γ-field over periods of 1 and 2 years. Dose rates between 10 and 60 R/day produced mutations of potential value, with changes in ripening time, and firmness and texture of the fruit flesh as well as for a freestone character. A late mutant of cv. Elberta had particularly great promise. Later, peach bud sticks were acutely irradiated with 1—6 kR X-rays and thermal neutrons and T-budded on seedling stock (Hough et al., 1965). Satisfactory survival was obtained between 1 and 4 kR X-rays; the mutants produced were all chimeral and included one with non-blooded flesh and late ripening which are linked characters. In some mutants, however, that linkage was broken.

Donini (1975, 1976b,c) and Roselli and Donini (1976) reported the irradiation of 2 300 summer buds of the cultivars Favorite and Fertilia with 3 krad γ-rays. They were grafted onto rootstocks and pruned in each of the following years. Growing plants were exposed in a γ-field to 20 or 40 R/day for several years. Morphological and chlorophyll mutations were produced by both dose rates, mostly in the form of "sectorial" chimeras. Among the mutants obtained were nectarine and early ripening.

Despite the activities discussed above, very few commercial results have been achieved. In fact, only one commercial mutant was reported in a list published by the IAEA, Vienna, and produced by De Terraciano in Castelar, Argentina. It was obtained by chronic γ-irradiation and has bigger fruits, deeper red skin colour and ripens earlier than normal fruit (Table 27).

Literature peach

Dermen, H., 1947c. Inducing polyploidy in peach varieties. J. Hered., 38(3): 77—82.

Dermen, H., 1953. Periclinal cytochimeras and origin of tissues in stem and leaf of peach. Am. J. Bot., 40: 154—168.

Dermen, H., 1954a. Histogenetic factors in color and nectarine sports of peach. Genetics, 39: 964.

Dermen, H., 1956. Histogenetic factors in color and fuzzless peach sports. J. Hered., 47(2): 64—76.

Donini, B., 1975. Induction and isolation of somatic mutations in vegetatively propagated plants. In: Improvement of Vegetatively Propagated Plants through Induced Mutations, Tokai, 1974. IAEA, Vienna, pp. 35—51.

Donini, B., 1976b. The use of radiation to induce useful mutations in fruit trees. In: Improvement of Vegetatively Propagated Plants and Tree Crops through Induced Mutations, Wageningen, 1976. IAEA, Vienna, 55—67.

Donini, B., 1976c. Breeding methods and applied mutagenesis in fruit plants. In: Proc. Workshop Eur. Comm. — Israel, The Use of Ionizing Radiation, Wageningen, 1976. Assoc. Euratom—ITAL, Wageningen, pp. 445—478.

Hesse, C.O., 1975. Peaches. In: J. Janick and J.N. Moore (Editors), Advances in Fruit Breeding. Purdue University Press, West Lafayette, Ind., pp. 285—335.

Hough, L.F. and Weaver, G.M., 1959. Irradiation as an aid in fruit variety improvement, I. Mutations in the peach. J. Hered., 50(2): 59—62.

Hough, L.F., Moore, J.N. and Bailey, C.H., 1965. Irradiation as an aid in fruit variety improvement. II. Methods for acute irradiation of vegetative growing points of the peach, *Prunus persica* (L.) Batsch. In: The Use of Induced Mutations in Plant Breeding. Radiat. Bot., 5 (Suppl.): 679—686.

IAEA, 1972. Induced Mutations and Plant Improvement. IAEA, Vienna, pp. 526—544.

Lapins, K.O. and Hough, L.F., 1970. Effects of gamma rays on apple and peach leaf buds at different stages of development. II. Injury to apical and axillary meristems and regeneration of shoot apices. Radiat. Bot., 10: 59—68.

Lapins, K.O., Bailey, C.H. and Hough, L.F., 1969. Effects of gamma rays on apple and peach leaf buds at different stages of development I. Survival, growth and mutation frequencies. Radiat. Bot., 9: 379—389.

Monet, R. and Salesses, G., 1975. Un nouveau mutant de nanisme chez le pêcher. Ann. Amélior. Plant., 25(3): 353—359.

Oberle, G.D. and Nicholson, J.O., 1953. Implications suggested by a peach to nectarine sport. Proc. Am. Soc. Hortic. Sci., 62: 323—326.

Pejkic, B., 1972/1973. Biological and economic characteristics of a mutant of 'J.H. Hale' (*Prunus persica* L.) (with English summary). Plant Breed. Abstr., 46: No. 1781.

Roselli, G. and Donini, B., 1976. Mutazioni radioindotte in ciliegio, pesco e olivo. In: Tecniche Nucleari per il Miglioramento e la Difesa dei Fruttifera, I. Com. Naz. Energ. Nucl., Cent. Stud. Nucl., Casaccia, 1974, pp. 71—83.

Thompson, A.H., Vierheller, A.F. and Dunbar, C.O., 1953. An unstable mutation of the shippers late red peach. Fruit Var. Hortic. Dig., 8(1): 39—40.

Yeager, A.F. and Meader, E.M., 1956. A flesh-color chimera in the peach. J. Hered., 47(2): 77—78.

VARIOUS OTHER TREE FRUITS

A small number of publications are concerned with spontaneous or induced mutations in tree fruit crops, other than those discussed before.

In almond, *Amygdalus communis* L., family Rosaceae, mainly grown in southern Europe and California, bud sports have been found and some have been put on the market as commercial cultivars, such as cv. Tardy Nonpareil, a late-blooming mutant of cv. Nonpareil (Brooks and Olmo, 1972). Other mutations have been reported by the same authors as well as by Kester and Asay (1975, p. 399). Therefore, it seems possible that induced mutations could contribute to improving important characteristics of almond, such as ability to adapt to particular environmental conditions, productivity, quality and self-fertility.

Avocado, *Persea americana* Mill., family Lauraceae, was originally mainly grown in Mexico and Central America but has now spread throughout the tropics and subtropics. Spontaneous mutations of several of its characters, such as those governing tree shape, colour, size and shape of fruit, have been found but, so far, no mutant has been put on the market. No references on mutation breeding were found, but it appears that a programme was started at the University of California, applying fast neutron irradiation to cv. Duke (Bergh, 1975).

In figs, *Ficus carica* L., family Moraceae, usually grown in southern Asia, Asia Minor and Mediterranean countries, the only known reference is by Storey (1975) who reported mutation induction by colchicine treatment of flushing axillary buds and seeds of a few cultivars. The results were poor aberrant plants, which may be partly polyploid (Storey, 1975).

Scions of the cultivar Champion of quince, *Cydonia oblonga* Mill. (fam. Rosaceae), mainly grown in southern Russia, were irradiated with 2—6 kR X-rays and grafted onto rootstock. Four of the 75 plants were observed to be mutants with altered growth, productivity, flowering time and all kinds of fruit characteristics. A dose of 2 krad seems closest to the optimum dose (Roby, 1972c).

Seedlings of the pecan nut tree, *Carya illinoensis*, family Juglandaceae, mainly grown in the southern U.S.A., were treated with the chemical mutagens N-nitroso-N-ethylurea and dimethyl sulphate to induce variability. Chlorophyll mutations and morphological changes were observed (Lebedinets, 1973).

Seeds of American chestnut, *Castanea dentata*, were irradiated by MacDonald et al. (1962) as reported by Jaynes (1975). As *C. dentata* was virtually exterminated by chestnut blight (Japanese chestnut, *C. crenata*, is resistant), the aim was to induce disease resistance. Since three generations are required to stabilize a mutation, following the method of Singleton (1969), it will take 15—20 years before it can be determined whether or not blight resistance has been induced by this irradiation.

Literature various other tree fruits

Bergh, B.O., 1975. Avocados. In: J. Janick and J.N. Moore (Editors), Advances in Fruit Breeding. Purdue University Press, West Lafayette, Ind., pp. 541—567.

Brooks, R.M. and Olmo, H.P., 1972. Register of New Fruit and Nut Varieties. University of California Press, Berkeley, Calif., 2nd edn.

Jaynes, R.A., 1975. Chestnuts. In: J. Janick and J.N. Moore (Editors), Advances in Fruit Breeding. Purdue University Press, West Lafayette, Ind., pp. 490—503.

Kester, D.E. and Asay, R., 1975. Almonds. In: J. Janick and J.N. Moore (Editors), Advances in Fruit Breeding. Purdue University Press, West Lafayette, Ind., pp. 387—419.

Lebedinets, L.N., 1973. Variability induced in seedlings of pecan (*Carya illinoensis* Wangh) in the M_0 by chemical mutagens. Isitol. Genet., 7(5): 467—468 (in Russian with English summary); Plant Breed. Abstr., 46: No. 1839.

MacDonald, R.D., Thor, E. and Andes, J.O., 1962. American chestnut breeding program at the University of Tennessee. Annu. Rep. No. Nut Grow. Assoc., 53: 19—21.

Roby, F., 1972c. Mutaciones en membrillero (*Cydonia oblonga* Mill.) producidas por Rayos X. Rev. Invest. Agropecu. Ser. 2, 9(1): 9—17 (with English summary).

Singleton, W.R., 1969. Mutations induced by treating maize seeds with thermal neutrons. In: Induced Mutations in Plants. IAEA, Vienna, pp. 479—483.

Storey, W.B., 1975. Figs. In: J. Janick and J.N. Moore (Editors), Advances in Fruit Breeding. Purdue University Press, West Lafayette, Ind., pp. 568—588.

ROOTSTOCKS

Rootstocks of several plants have been subjected to mutagenic treatments for various reasons.

Donini (1976b) and Roselli and Donini (1976) irradiated scions of the cherry rootstock F12/1 with 3.5 and 4.5 kR X-rays with the aim of obtaining dwarfing rootstock mutants. The frequency of morphological and chlorophyll mutants observed was approximately 4 and 5% after these doses, respectively.

Lacey (1975) reported about the irradiation of M9 apple rootstock scions hoping to find better rooting clones.

Przybyła and Zagaja (1974) irradiated root cuttings of various apple rootstocks and the cherry rootstock F12/1, and chemical mutagens were applied as well. Their main interest was to study whether mutants with larger mutated sectors or even completely mutated plantlets could be obtained. Their preliminary results indicated that this would be possible if the root cuttings were treated very early or stored at low temperatures to prevent tissue differentiation.

DeVay et al. (1965) irradiated seeds of several *Prunus* rootstocks, including *P. persica*, *P. amygdalus*, *P. mahaleb* and *P. cerasifera*, with thermal neutrons. No significant difference in frequency of resistance to crown gall,

Agrobacterium tumefaciens, or bacterial canker, *Pseudomonas syringae*, was found between the control and the irradiated population.

With regard to the interest in mutation induction in rootstocks and the problems connected with chimera formation, it could be of interest to develop in vitro propagation methods, such as those reported by Monsion and Dunez (1971).

Literature rootstocks

DeVay, J.E., Nyland, G., English, W.H., Schick, F.J. and Barbe, G.D., 1965. Effects of thermal neutron irradiation on the frequency of crown gall and bacterial canker resistance in seedlings of *Prunus* rootstocks. Radiat. Bot., 5(3): 197—204.

Donini, B., 1976b. The use of radiation to induce useful mutations in fruit trees. In: Improvement of Vegetatively Propagated Plants and Tree Crops through Induced Mutations, Wageningen, 1976. IAEA, Vienna, pp. 55—67.

Lacey, C.N.D., 1975. Induction and selection of mutant forms of fruit trees. Long Ashton Rep., 1975: 22—24.

Monsion, M. and Dunez, M.J., 1971. Obtention de jeunes plants de *Prunus mariana* à partir de boutures cultivées in vitro. C. R. Acad. Sci., 272: 1861—1864.

Przybyła, A. and Zagaja, S.W., 1974. Reaction of apple and cherry root cuttings to mutagen treatment. Fruit Sci. Rep., 1(1): 60—66.

Roselli, G. and Donini, B., 1976. Mutazioni radioindotte in ciliegio, pesco e olivo. In: Tecniche Nucleari per il Miglioramento e la Difesa dei Fruttifera I. Com. Naz. Energ. Nucl., Cent. Stud. Nucl., Casaccia, 1974, pp. 71—83 (with English summary).

Small fruits

In comparison to other crops, the application of artificial mutation induction does not play an important role in achieving breeding objectives in small fruit, as can be deduced from the relatively few literature references found. This situation probably exists because the high breeding potential in the genera has hardly been used.

STRAWBERRY

The breeding objectives in strawberry, *Fragaria* × *ananassa* Duch., family Rosaceae, are improvement of yield, vigour, fruiting habit, time of ripening, hardiness, disease resistance and many other characteristics. The breeder mainly tries to achieve these objectives by cross-breeding methods (Scott and Lawrence, 1975). Although spontaneous mutants are known (Staudt, 1959), very few breeders have taken the possibilities of mutation induction into consideration. De Ranieri and Scarascia Venezian (1964) and Scarascia Venezian and Esposita Seu (1965) compared the results of cross breeding and γ-irradiation. They reported higher yields in both groups following analysis of the offspring after sexual or vegetative propagation.

Abdullaev and Mekhtieva (1974) treated several cultivars with chemical mutagens and found mutations in various morphological characters and processing characteristics, a few of which were classified as promising.

Rybakov (1966) treated clones of strawberry with different γ-ray doses (up to 8 kR) and with ethylene imine. The sensitivity of the material depended on the conditions as well as on the cultivar used. He found mutations of leaf form, time of ripening and fruit quality. Since chimeras are formed, it required several cycles of vegetative propagations before the mutants' characters were stable.

The use of a fast in vitro propagation technique, as reported by Boxus (1974), Boxus et al. (1977) and Watkins (1975), may avoid some of the difficulties connected with chimera formation and thus lead to a larger role for mutation breeding in the future than it has today.

Literature strawberry

Abdullaev, I.K. and Mekhtieva, T.D., 1974. Useful mutant forms of strawberry obtained by treatment with chemical mutagens. Eksp. Mutagenez. Rast., 2: 147–148 (in Russian). Plant Breed. Abstr., 47: No. 1636.

Boxus, P., 1974. La micropropagation 'in vitro' du fraisier. Proc. Int. Hortic. Congr., 19(1A): 65.

Boxus, P., Quoirin, M. and Laine, J.M., 1977. Large scale propagation of strawberry plants from tissue culture. In: J. Reinert and Y.P.S. Bajaj (Editors), Plant Cell, Tissue and Organ Culture. Springer Verlag, Berlin, pp. 130–143.

De Ranieri, M. and Scarascia Venezian, M.E., 1964. Miglioramento genetico e colturale. Ital. Agric., 101 (Spec. No.): 773—775.

Rybakov, M.N., 1966. Variability of garden strawberry clones induced by gamma rays and ethylene imine. Timirjazevsk. S-kh. Akad., IXV(2): 36—44 (in Russian with English summary).

Scarascia Venezian, M.E. and Esposito Seu, M., 1965. Fioritura, epoca di fruttificazione e produzione di piante di fragole (varietà Elite Climax, Cambridge e Huxley) sottoposte ad irraggiamento cronico gamma. Agric. Ital., XI(7): 3—14.

Scott, D.H. and Lawrence, F.J., 1975. Strawberries. In: J. Janick and J.N. Moore (Editors), Advances in Fruit Breeding. Purdue University Press, West Lafayette, Ind., pp. 71—97.

Staudt, G., 1959. Eine spontan aufgetretene Grossmutation bei *Fragaria vesca* L. Naturwissenschaften, 46: 23—24.

Watkins, P.A.C., 1975. Strawberry meristem culture. John Innes Inst. Annu. Rep., 66: 18—19.

BRAMBLES

Brambles is the pomologist's name for members of the genus *Rubus*, family Rosaceae, and applies to raspberries, blackberries and dewberries.

Raspberry, subgenus *Idaeobatus*, includes a large number of diploid species (2n = 14) which produce edible fruits, such as the European red raspberry (*R. idaeus* L. var. *vulgatus*), black raspberry, thimbleberry, salmonberry and red raspberry (Ourecky, 1975).

Polyploidy, spontaneous or colchicine induced, is fairly common (Haskell, 1968; Kichina and Ogoltzova, 1973), while the occurrence of mutations seems rather rare. Ourecky (1975) reports an amber-fruited sport, cv. Golden Queen, of cv. Cathbert. Jennings (1961, 1966) describes a bud sport with larger fruit, believed to be based on a dominant gene; the sport has successfully been used in breeding.

The only report of the deliberate induction of mutants comes from Lacey (1975), who irradiated raspberry roots with γ-rays. The optimum dose was found to be 5—7 kR, but no further details were given.

Similarly, very few reports dealing with the occurrence of spontaneous mutants in blackberry were found. Thornless sports seem to be fairly common in blackberry and dewberry, several of which have been introduced as commercial cultivars (Darrow, 1928, 1929, 1931, 1955; Ourecky, 1975). A drawback to transferring the thornless character by cross breeding is that they are periclinal chimeras. Only L-I carries the mutation which is proven by the fact that neither seedlings nor plants grown from root cuttings are thorny. The artificial production of thornless mutants using an in vivo or in vitro vegetative propagation technique, or the in vitro production of plantlets from exclusively L-I tissue, and the subsequent production of solid, thornless mutants could solve this problem. Another possibility is to use radiation for tissue rearrangement.

The only in vitro propagation method of blackberry, *Rubus procerus* was described by Richardson and Amor (1975).

Literature brambles

Darrow, G.M., 1928. Notes on thornless blackberries. Their chromosome number and their breeding. J. Hered., 19(3): 139—142.
Darrow, G.M., 1929. Thornless sports of the young dewberry. J. Hered., 20(12): 567—569.
Darrow, G.M., 1931. A productive thornless sport of the evergreen blackberry. J. Hered., 22: 405—406.
Darrow, G.M., 1955. Nature of thornless blackberry sports. Fruit Var. Hortic. Dig., 10(1): 14—15.
Haskell, G., 1968. Biochemical differences between spontaneous and colchicine induced auto-tetraploids. Hereditas, 23: 139—141.
Jennings, D.L., 1961. Mutation for larger fruit in the raspberry. Nature (London), 191 (4785): 302—303.
Jennings, D.L., 1966. The manifold effects of genes affecting fruit size and vegetative growth in the raspberry. I. Gene L_1. New Phytol., 65: 176—187.
Kichina, V.V. and Ogoltzova, T.P., 1973. Spontaneous mutagenesis in red raspberry (*Rubus idaeus* L.). Genetika, 9(6): 165—167 (in Russian with English summary).
Lacey, C.N.D., 1975. Induction and selection of mutant forms of fruit trees. Long Ashton Annu. Rep., 1975: 22—24.
Ourecky, D.K., 1975. Brambles. In: J. Janick and J.N. Moore (Editors), Advances in Fruit Breeding. Purdue University Press, West Lafayette, Ind., pp. 98—129.
Richardson, R.G. and Amor, R.L., 1975. Effect of 2,4,5-T and picloram on the regeneration of blackberry (*Rubus procerus* P.J. Muell) from root segments. Weed Res., 15: 227—231.

BLUEBERRY AND CRANBERRY

The breeders of blueberry and cranberry, members of the genus *Vaccinium*, family Ericaceae, seem to have very little interest in the possibilities of artificial mutation induction as demonstrated by the almost complete absence of literature in this field (Galletta, 1975). Except for one general article on the use of irradiation (Boehm, 1955), the publications found are concerned with the phenomenon of chimerism, using polyploidy as a means to study histogenesis in the cranberry (Bain and Dermen, 1944; Dermen, 1947a).

It may be of interest for future work that cranberry is able to produce adventitious shoots, as was reported by Bain (1940).

Literature blueberry and cranberry

Bain, H.F., 1940. Origin of adventitious shoots in decapitated cranberry seedlings. Bot. Gaz. (Chicago), 101: 872—880.
Bain, H.F. and Dermen, 1944. Sectorial polyploidy and phyllotaxy in the cranberry (*Vaccinium macrocarpon* Ait.). Am. J. Bot., 31(9): 581—587.
Boehm, R.F., 1955. Atomic irradiation of blueberries. Annual Blueberry Open House. N. Jersey Agric. Exp. Stn, Rutgers Univ., New Brunswick, N.J., 23: 9—10.
Dermen, H., 1947a. Periclinal cytochimeras and histogenesis in cranberry. Am. J. Bot., 34: 32—43.

Galletta, G.J., 1975. Blueberries and cranberries. In: J. Janick and J.N. Moore (Editors), Advances in Fruit Breeding. Purdue University Press, West Lafayette, Ind., pp. 154–196.

CURRANTS

The currants and gooseberries are grouped in the genus *Ribes*, family Grossulariaceae, because of their crossing relationship. The genus includes 150 species, most of them native to temperate regions. Most species and commercial cultivars are diploid (2n = 16) (Keep, 1975).

Though rather rarely, spontaneous mutations have been reported: a mutant of red currant with early ripening by Knight and Keep (1958) and one of black currant with changed ripening time by Hughes (1963).

Colchicine has been widely used to induce tetraploidy, more specifically, to produce allotetraploids from wide species crosses. Mutagenic effects of colchicine have been observed as well, for example in red currants (Zeilinga, 1963).

Artificially induced mutations have been reported by several authors, such as Donini (1976c). The publications of a number of East European mutation breeders are, unfortunately, almost inaccessible because of language barriers (Boda, 1975; Melekhina, 1966; Ravkin, 1972, 1974), except for one by Privalov, who discussed the possibilities of mutation induction in woody plants, though he did so in general terms (Privalov, 1967).

Black currant became recognized as suitable material for mutation studies mainly through the more or less classical work of Bauer (1957). Bauer intensively studied the usefulness of artificial mutation induction in a vegetatively propagated species of black currant since, as he stated, black currant is very hard to improve by cross breeding. He irradiated several groups of cultivars with X-rays (optimum dose approximately 3 kR). The lower part of the cutting was protected by lead. The merit of his work was that he demonstrated very clearly that, by repeatedly cutting shoots back, the chimera problem can be solved. Numerous mutants were obtained. A distinction was made between large mutations (many characters changed), small mutations (few characters changed) and uncertain mutations (not clearly visibly different to the parent genotype, but characteristics, such as fruiting, may be favourably changed). Large mutations are of no value, but the two other categories may be useful, though most changes are either unfavourable themselves or accompanied by unfavourable changes, only approximately 1% being positive.

One mutant was eventually commercialized, namely cv. Westra, an X-ray induced mutant with erect growth (Table 28) (Bauer, 1974).

Gröber (1967) similarly irradiated thousands of cuttings with 2.5–3 kR X-rays and produced many mutants after repeatedly cutting shoots back to induce the basal buds to develop. Apart from mutations in leaf form and

TABLE 28

The commercial mutant of black currant

Name of new cultivar	Place and date of release (or approval) and name of principal worker(s) and institute	Conditions and date of mutagenic treatment [Parent cultivar]	Main improved attributes of cultivar
Westra	German Federal Republic, 1968 R. Bauer, Max Planck Institut für Züchtungsforschung, Köln-Vogelsang	1.5 kR X-rays, 1949 [Westwick Choice]	Strong erect habit

Sources: IAEA (1972, pp. 526—544) and *Mutation Breeding Newsletter*, 1 (1972)—9 (1977).

internode length, mutants with increased ascorbic acid content were obtained.

Other reports of mutation induction have been made by Kaplan (1953), who obtained variation in leaf characteristics, growth habit, yield, flavour etc., and Nybom and Bergendal, who used γ-irradiation as well as chemicals like ethyl methanesulphonate (EMS) and N-nitroso-N-methyl urethane (NMU). The results are described in the annual reports of the Balsgård Fruit Breeding Institute for the years 1959—1970. Mutations affecting several characteristics were obtained; however, none was outstanding.

Literature currants

Bauer, R., 1957. The induction of vegetative mutations in *Ribes nigrum*. Hereditas, 43: 323—337.

Bauer, R., 1974. Westra, an X-ray induced erect-growing black currant variety, and its use in breeding. In: Polyploidy and Induced Mutations in Plant Breeding, Bari, 1972. IAEA, Vienna, pp. 13—20.

Boda, J., 1975. Effect of gamma irradiation on red currant. Kertgazd, 7(2): 27—36 (in Hungarian with English summary).

Donini, B., 1976c. Breeding methods and applied mutagenesis in fruit plants. In: Proc. Workshop Eur. Comm. — Israel, The Use of Ionizing Radiation, Wageningen. Assoc. Euratom—ITAL, Wageningen, pp. 445—478.

Gröber, K., 1967. Some results of mutation experiments in apples and black currant. In: H. Stubbe (Editor), Induzierte Mutationen und ihre Nützung. Erwin Baur Gedächtnisvorlesungen 1966, IV. Akademie-Verlag, Berlin, pp. 377—382.

Hughes, H.M., 1963. A study of two blackcurrant chimaeras. J. Hortic. Sci., 38: 286—296.

IAEA, 1972. Induced Mutations and Plant Improvement. IAEA, Vienna, pp. 526—544.

Kaplan, R.W., 1953. Über Möglichkeiten der Mutationsauslösung in der Pflanzenzüchtung. Z. Pflanzenzücht., 32(2): 121—131.

Keep, E., 1975. Currants and Gooseberries. In: J. Janick and J.N. Moore (Editors), Advances in Fruit Breeding. Purdue University Press, West Lafayette, Ind., pp. 197—268.

Knight, R.L. and Keep, E., 1958. An early sport of a red currant. Annu. Rep. 1957, East Malling Res. Stn, p. 74.

Melekhina, A.A., 1966. Variability of black currant under the influence of ionizing irradiation (in Russian); Nucl. Sci. Abstr., 21(15) (1967): No. 26669.

Privalov, G.F., 1967. Experimental mutations in woody plants. In: H. Stubbe (Editor), Induced Mutations and their Utilization. Erwin Baur Memorial Lectures, 1966, IV. Akademie-Verlag, Berlin, pp. 383—386.

Ravkin, A.S., 1972. The frequency of somatic mutations in black currant after induced mutagenesis. Dokl. TSKh A, 186: 175—180 (in Russian).

Ravkin, A.S., 1974. Reaction of black currant cultivars to different ways of irradiation in the gamma-field. Biofiz. Fiziol. Biokhim. Issled. Plodovykh lagodnykh Kult., 1974: 191—198 (in Russian).

Zeilinga, A.E., 1963. Toepassing van colchicine ter verkrijging van polyploïde vormen van de rode bes. In: Jaarverslag Tuinbouwkundig Onderzoek. Minist. Agric., The Hague, p. 68.

Grapes

Grapes, *Vitis vinifera* L. and other genera of the family Vitaceae, are widely distributed in the tropics and subtropics with ranges extending into the temperate regions. Most of the breeding has been of *Vitis vinifera*, but other sources of germ plasm are used as well to incorporate resistance to pests and diseases in *V. vinifera* (Einset and Pratt, 1975).

According to Einset and Pratt, the genus *Vitis* is divided into two subgenera, *Muscadinia* Planch. (diploid; 2n = 40) and *Euvitis* Planch. (diploid; 2n = 38), but other classifications have also been presented (Olmo, 1976).

Spontaneous mutations are frequently observed, especially in old cultivars (Breider, 1964), and are an important source of variation in grapes. They involve all kinds of characteristics, including leaf colour (Boubals, 1976), yield (e.g. Woodham and Alexander, 1966), seedlessness (e.g. Nitsch et al., 1960; Olmo, 1940) and change in the colour of berries (Breider, 1953). Some mutants have been of economic importance and released as a new cultivar after selection and testing (e.g. Olmo, 1964). (For more information on the subject the reader is referred to Breider (1964) and Einset and Pratt, 1975, p. 140.)

Mutations have been induced by many investigators (Breider, 1952, 1953, 1956, 1959, 1964; Das and Mukherjee, 1968; Donini, 1975, 1976a,b,c; Fry, 1963; Golodriga and Kireeva, 1974, 1975; Milosavljević and Mijajlović, 1965; Olmo, 1960; Pratt, 1959; Reichardt, 1955; Romisondo et al., 1976; Sharma and Mukherjee, 1972; Shimotsuma, 1962). The mutations reported affected all kinds of characteristics, including yield, earliness, size, number and colour of berries, hardiness, resistance to diseases (e.g. downy mildew, *Plasmopara viticola* (Golodriga and Kireeva, 1975), and form and size of leaves.

Generally, physical mutagens were used, either X-rays, ^{60}Co γ-rays or thermal neutrons. The optimum dose for X-rays or γ-rays is approximately 2—6 kR if dormant (rooted) cuttings are irradiated (Breider, 1964; Das and Mukherjee, 1968; Donini, 1976a,b,c; Fry, 1963; Reichardt, 1955), although lower optimum doses have also been reported, for instance 1—2 kR (Shimotsuma, 1962). For neutrons, an optimum irradiation time is hard to "translate" because the physical characteristics of the reactors used varied.

Several authors have used chemical mutagens, for example, ethyl methanesulphonate (EMS) (Das and Mukherjee, 1968), diazomethane (DAB) and EMS (Golodriga and Kireeva (1975) who obtained resistance to downy mildew) and N-nitroso-N-methyl urethane (NMU) (Sharma and Mukherjee (1972) who induced chlorophyll and other mutations affecting leaf characteristics).

Chimeras are usually formed in grapes as in all vegetatively propagated plants if multi-cellular apices have been treated with physical or chemical mutagens. Spontaneous tetraploids or mixoploids (cyto-chimeras) have been used to study the phenomenon of chimerism, for example, by Einset and

Lamb (1951), Einset and Pratt (1954) and Thompson and Olmo (1963). The latter carried out cyto-histological studies of 62 tetraploid or mixoploid clones of grape and described the characteristics of 4n grapes (larger berry size, earlier ripening but generally having a poor growth habit, irregular setting of fruit and reduced yield) (Olmo, 1935, 1942). Colchicine and radiation-induced tetraploids (mixoploids) were used as well, to study chimeras or to produce tetraploids for the production of fertile tetraploid hybrids or to ensure fertility in tetraploids of sterile diploid hybrids (Abdullaev et al., 1974; Dermen, 1954b; Fry, 1963). In addition, chimerism has been studied by the use of radiation-induced mutations (Breider, 1953, 1956, 1964; Reichardt, 1955).

To avoid or restrict the consequences of chimerism and to arrive at stable periclinal chimeras or solid mutants as quickly as possible the post-radiation treatment of material is of great importance. Donini (1975, 1976a,b,c) and Romisondo et al. (1976) irradiated scions of six cultivars with 3—4 kR γ-rays or with thermal neutrons. The main buds of the scions were grafted onto rootstocks and the vM_1 shoots were either pruned during growth (axillary buds grow out) or cut back in winter time (only two buds were allowed to develop). In another experiment, irradiated scions were rooted and cut back

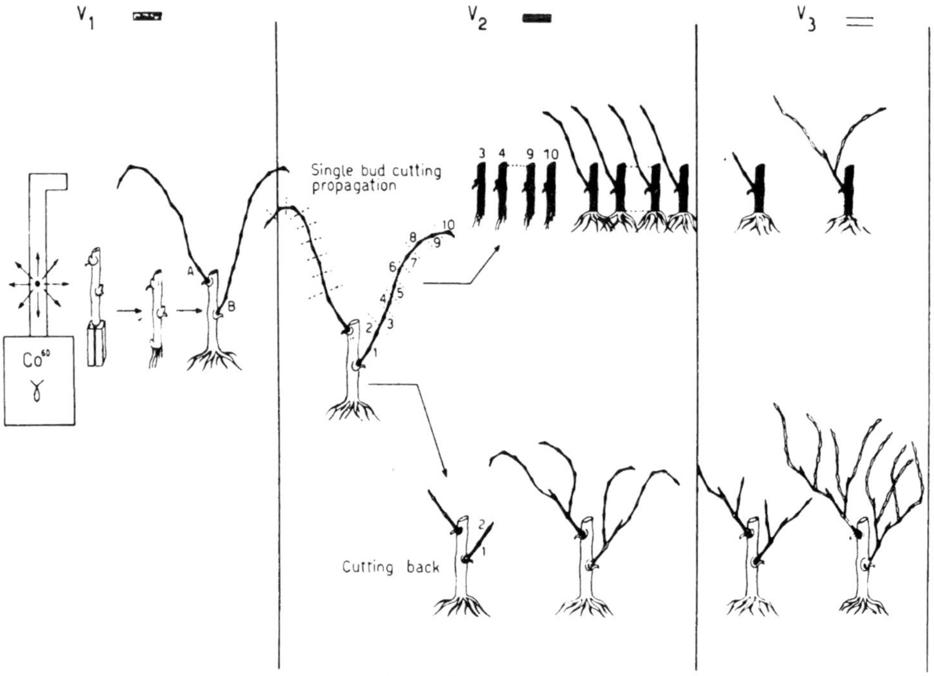

Fig. 26. Methodology scheme used for isolation of somatic mutations in grapes. (Donini, 1975, Fig. 3.)

at the end of vM_1 growth (two basal buds were allowed to develop and the other buds were singly propagated) (Fig. 26). Chlorophyll mutations as well as morphological mutations affecting leaf, shoot and fruit characteristics, were studied in vM_2 and vM_3. Mutants that were interesting as regards their fruit ripening period or the self-thinning character were propagated for practical evaluation. No definite conclusions as to the optimum treatment are available at present.

Despite the numerous investigations concerned with the improvement of grapes by artificial mutation induction, very few commercial results have been reported. The only author to mention a mutant being used in practice is Breider (1964), who described a radiation-induced mutant of cv. Perle, cultivated on a small scale in Germany as it tolerates lower temperatures than most grape cultivars (no further details are available).

Literature grapes

Abdullaev, I.K., Mekhtieva, T.D., Pirieva, G.G. and Akhmedova, S.M., 1974. Promising mutant forms of grape obtained from the variety Marandi. Eksp. Mutagenez Rast., 2: 126—129 (in Russian). Plant Breed. Abstr., 47: No. 1661.

Boubals, D., 1976. Une mutation génétique de l'Alicante Bouschet. Perte du caractère teinturier chez la baie. Prog. Agric. Vitic., 93(3): 79.

Breider, H., 1952. Röntgenstrahlen im Weinbau. Umsch. Wiss. Tech., 52: 715—717.

Breider, H., 1953. Entwicklungsgeschichtlich-genetische Studien über somatische Mutationen bei der Rebe. Züchter, 23: 208—222 (41 references).

Breider, H., 1956. Über strahleninduzierte somatische Mutationen bei Reben. Bayer. Landwirtsch. Jahrb., 33: 515—533.

Breider, H., 1959. Röntgeninduzierte Mutationen bei *Vitis, Malus* und *Cyclamen*. Bayer Landwirtsch. Jaarb., 36: 396—401.

Breider, H., 1964. Über die züchterische Auswertung und über die praktische Verwertung röntgeninduzierter somatische Mutationen bei langlebigen und vegetativ vermehrbaren Kulturpflanzen. Mitt. Klosterneuburg, Ser. A, 14: 165—171.

Das, P.K. and Mukherjee, S.K., 1968. Effect of gamma radiation and ethyl methane sulphonate on seeds, cuttings and pollen in grapes. Indian J. Genet. Plant Breed., 28(3): 347—351.

Dermen, H., 1954b. Colchiploidy in grapes. J. Hered., 45(4): 159—172.

Donini, B., 1975. Induction and isolation of somatic mutations in vegetatively propagated plants. In: Improvement of Vegetatively Propagated Plants through Induced Mutations, Tokai, 1974. IAEA, Vienna, pp. 35—51.

Donini, B., 1976a. Use of radiations to induce useful mutations in fruit trees. Mutat. Breed. Newsl., 8: 7—8.

Donini, B., 1976b. The use of radiations to induce useful mutations in fruit trees. In: Improvement of Vegetatively Propagated Plants and Tree Crops through Induced Mutations, Wageningen. IAEA, Vienna, pp. 55—67.

Donini, B., 1976c. Breeding methods and applied mutagenesis in fruit plants. In: Proc. Workshop Eur. Comm. — Israel, The Use of Ionizing Radiation, Wageningen. Assoc. Euratom—ITAL, Wageningen, pp. 445—478.

Einset, J. and Lamb, B., 1951. Chimeral sports of grapes. Alleged tetraploid varieties have diploid "skin". J. Hered., 42(4): 158—162.

Einset, J. and Pratt, C., 1954. "Giant" sports of grapes. Proc. Am. Soc. Hortic. Sci., 63: 251—256.

Einset, J. and Pratt, C., 1975. Grapes. In: J. Janick and J.N. Moore (Editors), Advances in Fruit Breeding. Purdue University Press, West Lafayette, Ind., pp. 130—153.
Fry, B.O., 1963. Production of tetraploid muscadine (*Vitis rotundifolia*) grapes by gamma radiation. Proc. Am. Soc. Hortic. Sci., 83: 388—394.
Golodriga, P.Y. and Kireeva, L.K., 1974. Induced mutations in grapes using physical mutagenic factors. Biofiz. Rast., 10—11 (in Russian); Plant Breed. Abstr., 46(12): No. 11493.
Golodriga, P.Y. and Kireeva, L.K., 1975. Using methods of induced mutagenesis in breeding grape. Tr. Prikl. Bot., Genet. Sel. Ser. A, 54(2): 142—150 (in Russian with English summary). Plant Breed. Abstr., 46: No. 4724.
Milosavljević, M. and Mijajlović, R., 1965. Untersuchungen über die Strahlungsempfindlichkeit von Rebenknospen. Vitis, 5: 88—93.
Nitsch, J.P., Pratt, C., Nitsch, C. and Shaulis, N.J., 1960. Natural growth substances in Concord and Concord Seedless grapes in relation to berry development. Am. J. Bot., 47: 566—576.
Olmo, H.P., 1935. Bud mutation in the vinefera grape II. *Sultania gigas*. Proc. Am. Soc. Hortic. Sci., 33: 437—439.
Olmo, H.P., 1940. Somatic mutation in the vinifera grape III. The Seedless Emperor. J. Hered., 31: 211—213.
Olmo, H.P., 1942. Breeding new tetraploid grape varieties. Proc. Am. Soc. Hortic. Sci., 41: 225—227.
Olmo, H.P., 1960. Plant breeding program aided by radiation treatment. Calif. Agric., 14(7): 4.
Olmo, H.P., 1964. Improvement in grape varieties. Wines Vines, 45(2): 23, 25.
Olmo, H.P., 1976. Grapes. *Vitis, Muscadinia* (Vitaceae). In: N.W. Simmonds (Editor), Evolution of Crop Plants. Longman, London, pp. 294—298.
Pratt, C., 1959. Radiation damage in shoot apices of Concord grape. Am. J. Bot., 46: 103—109.
Reichardt, A., 1955. Experimentelle Untersuchungen über den Effekt von Röntgenstrahlen in der vegetativen Vermehrung einer alten Rebensorten. Gartenbauwissenschaft, 2: 355—413.
Romisondo, P., Donini, B. and Me, G., 1976. Induzione di mutazioni gemmarie nella vite mediante radiazoni ionizzanti. In: Tecniche Nucleari per il Miglioramento e la Difesa dei Fruttifera, 1. Com. Naz. Energ. Nucl., Cent. Stud. Nucl. Cassacia, 1974, pp. 85—94.
Sharma, R.L. and Mukherjee, S.K., 1972. Morphological descriptions of some induced systematic mutants of grapes (*Vitis vinifera* L.). Vitis, 11(3): 177—188.
Shimotsuma, M., 1962. Irradiation experiments with grapes. Seiken Jihô, 14: 102—103.
Thompson, M.M. and Olmo, H.P., 1963. Cytohistological studies of cytochimeric and tetraploid grapes. Am. J. Bot., 50(9): 901—906.
Woodham, R.C. and Alexander, D.M., 1966. Reproducible differences in yield between sultana vines. Vitis, 5: 257—264.

BANANA

The importance of banana cultivation for tropical countries is sufficiently demonstrated by some economic data. World production of bananas (*Musa sapientum* L., fam. Musaceae), together with plantains (*Musa paradisiaca* L.) is estimated at over 20×10^6 metric tons. Africa, with the bulk of its production in East Africa, contributes about 50%. Some 5×10^6 metric tons are traded on the world market, while the remainder, a great deal of it being plantains, is used for local consumption. Bananas are quite large herbaceous perennials. Their aerial shoots develop from lateral buds on short underground rhizomes. Cultivated bananas are propagated vegetatively by peepers, suckers or corms. The majority of fruits, which develop parthenocarpically, are sterile (i.e. seedless).

Approximately 200—300 cultivars (clones) exist but only very few, for example, cultivars Gros Michel, Lacatan and Dwarf Cavendish, are important for the international market. All significant cultivars are triploid and this makes breeding by conventional methods difficult, apart from some other intricate problems.

The agricultural importance of somatic mutations in bananas is very great. Darwin mentioned a case of "bud sport variation" in banana, and Simmonds (1964) provides a list of 18 different somatic mutations involving stature, habit and bunch or fruit characteristics. Simmonds also states that many less obvious mutants have not been recognized. Since then, many more examples of other spontaneous mutations have been published.

Although the use of induced mutations in banana breeding has been suggested on several occasions (Champion, 1963; De Langhe, 1969) there are only a few publications on the use of mutagenic agents for the improvement of certain genetic characters. Preliminary studies were performed, for example, by Stotzky et al. (1964), Moh and Alan (1965) and Azzam and Linden (1965). Stotzky et al. (1964) reported the effects of γ-irradiation on seeds of the wild, diploid *Musa balbisiana* and rhizomes of *Musa sapientum* cultivar Gros Michel. Some figures concerning reduction of survival etc. were given. Of course, true seeds and rhizomes do not have comparable reactions to mutagenic treatments. Moreover, both species mentioned have a different ploidy level.

In an annual report, Moh and Alan (1965) briefly outlined the behaviour of banana plants in a γ-field and studies of the radiosensitivity of this crop. Azzam and Linden (1965) performed some preliminary investigations with cultivar Gros Michel in Puerto Rico. After the death of Azzam, the work was continued by Velez Fortuño and Cedeño Maldonado (1972). The cultivar Gros Michel is a banana with outstanding fruit quality but, owing to its high susceptibility to *Fusarium oxysporum cubensis* and *Cercospora musae*, it was rapidly replaced by less susceptible Cavendish types. However, from the commercial point of view, a resistant clone of the more attractive 'Gros Michel' would be highly desirable. For this reason, rhizomes of cultivar Gros Michel were irradiated with 2.5—40 kR of γ-rays. There was no survival after doses higher than 5 kR. Two mutants, one with drastic leaf aberrations and the other with more intense pigmentation and other smaller morphological changes, were selected in vM_2 for further testing. The results of the tests are not yet known. Because of the economic significance of 'Gros Michel' and the great difficulty experienced in attempts to obtain good genotypes which combine fruit quality similar to that of 'Gros Michel' and the desired disease resistance, this work is important and deserves to be continued.

Contributions have recently been made by Panton and Menendez (1972) and Menendez (1973). They applied ethyl methanesulphonate (EMS) to seeds of a diploid breeding line of *Musa acuminata* in Jamaica in an attempt to obtain plants with decreased plant height. Soaking seeds in a 0.1 M EMS solution for 48 h at a temperature of 20°C reduced germination by 50%. In addition, the authors expressed the opinion that induction of resistance to several diseases using mutation methods appears to be quite feasible. They also suggested the use of adventitious buds obtained from callus as starting material. This approach has also been suggested by several later authors and, indeed, might well be a key to efficient mutation procedures in banana (De Guzman, 1975; De Guzman et al., 1976; Hamilton, 1965; Ma and Shii, 1972, 1974).

Literature banana

Azzam, H. and Linden, D.B., 1965. Radiation effects on banana corms, *Musa sapientum*. J. Agric. Univ. P. R., 4: 270—271.

Champion, J., 1963. Le Bananier. Ed. G.P. Maisonneuve et Larouse, Paris, 23 pp.

De Guzman, E.V., 1975. Project on production of mutants by irradiation of in vitro cultured tissues of coconut and banana and their mass propagation by the tissue culture technique. In: Improvement of Vegetatively Propagated Plants through Induced Mutations, Tokai, 1974. IAEA, Vienna, pp. 53—76.

De Guzman, E.V., Ubalde, E.M. and Del Rosario, A.G., 1976. Banana and coconut in vitro cultures for induced mutations study. In: Improvement of Vegetatively Propagated Plants and Tree Crops through Induced Mutations, Wageningen, 1976. IAEA, Vienna, pp. 33—54.

De Langhe, E., 1969. Bananas, In: F.P. Ferwerda and F. Wit (Editors), Outlines of Perennial Crop Breeding in the Tropics. Misc. Pap. 4. Landbouwhogeschool, Veenman, Wageningen, pp. 53—78.

Hamilton, K.S., 1965. Reproduction of banana from adventitious buds. Trop. Agric. (Trinidad), 40(1): 69—73.

Ma, S. and Shii, C., 1972. In vitro formation of adventitious buds in banana shoot apex following decapitation. J. Hortic. Soc. China, 18(3): 1—8.

Ma, S. and Shii, C., 1974. Growing banana plantlets from adventitious buds. J. Hortic. Soc. China, 20(1): 1—7.

Menendez, T., 1973. Application of mutation methods to banana breeding. In: Induced

Mutations in Vegetatively Propagated Plants. IAEA, Vienna, pp. 75—83.
Moh, C.C. and Alan, J.J., 1965. Radiosensitivity of tropical plants. In: The Application of Nuclear Energy. Annu. Rep., Turrialba, Costa Rica, pp. 33—36.
Simmonds, N.W., 1964. Bananas. Longmans, London, 3rd edn, 466 pp.
Stotzky, G., Cox, E.A., Goos, R.D., Wornick, R.C. and Badger, A.M., 1964. Some effects of gamma irradiation of seeds and rhizomes of *Musa*. Am. J. Bot., 51(7): 724—729.
Panton, C.A. and Menendez, T., 1972. Possibilities and implications of mutation breeding in Jamaica. In: Induced Mutations and Plant Improvement. IAEA, Vienna, pp. 61—66.
Velez Fortuño, J. and Cedeño Maldonado, A., 1972. The use of radiation in breeding banana (*Musa sapientum* L.). In: Induced Mutations and Plant Improvement, FAO/IAEA, Buenos Aires, 1970. IAEA, Vienna, pp. 485—489 (in Spanish).

CITRUS

Citrus, family Rutaceae, is grown throughout the world in tropical and subtropical climates and is economically a very important crop with large-scale production of both the fresh fruit and processed products.

The taxonomy of the genus *Citrus* seems to be a matter of contention. It consists of several tens of species, including the sour orange (*C. aurantium*), the sweet orange (*C. sinensis*), the limes (*C. aurantifolia*), lemons, citrons, grapefruits, mandarins, pummelos and kumquats as well as hybrids between these and with related species like *Poncirus* (Soost and Cameron, 1975).

Breeding objectives are manifold and vary from improved tree vigour and longevity, reduced tree size, better tolerance of low temperatures, improvement of fruit size, fruit quality, colour of the rind and ease of peeling, seedlessness, and season of ripening, to disease and pest resistance.

Somatic mutations have played an important role in the development of new cultivars, partly because apomixis is common in most *Citrus* species through the formation of nucellar embryos which derive from somatic cells of the nucellus. Seedlings may be of zygotic or non-zygotic (nucellar) origin. The percentage of seedlings that are of non-zygotic origin varies greatly, often being high to very high and very rarely zero (Soost and Cameron, 1975, Table 2, p. 522).

Spontaneous mutations have been reported in various *Citrus* fruits (e.g. oranges, lemons, mandarins) by, for example, Anonymous (1968; variegated leaves, variation in size, shape and colour of fruit, and various other abnormalities), Kerkadze (1974; high percentages in old trees of Unshiu mandarin and Orange Washington Navel), Olson et al. (1966; the grapefruit Burgundy sport), Shamel (1943; somatic mutations in commercial citrus cultivars, mostly unfavourable, though several had useful characters) and Tanaka (1932). Tanaka wrote a monograph on the Satsuma orange and paid special attention to the occurrence of new varieties from bud mutations. Soost and Cameron (1965) and Nishiura (1965, 1968) listed a long series of sports in many citrus fruits and the latter even stated "that most of the present commercial varieties cultivated are said to have arisen through some kind of spontaneous mutation".

Many of the sports are periclinal chimeras, resulting from unstable mericlinal chimeras (Cameron et al., 1964). Cyto-chimeras are also known to occur. Frost and Krug (1942) vegetatively propagated a bud variant of a hybrid mandarin and obtained different cyto-chimeras. In fact, chimerism is *the* problem in mutation breeding of *Citrus*, as it is in most vegetatively propagated plants and, since the interest in commercial mutation induction has grown considerably recently, it is not surprising that several breeders attempt to develop and use methods which avoid or limit the unfavourable consequences of chimera formation.

To limit chimerism, nucellar material is irradiated at a very early developmental (proembryonic) stage or nucellar seedlings are produced which are decapitated immediately after or before irradiation. Ikeda (1976b) irradiated floral buds at three different stages of development with 2 or 4 kR γ-rays. The zygotic seedlings could easily be recognized and discarded since the Sunki mandarin plants were pollinated with *Poncirus trifoliata* pollen, which carries a dominant gene for the *trifoliata* character. Solid mutants as well as chimeras were observed and the author thus concluded that the nucellar embryo develops from more than one cell. Decapitated nucellar seedlings, cut off below the first leaf, produced new shoots on the cut surface approximately 2 months after decapitation. This method, however, produced chimeras and solid mutants as well. The final conclusion was that the irradiation of young flower buds with 2 krad γ-rays is preferable.

Kim et al. (1972) used similar techniques and obtained four mutants selected for tolerance to low temperatures, compact habit and small leaves. The same technique was mentioned by Spiegel-Roy and Kochba (1973, 1975) and Spiegel-Roy and Padova (1973), who produced chlorophyll mutants using the method. Whether these were chimeral or solid was not clear (Spiegel-Roy and Kochba, 1976).

Entire leaves or leaf fragments of various citrus species were successfully rooted and, in the case of lemon, shoots were produced 7 months after rooting. Whether this method could be used to produce solid mutants has not yet been studied (Salomon and Mendel, 1965).

Much attention is paid to the use of in vitro techniques since they can be used to propagate several *Citrus* species (Button and Kochba, 1977; Chaturvedi and Mitra, 1974; Grinblat, 1972; Murashige and Tucker, 1969; Rangan et al., 1968). For mutation breeding purposes, callus from unfertilized ovules, irradiated with 12—16 kR γ-rays, is generally used (Kochba and Spiegel-Roy, 1973, 1976; Mitra and Chaturvedi, 1972). Most studies have been of the Shamouti orange from which established plants could be obtained using callus from ovules and the respective formation of embryoids, shoots and plantlets (Spiegel-Roy and Kochba, 1975). This procedure is also possible in grapefruit, *C. paradisi* cv. Marsh Seedless (Kochba et al., 1972).

Scions have also been used, for example, to produce mutants of sweet orange more tolerant to low temperatures (Anonymous, 1970c; Gregory,

TABLE 29

The commercial mutant of grapefruit

Name of new cultivar	Place and date of release (or approval) and name of principal worker(s) and institute	Conditions and date of mutagenic treatment [Parent cultivar]	Main improved attributes of cultivar
Star Ruby	U.S.A.(Patent applied for) 1970 R.A. Hensz, Texas A and I University, Citrus Centre, Weslaco, Texas	Seeds, thermal neutrons, 1959 [Hudson]	Red flesh like parent variety, but almost seedless (0—9 seeds instead of 40—60)

Sources: IAEA (1972, pp. 526—544) and *Mutation Breeding Newsletter*, 1 (1972)—9 (1977).

1972; Gregory and Gregory, 1965). X-irradiated scions were tested outdoors for three winters (in Raleigh, N.C.) after successive graftings onto the winter-hardy *Poncirus* rootstock. Of over 1 000 initially irradiated buds and approximately 4 000 secondary buds, 1% survived and replicates showed a significant difference in resistance to cold from the original clone.

Attempts to induce virus disease resistance in *Citrus* species are made as well (Ryan, 1976; preliminary report). Kerkadze (1970) described different types of mutants induced by various agents.

Citrus is an instructive example of the various radiosensitivities of the various starting materials used for irradiation. The LD_{50} (approximately the optimum dose) of *Citrus sinensis* (Shamouti orange) for (nucellar) seeds is close to 10 krad (Spiegel-Roy and Padova, 1973); for bud wood, approximately 4—8 krad is usually used (Spiegel-Roy and Kochba, 1975; Spiegel-Roy and Padova, 1973), although 10 krad has also been reported to be the optimum (Tachikawa et al., 1962). In vitro material is much less sensitive and the best dose seems to be around 20 krad, whereas decapitated young seedlings are very sensitive, having an optimum dose of 2—4 krad (Spiegel-Roy and Kochba, 1973, Table VIII).

Most mutation breeding projects have only recently begun and have not yet produced useful mutants. Only one commercial mutant has been reported: Hensz (1960) found a grapefruit with few seeds and red flesh among irradiated seedlings (Table 29). In particular, the development of new seedless cultivars, as well as better coloured varieties and some with different ripening times are of interest to *Citrus* breeders (Spiegel-Roy and Kochba, 1976).

Literature Citrus

Anonymous, 1968. Mutations or sports in citrus. Agric. Gaz. N.S.W., 79: 204—208.

Anonymous, 1970c. Northward Migration of Citrus by Means of Radiation-induced Mutation. Progr. Rep., N.C. State Univ., Raleigh, N.C., 20 pp.

Button, J. and Kochba, J., 1977. Tissue culture in the citrus industry. In: J. Reinert and Y.P.S. Bajaj (Editors), Plant Cell, Tissue, and Organ Culture. Springer Verlag, Berlin, pp. 70—92.

Cameron, J.W., Soost, R.K. and Olson, E.O., 1964. Chimeral basis for color in pink and red grapefruit. J. Hered., 55(1): 23—28.

Chaturvedi, H.C. and Mitra, G.C., 1974. Clonal propagation of *Citrus* from somatic callus cultures. HortScience, 9(2): 118—120.

Frost, H.B. and Krug, C.A., 1942. Diploid-tetraploid periclinal chimeras as bud variants in citrus. Genetics, 27: 619—634.

Gregory, W.C., 1972. A test of the mutation breeding hypothesis with citrus. In: M.J. Constantin (Editor), Mutat. Breed. Workshop, Knoxville, Tenn. University of Tennessee, Knoxville, Tenn.

Gregory, W.C. and Gregory, M.P., 1965. Induced mutations in quantitative characters: experimental basis for mutations to hardiness in citrus. Proc. Soil Crop Sci. Soc. Fla., 25: 372—396.

Grinblat, U., 1972. Differentiation of citrus stem in vitro. J. Am. Soc. Hortic. Sci., 97(5): 599—603.
Hensz, R.A., 1960. Effect of X-ray and thermal neutrons on *Citrus* propagating material. J. Rio Grande Val. Hortic. Soc., 14: 21—25.
IAEA, 1972. Induced Mutations and Plant Improvement. IAEA, Vienna, pp. 526—544.
Ikeda, F., 1976b. Induced bud sports in *Citrus sunki* from nucellar seedlings. In: Improvement of Vegetatively Propagated Plants and Tree Crops through Induced Mutations, Wageningen, 1976. IAEA, Vienna, pp. 95—96.
Kerkadze, I.G., 1970. Type of obtained mutants in *Citrus* by induced mutagenesis. Genetika, 6: 26—32 (in Russian with English summary).
Kerkadze, I.G., 1974. Mutation and somatic drift in *Citrus*. Subtrop. Kult., 3: 36—40 (in Russian); Plant Breed. Abstr., 46: No. 10480.
Kim, C.M., Kim, J.K., Kim, H.W. and Moon, J.D., 1972. Studies on the induction and selection of bud sport by chronic gamma irradiation on *Citrus* tree. Korean J. Breed., 4(2): 132—140 (in Korean). Plant Breed. Abstr., 44(8): No. 5430 (August, 1974).
Kochba, J. and Spiegel-Roy, P., 1973. Effect of culture media on embryoid formation from ovular callus of 'Shamouti' orange (*Citrus sinensis*). Z. Pflanzenzücht., 69: 156—162.
Kochba, J. and Spiegel-Roy, P., 1976. The use of *Citrus* tissue culture for mutation breeding: Effects of plant growth substances and gamma irradiation on embryogenesis. In: Improvement of Vegetatively Propagated Plants and Tree Crops through Induced Mutations, Wageningen, 1976. IAEA, Vienna, pp. 83—92.
Kochba, J., Spiegel-Roy, P. and Safran, H., 1972. Adventive plants from ovules and nucelli in *Citrus*. Planta, 106: 237—245.
Mitra, G.C. and Chaturvedi, H.C., 1972. Embryoids and complete plants from unpollinated ovaries and from ovules of in vivo-grown emasculated flower buds in *Citrus* spp. Bull. Torrey Bot. Club, 99(4): 184—189.
Murashige, T. and Tucker, D.P.H., 1969. Growth factor requirement of *Citrus* tissue culture. In: Proc. First Int. Citrus Symp., Vol. 3. pp. 1155—1161.
Nishiura, M., 1965. Natural mutation and its utilization in the selection of citrus fruits. Gamma Field Symp., 4: 27—42.
Nishiura, M., 1968. Mutation in citrus. Jpn. Agric. Res. Q., 3(2): 10—14.
Olson, E.O., Cameron, J.W. and Soost, R.K., 1966. The Burgundy sport: further evidence of the chimeral nature of pigmented grapefruits. HortScience, 1(2): 57—58.
Rangan, T.S., Murashige, T. and Bitters, W.P., 1968. In vitro initiation of nucellar embryos in monoembryonic *Citrus*. HortScience, 3: 226—227.
Ryan, G.S., 1976. Attempts to induce mutations for virus disease resistance in citrus. In: Improvement of Vegetatively Propagated Plants and Tree Crops through Induced Mutations, Wageningen. IAEA, Vienna, pp. 149—150.
Salomon, E. and Mendel, K., 1965. Rooting of citrus leaf cuttings. Am. Soc. Hortic. Sci., 86: 213—219.
Shamel, A.D., 1943. Bud variation and bud selection. In: H.J. Webber and L.D. Bachelor (Editors), The *Citrus* Industry, Vol. 1. University California Press, Berkeley, Calif., pp. 915—952.
Soost, R.K. and Cameron, J.W., 1975. Citrus. In: J. Janick and J.N. Moore (Editors), Advances in Fruit Breeding. Purdue University Press, West Lafayette, Ind., pp. 507—540.
Spiegel-Roy, P. and Kochba, J., 1973. Mutation breeding in citrus. In: Induced Mutations in Vegetatively Propagated Plants. IAEA, Vienna, pp. 91—105.
Spiegel-Roy, P. and Kochba, J., 1975. Production of solid mutants in citrus, utilizing new approaches and techniques. In: Improvement of Vegetatively Propagated Plants through Induced Mutations, Tokai, 1974. IAEA, Vienna, pp. 113—127.
Spiegel-Roy, P. and Kochba, J., 1976. Use of ionizing radiation for mutagenesis in tissue

culture. In: Proc. Workshop Europ. Comm. — Israel, The Use of Ionizing Radiation, Wageningen, 1976. Assoc. Euratom—ITAL, Wageningen, pp. 434—444.

Spiegel-Roy, P. and Padova, R., 1973. Radiosensitivity of Shamouti orange (*Citrus sinensis*) seeds and buds. Radiat. Bot., 13: 105—110.

Tachikawa, T., Hara, S. and Taniguchi, T., 1962. Radiation breeding in *Citrus*. Seiken Jihô, 11: 103—106.

Tanaka, T., 1932. A monograph of the Satsuma orange, with special reference to the occurrence of new varieties through bud variation. Mem. Taihoku Imp. Univ. Fac. Sci. Agric., 4: 1—626.

MANGO

Mango (*Mangifera indica* L., fam. Anacardiaceae) is a very important fruit all over the tropics but especially in India, where 60% of the total fruit growing area is devoted to mango with a yearly production of some 5×10^6 metric tons of the fruit (Purseglove, 1974). Little has been done to improve the crop except by selection of clones from a seedling population for better fruit colour and flavour and, to a lesser extent, for yield and earliness. The mango is an outbreeder and hence very heterozygous. Many types, moreover, are polyembryonic. Improved cultivars normally originate from seedlings which are propagated vegetatively.

According to Singh (1969) several examples of spontaneous mutations exist, and some clonally propagated cultivars were obtained from such mutants (e.g. the variety Davis—Haden in Florida).

One case of radiation work was reported by Siddiqui et al. (1966) in Pakistan. Bud wood was subjected to acute radiation doses of 1—5 kR of X-rays and then budded onto a parent tree. One mutant shoot had different sized leaves, branching habit and delayed floral bud initiation and anthesis to those of the parent. According to the authors, this mutant could be useful for extending the fruiting season. However, it is not known to us whether any practical use was made of this mutant afterwards.

Literature mango

Purseglove, J.W., 1974. Tropical Crops: Dicotyledons. Longman, London, pp. 24—32.

Siddiqui, S.H., Mujeeb, K.A. and Vasti, S.M., 1966. Evolution of new varieties by ionizing radiation. In: Proc. Agric. Symp., May 1966. Atomic Energy Centre, Dacca, pp. 34—37.

Singh, L.B., 1969. Mango (*Mangifera indica* L.). In: F.P. Ferwerda and F. Wit (Editors), Outlines of Perennial Crop Breeding in the Tropics. Misc. Pap. 4, Landbouwhogeschool, Wageningen, pp. 309—327.

PINEAPPLE

Ananas comosus L. (fam. Bromeliaceae) is a tropical fruit with a world production of more than 3×10^6 metric tons, one third of which is produced in Hawaii. Other important producing countries are Thailand, Taiwan, the Philippines, Malaysia, South Africa, Mexico and Brazil. Most pineapples are canned but consumption of the fresh fruit is also considerable (Purseglove, 1975). Cv. Cayenne has been the most important cultivar in commercial production for almost 150 years.

Cv. Cayenne started as a single genotype but, because of asexual propagation and the occurrence of somatic mutations, a considerable number of different clones within this cultivar can now be distinguished.

Collins and Kerns (1938) mentioned that the more extreme mutant types are eliminated through natural and artificial selection. The spontaneous mutation rate is said to be rather low but, as a very high number of plants of one clone is grown, even less frequently occurring spontaneous mutations can be found. The same authors described 30 mutant types of the cultivar Cayenne. It was reported that some of them reproduce sexually. The mutations observed affected very different characters of the plant, the inflorescence and the fruit and some were reported to be dominant.

Breeding work usually starts from cv. Cayenne and concentrates on obtaining types with, among other things, better root systems, better fruit shape, more uniform ripening, deeper yellow flesh colour and resistance to several diseases. In order to achieve these goals, crosses are made generally with other cultivars and sometimes even with other species, and this is followed by back-crossing with cv. Cayenne.

Only a few irradiation experiments have been reported. In Kerala, India, irradiation of plants of the cultivars Kew and Mauritius led to growth retardation and, in one plant, to premature suckers (Anonymous, 1964). Further details are not known.

Marz (1964) reported the induction of self-fertile mutants by X-irradiation of pollen during meiosis. Recently, Singh and Iyer (1974) described a suitable technique for applying chemical mutagens like ethyl imine (EI), N-nitroso-N-methyl urethane (NMU) and diethyl sulphate (DES) to detached slips of 1—1.5 months old. In this preliminary work, several morphological mutations were found, one of which produced spineless plants from cv. Queen and was economically significant.

In conclusion, it can be said that utilization of mutation induction in this crop seems to be quite feasible. However, natural variation is still so large that most breeders do not yet see any direct need for induced mutations. Whether induced mutations become important in pineapple breeding in the future will mainly be determined by whether it is the most economic breeding method for each specific breeding objective. If mutation breeding is chosen then rapid vegetative reproduction seems necessary. For guidance on this aspect one may refer to Ma (1968), who described various methods of rapid multiplication based on the principle of decapitation, and Ma and Wang (1977), who developed a rapid in vitro propagation technique.

Literature pineapple

Anonymous, 1964. Pineapple Research Station, Kannara, Kerala. Agric. Res. Indian Counc. Agric. Res., 4: 141—142.

Collins, J.L. and Kerns, K R , 1938. Mutations in the pineapple. A study of thirty inherited abnormalities in the Cayenne variety. J. Hered., 29: 162—172.

Ma, S., 1968. The decapitation method for rapid multiplication of pineapples. J. Hortic. Soc. China, 14(1,2): 1—5.

Ma, S. and Wang, S., 1977. Tissue culture propagation of pineapple. J. Hortic. Soc. China, 23(3): 107—113 (in Chinese with English summary).

Marz, G.S., 1964. Study of self-incompatibility in the pineapple. Agric. Res., Pretoria, 2-6: 561—562 (Abstract).

Purseglove, J.W., 1975. Tropical Crops: Monocotyledons. Longman, London, pp. 76—91.

Singh, R. and Iyer, C.P.A., 1974. Chemical mutagenesis in pineapples (*Ananas comosus*). Proc. Int. Hortic. Congr., 19(1A): 108.

Chapter 8

OTHER CROPS

PEPPERMINT

An interesting case of the successful use of irradiation to induce resistance is that of *Mentha* × *piperita* L. (fam. Labiateae). In the U.S.A., Murray (1969, 1971) was able to produce strains resistant to the very serious and economically important disease *Verticillium albo-atrum* R and B var. *Menthae* Nelson, better known as *Verticillium*-wilt.

Peppermint is mainly grown in the U.S.A., the U.S.S.R., Moldavia, the Ukraine and the Balkans for the production of peppermint oil, which is obtained by steam distillation. In the U.S.A. approximately 30 000 ha was used to produce about 2.1×10^6 kg around 1970. This amount of peppermint had a value of over US \$ 20×10^6. *Mentha* is a highly sterile allohexaploid (2n = 72). Since 1890 only one clonal strain, known as the Mitcham variety, has been used; it is propagated vegetatively from stolons. The above disease has been known since 1924 but did not become a serious problem until 1940.

Despite tremendous efforts, no attempt using genetic techniques to obtain a resistant strain without diminishing the quality of the oil was successful. The only possibility left that might induce disease resistance without altering the genes responsible for the quality and composition of the complex peppermint oil seemed to be irradiation. Although the objective was to induce a dominant mutation for resistance in the Mitcham clone (a far from eviable task for a mutation breeder), the project was successful for two very important reasons: the scale at which it was possible to work, and the availability of a selective screening method.

Small-scale tests in which dormant stolons were irradiated were started by S. Shapiro in Brookhaven in 1955, both with X-rays and thermal neutrons. The experiments, which went on till 1959, indicated that the optimum neutron dosage is around 22×10^{12} N_{th} per cm^2 per sec and that of X-rays 6 kR. Over 100 000 irradiated plants were planted in heavily wilt-infested soil, from which more than 6×10^6 plants were obtained in the second year. In the subsequent 4 years this number was reduced by 99%, because of severe attacks of wilt. From a testing programme with almost 60 000 unattacked selections, seven highly resistant and five moderately resistant strains were obtained, though these strains also showed some morphological changes. The chemical composition and quality of the oil from the wilt-resistant material was next tested. It was found that the selected strains did not differ greatly from the Mitcham clone in these two important characteristics and were acceptable from both an industrial and an agricultural point of view.

TABLE 30

Commercial mutants of peppermint

Name of new cultivar	Place and date of release (or approval) and name of principal worker(s) and institute	Conditions and date of mutagenic treatment [Parent cultivar]	Main improved attributes of cultivar
Todd's Mitcham Peppermint	U.S.A., 1971 M.J. Murray, A.M. Todd Co., Kalamazoo, Mich.	X-rays and neutrons, 1955—1959 [Mitcham]	Darker green foliage colour, smaller leaves, more erect and less branched plant habit; earlier maturity (5—10 days); wilt resistant
Murray Mitcham	U.S.A., 1976 M.J. Murray, A.M. Todd Co., Kalamazoo, Mich., Purdue, Oregon and Washington Agricultural Experiment Station and Agricultural Research Service USDA	Neutrons [Mitcham]	*Verticillium* wilt resistant; improved first year yield over Todd's Mitcham

Sources: IAEA (1972, pp. 526—544) and *Mutation Breeding Newsletter*, 1 (1972)—9 (1977).

A new strain 'Todd's Mitcham Peppermint' was finally registered by the Crop Science Society of America in 1972. The oil from this wilt-resistant strain (selection 58) is quantitatively as well as qualitatively about the same as the oil from the old Mitcham variety. The strain is also ready for harvest 5—10 days earlier than the Mitcham variety (Murray, 1972). A second mutant was put on the market in 1976 (Table 30).

This is one of the best examples of successful mutation breeding, which could only be successful because of the scale, the efficient screening technique and the simple way in which peppermint propagates (from pieces of stolons) that reduced or even avoided the disadvantages of chimera formation.

Mutation breeding of *Mentha piperita* was also reported by Gupta (1976).

In an irradiation programme with *Mentha arvensis* L. var. *piperascens*, dormant runners were treated with γ-rays and several chemical mutagens (Kaul and Kak, 1973, 1975). The chemicals produced a higher genetic variability in leaf/stem ratio, growth habit and percentage and quality of oil that did γ-rays. The effect of γ-rays upon the herbage yield and oil content was studied by Mital et al. (1972). Ono (1971, 1972a,b) used various *Mentha* species with different ploidy levels to study the effect of acute and chronic irradiation with X- or γ-rays on, amongst other things, chlorophyll mutation frequency.

Literature peppermint

Gupta, M.N., 1976. Mutation breeding of peppermint (*Mentha piperita* Linn.). I.D.M.A. Bull., 7(26): 370—377.

IAEA, 1972. Induced Mutations and Plant Improvement. IAEA, Vienna, pp. 526—544.

Kaul, B.L. and Kak, S.N., 1973. Improvement of *Mentha arvensis* L. through induced mutations. In: Advancing Frontiers in Cytogenetics. Hindustan Publishing Corporation (India), Delhi, pp. 189—195.

Kaul, B.L. and Kak, S.N., 1975. Use of radiations and radiomimetic chemicals in the breeding of vegetatively propagated plants. I. *Mentha arvensis*. Reg. Res. Lab. Jammu, 41; Plant Breed. Abstr.: No. 1009.

Mital, S.P., Issar, S.C., Kidwai, M.A. and Saxena, D.B., 1972. Improvement in Japanese mint (*Mentha arvensis* L. var. *piperascens* Holmes) through gamma-irradiation. Indian J. Agric. Sci., 42(7): 550—553.

Murray, M.J., 1969. Successful use of irradiation breeding to obtain *Verticillium*-resistant strains of peppermint, *Mentha piperita* L. In: Induced Mutations in Plants (Pullman, Wash.). IAEA, Vienna, pp. 345—370.

Murray, M.J., 1971. Additional observations on mutation breeding to obtain *Verticillium*-resistant strains of peppermint. In: Mutation Breeding for Disease Resistance. IAEA, Vienna (1970), pp. 171—195.

Murray, M.J., 1972. Mutation breeding in *Mentha*. In: M.J. Constantin (Editor), Mutation Breeding Workshop, Knoxville, Tenn. University of Tennessee, Knoxville, Tenn.

Ono, S., 1971. Studies on the radiation breeding in the genus *Mentha* IX. Effective irradiation techniques to induce mutation in mint. Sci. Rep. Fac. Agric. Okayama Univ., (37): 9—15.

Ono, S., 1972a. Studies on the radiation breeding in the genus *Mentha* XIII. Effects of irradiation on pollen. Sci. Rep. Fac. Agric. Okayama Univ., (39): 1—8.

Ono, S., 1972b. Studies on the radiation breeding in the genus *Mentha* XV. A comparison of biological effects between γ-rays and X-rays on mint seeds. Sci. Rep. Fac. Agric. Okayama Univ., (40): 1—7.

SISAL

Sisal, *Agave sisalana* Perz., and, to a lesser degree, henequen, *Agave fourcroydes* Lem., are the most important "hard fibres". Both species are grown in arid and semi-arid areas of the tropics.

Sisal is a pentaploid (5x = ca 138—149) and native to Central America, with Brazil and, in particular, Tanzania as major world producers. Because of the synthetic fibres, interest in the crop is considerably declining. Propagation normally takes place mainly from bulbils but also from suckers. Somatic mutations have been reported, but up to now no useful types have been obtained in this way. It is difficult to improve the crop sexually. The plants seldom set seed and, in addition, the vegetative life cycle is a very long one. However, Wienk (1969) points out that breeding can improve fibre agaves.

So far, the use of irradiation has only once been reported in a sisal improvement programme (Anonymous, 1963); however, the results were never published. It seems highly improbable that any more mutation work will be undertaken for practical purposes in this crop.

Literature sisal

Anonymous, 1963. Annual report of the Tanganyika Sisal Growers Association, Sisal Research Station, Mlingano, Ngomeni for the year 1961—1963. Sisal Research Stn, Mlingano, Ngomeni, 71 pp.

Wienk, J.F., 1969. Long fibre agaves (*Agave sisalana* Perz. and *A. fourcroydes* Lem.). In: F.P. Ferwerda and F. Wit (Editors), Outlines of Perennial Crop Breeding in the Tropics. Misc. Pap. 4, Landbouwhogeschool, Wageningen, pp. 1—21.

SUGARCANE

Sugarcane is undoubtedly one of the oldest cultivated crops in the world. It is grown between the latitudes 35° N and 35° S, in different climates, on various soils and at various altitudes. Its considerable economic importance is sufficiently indicated by the fact that over 10^7 ha are used to produce more than 45×10^6 metric tons/year of sugar. At present, the principal producing countries are Cuba, India and Brazil. Cultivars of the botanical species *Saccharum officinarum* L., usually referred to as noble cane, were the main types that used to be grown. Nowadays, all cultivars are interspecific hybrids of different species of the genus *Saccharum* (fam. Graminae). In his standard work on the breeding of sugarcane, Stevenson (1965) provides a key to the species of this genus, including *S. spontaneum*, *S. robustum*, *S. sinense*, *S. barberi*, *S. edule* and *S. officinarum*. Strictly speaking, only the first two wild species are true botanical species.

Commercially used plants are clones grown from stem cuttings with two or more nodes from which buds start to develop. Each cutting grows into a "stool" with perhaps several dozens of individual canes. Sugar is stored in the stems' tissues. Man has been looking for better, i.e. sweeter or less fibrous types for as long as the crop has been cultivated, but deliberate breeding did not begin until the end of the 19th century.

Sugarcane is a difficult crop from a breeding point of view; its genetic make-up is not well understood. Sugarcane is highly polyploid and very heterozygous, with many, small chromosomes. *S. officinarum* cultivars are octoploids and contain 80 chromosomes. Most commercial cultivars have chromosome numbers of 100—120 or more. Due to meiotic irregularities, many cultivars are aneuploids. The famous old clone POJ 2878, for example, is reported to have 121 chromosomes. To begin with, breeding mainly consisted of selecting from seedlings of openly pollinated plants of *S. officinarum*. At present, seedlings are selected from the progeny obtained from controlled crosses which, as mentioned above, are normally interspecific.

Another way to obtain genetic variation is from bud mutations which have been known to occur fairly frequently for a long time. According to Purseglove (1975), however, they have had little influence on crop improvement. Nevertheless, several useful, or potentially useful spontaneous bud mutations have been reported in the literature. Cross (1933a,b), for instance, described two valuable somatic mutants of POJ 36 — at that time the leading cultivar in Argentina — that had higher sugar contents and were much more vigorous than the parent clone. Mutations affecting characteristics such as stature, waxy bloom, stem colour, striped appearance and chlorophyll content have also been reported. An important economic character for which mutations are most desirable, especially in tropical areas (and have been reported too), is that which makes a clone non-flowering. (It must be kept in mind, however, that for cross-breeding work a certain amount of flowering is essential.)

Much information about induced mutations is derived from annual reports of various research institutes. These reports are not always easily available.

As early as 1927, sugarcane was irradiated with X-rays with the intention of inducing mutations (Heinz, 1973). This experiment did not lead to useful mutations, but to the occurrence of coloured stripes on the stem. Heinz expresses the opinion that mutation breeding is relatively promising in countries with a limited or no sexual breeding programme, and especially useful in cases of sudden serious disease attacks. Because of necessarily lengthy quarantine periods, replacement of diseased plants by resistant clones from outside may take too long.

Early work with X-rays was also reported from India (Dorosami and Venkoba, 1947; Panje and Prasad, 1959; Rao, 1954). Stevenson (1957) unsuccessfully treated germinated seeds of an inbred line with 50—1 000 R of X-rays. According to Urata and Heinz (1971), γ-rays were used as early as 1956, but no further information was given. Price and Warner (1959) suggested the transfer of desirable genes from otherwise undesirable chromosomes through irradiation as used by Sears (1956) to obtain his famous *Triticum* X *Aegilops* hybrid. This was tried in Hawaii in 1962 (Anonymous, 1962), but without success. At the Sugarcane Breeding Institute, Coimbatore, India, several cultivars were irradiated around 1960 in an attempt to induce, among other things, resistance to red rot or *Physalospora tucumanensis* (Anonymous, 1962). Two publications (Anonymous, 1965; Rao et al., 1966) report that cv. Co 499, having been treated with 500 R of γ-rays produced a plant with five stalks and a high degree of resistance to red rot strain D. The use of X-rays, γ-rays and chemical mutagens resulted in further mutants that had, for example, no spines and profuse tillers. Rao et al. (1966) mentioned that doses of 500 and 300 R of γ-rays induced resistance to *Glomerella tucumanensis* (N.B. *Physalospora*, *Glomerella* and *Colletotrichum* all cause red rot disease). The 1966/1967 Annual Report of Coimbatore (Anonymous, 1966a) also mentioned ethyl methanesulphonate (EMS)-induced resistance to *Proceras indicus*, a stem borer.

Several annual reports of the Sugarcane Breeding Station, Barbados, also refer to morphological mutants that were obtained from irradiation or treatments with EMS (Anonymous, 1966b; Anonymous, 1970a).

According to Hrishi and Marimuthammal (1968) and Hrishi et al. (1968a, b), the treatment of buds and growing meristems with different chemical mutagens produced morphological mutants. Mee et al. (1969) also used various chemicals, such as 8-ethoxy caffeine, methyl methanesulphonate (MMS) and EMS.

In the experiments discussed so far, X-rays, γ-rays as well as chemical agents were used but, according to Heinz (1973), γ-rays were most commonly applied. If stem cuttings are to be irradiated an acute dose of 2—6 kR seems to be optimal as it results in high rates of survival as well as a high frequency of mutations (Siddiqui et al., 1976). Chronic γ-irradiation was used by Sankaranarayanan and Babu (1970).

According to Heinz (1973), the occurrence of chimerism, a common problem in mutation breeding of vegetatively propagated plants, seems to be limited to two or three vegetative generations if proper selection is carried out, although Jagathesan (1976) emphasizes that the stabilization of a mutant is the main problem and that more vegetative propagations are needed to ensure that the mutant will never revert back to the original type.

In the 1969 annual report of the Hawaiian Sugar Planters Association (Anonymous, 1970b) it is stated that upper, middle and lower parts of young stalks were treated with up to 10 kR of γ-rays. Stalks irradiated at the top showed greatest survival and varietal differences in resistance to radiation were noted. A dose of 6 kR produced the highest rate of deviations, most of which were morphological. Jagathesan and Jalala (1969) treated pollen grains with both ultra-violet and X-rays.

Induction of the non-flowering character has been reported by Walker and Sisodia (1969), by Rao (1974) and in several annual reports. The influence of mutagenic treatment on juice quality and cane weight was studied by Mustafa Khan and Zakir (1970), while "thick-cane mutants" were reported by Shama Rao et al. (1977). Some preliminary work has been carried out by Anderez (1971) in Cuba.

Jagathesan and Sreenivasan (1970) treated single-budded sets (cuttings) of several Coimbatore cultivars with X- and γ-rays. The material was observed up to vM_3, and some potentially useful mutants were described. Later published details about this programme indicate that 52 mutants were isolated, including types with glabrous leaf sheaths, short types (Jagathesan, 1977; Jagathesan and Jebadhas, 1976), non-flowering types and plants with a high sugar content. Between 3 and 4 years were necessary for the mutants, which were selected on the basis of single stalks, to become stable. There were some promising results: induced resistance against red rot (*Colletotrichum falcatum*) and smut (*Ustilago sciiaminea*) were reported. Red rot resistant mutants of Co 449 and Co 997 were released several years ago (Rao et al., 1966). D. Jagathesan (personal communication, 1976) suggested the use of the decapitation technique and closer planting to increase the frequency of (solid) mutants.

In recent years, several other reports concerning the induction of resistance against red rot have appeared (Singh, 1970; Nair, 1973; Bari, 1974; Haq et al., 1970, 1974a,b; Jagathesan et al., 1974). Resistance to the toxin from *Helminthosporium sacchari* in γ-ray-induced mutants was reported by Strobel et al. (1975).

Recently, several potentially useful mutants, displaying, for example, disease resistance, have been reported (Darmodjo and Wirioatmodjo, 1976) or even released for use in the field, while one or two high yielding mutants soon will become available (Jagathesan, 1976, 1977). So far, however, we have not heard of any officially released commercial variety that is the result of mutagenic treatment, which seems to be contradictory to the relatively extensive literature. It appears that large-scale mutation breeding has been prevented by the size of the plant and the chimera problem. To solve the latter problem, three or more vegetative generations are required, and this takes (too) much time, space and labour. It may, therefore, be of importance that a number of in vitro propagation techniques have been described (Heinz, 1971, 1973; Heinz and Mee, 1969, 1970, 1971; Heinz et al., 1969, 1977; Krishnamurthi, 1975; Liu and Chen, 1976; Mee and Heinz, 1969; Mee et al., 1969). The use of these techniques will most probably restrict and may even avoid chimera formation. They may also facilitate large-scale selection for certain quantitative traits. Recently, Lyrene (1976) discussed the use of tissue culture techniques to obtain sugarcane mutants. He maintained that much higher mutation frequencies are thus obtained than is the case with spontaneous bud mutations in the field. Heinz (1973) reported that as many as 15—20% mutants are found without the application of mutagenic agents. It, therefore, appears to be only a matter of time before commercial mutants are released.

Literature sugarcane

Anderez, M., 1971. Preliminary report of studies on inducing mutations in sugarcane in Cuba with a Co^{60}-γ-source. Rev. Agric. Cuba, 4(2): 1—4 (in Spanish).

Anonymous, 1962. Hawaii. Sugar Plant. Assoc. Exp. Stn Annu. Rep. 1961, p. 70.

Anonymous, 1965. Sugar cane breeding at Coimbatore. Int. Sug. J., 68: 355—356.

Anonymous, 1966a. Sugarcane Breeding Institute, Coimbatore, India, Annu Rep. 1966—67, p. 117.

Anonymous, 1966b. Cent. Sugar Cane Breed. Stn, Barbados, 33rd Annu. Rep., 33: 1—58.

Anonymous, 1970a. Cent. Sugar Cane Breed. Stn, Barbados, 37th Annu. Rep., 37: 1—70.

Anonymous, 1970b. Hawaii. Sugar Plant. Assoc. Exp. Stn Annu. Rep. 1969, pp. 3—7.

Bari, G., 1974. Induction of somatic mutations for disease resistance in sugarcane. In: Induced Mutations for Disease Resistance in Crop Plants. Proc. Meet. FAO/IAEA/SIDA, Novi Sad, 1973. IAEA, Vienna, p. 149.

Cross, W.E., 1933a Una mutacion de la P.O.J. 36 de la mayor riquera sacarina. La P.O.J. 36M. Rev. Ind. Agric. Tucumán, 23: 47—52.

Cross, W.E., 1933b. Otra mutacion de la P.O.J. 36 de mayor productividad. La cana "Paz posse". Rev. Ind. Agric. Tucumán, 23: 104—108.

Darmodjo, S. and Wirioatmodjo, B., 1976. Resistance Mutation of Sugarcane Species (*Saccharum officinarum*) from Top Borer (*Scirpophaga nivella*) as the Effect of Irradiation with Gamma Rays. Lokakarya Pemberantasan Hama Dengan Radiasi (4). Jakarta, 10 pp. (in Indonesian with English summary).

Dorosami, L.S. and Venkoba, B., 1947. Effect of X-ray irradiation on sugarcane. Indian Sci. Congr. Assoc. Proc., 33(3): 111.

Haq, M.S., Maniruzzaman, A.F.M. and Ali, S.M., 1970. Evolution of high yielding disease resistant strains of sugarcane through gamma radiation. In: Proc. 21st—22nd Pak. Sci. Conf. Rajshatic 1970. Atomic Energy Centre, Dacca.

Haq, M.S., Tin, M.M. and Myint, T., 1974a. The prospect of breeding red-rot resistant sugarcane varieties through induced mutations in Burma. In: Induced Mutations for Disease Resistance in Crop Plants. Proc. Meet. FAO/IAEA/SIDA, Novi Sad, 1973. IAEA, Vienna, pp. 150—153.

Haq, M.S., Rahman, M.M., Mia, M.M. and Ahmed, H.U., 1974b. Disease resistance of some mutants induced by gamma rays. In: Induced Mutations for Disease Resistance in Crop Plants. Proc. Meet. FAO/IAEA/SIDA, Novi Sad, 1973. IAEA, Vienna, p. 150.

Heinz, D.J., 1971. New procedures for sugarcane breeders. Proc. Congr. Int. Sugar Cane Technol., 14.

Heinz, D.J., 1973. Sugar-cane improvement through induced mutations using vegetative propagules and cell culture techniques. In: Induced Mutations in Vegetatively Propagated Plants. IAEA, Vienna, pp. 53—61.

Heinz, D.J. and Mee, G.W.P., 1969. Plant differentiation from callus tissue of *Saccharum* species. Crop Sci., 9: 346—348.

Heinz, D.J. and Mee, G.W.P., 1970. Colchicine-induced polyploids from cell suspension cultures of sugarcane. Crop Sci., 10: 696—699.

Heinz, D.J. and Mee, G.W.P., 1971. Morphologic, cytogenetic and enzymatic variation in *Saccharum* species hybrid clones derived from callus tissue. Am. J. Bot., 58 (3). 257—262.

Heinz, D.J., Mee, G.W.P. and Nickell, L.G., 1969. Chromosome numbers of some *Saccharum* species hybrids and their cell suspension cultures. Am. J. Bot., 56(4): 450—456.

Heinz, D.J., Krishnamurthi, M., Nickell, L.G. and Maretzki, A., 1977. Cell, tissue and organ culture in sugarcane improvement. In: J. Reinert and Y.P.S. Bajaj (Editors), Plant Cell, Tissue, and Organ Culture. Springer Verlag, Berlin, pp. 3—17.

Hrishi, N. and Marimuthammal, S., 1968. Studies in mutagenesis in sugarcane. I. Effect of chemical mutagens. Proc. Indian Acad. Sci., Sect. B, 68: 131—142.

Hrishi, N., Krishnamurthy, T.N. and Marimuthammal, S., 1968a. Studies in mutagenesis in sugarcane II. Induction of visible and micromutations in sugarcane by chemical mutagens. Proc. Indian Acad. Sci., Sect. B, 68: 181—189.

Hrishi, N., Marimuthammal, S. and Selvanathan, J., 1968b. The use of chemical mutagens in sugarcane. Proc. Int. Soc. Sugar Cane Technol., 13: 1024—1033.

Jagathesan, D., 1976. Induction and isolation of mutants in sugarcane. In: Improvement of Vegetatively Propagated Plants and Tree Crops through Induced Mutations, Wageningen, 1976. IAEA, Vienna, pp. 69—82.

Jagathesan, D., 1977. Induction and isolation of mutants in sugarcane. Mut. Breed. Newsl., 9(1): 5—6.

Jagathesan, D. and Jalala, N.C., 1969. Effect of UV and X-rays on the pollen grains of *Saccharum* and *Erianthus*. In: Proc. Symp. Radiation and Radiomimetic Substances in Mutation Breeding, Bombay. pp. 103—111.

Jagathesan, D. and Jebadhas, A.W., 1976. A rare 'dwarf' mutant in sugarcane. Sci. Cult., 42(4): 233—234.

Jagathesan, D. and Sreenivasan, T.V., 1970. Induced mutations in sugarcane. Indian J. Agric. Sci., 40: 165—172.

Jagathesan, D., Balasundaram, N. and Alexander, K.C., 1974. Induced mutations for disease resistance in sugarcane. In: Induced Mutations for Disease Resistance in Crop Plants. Proc. Meet. FAO/IAEA/SIDA, Novi Sad, 1973. IAEA, Vienna, p. 151.

Krishnamurthi, M., 1975. Notes on disease resistance of tissue culture subclones and fusion of sugarcane protoplasts for the improvement of varieties. Sugarcane Breed. Newsl., 35: 24—26.

Liu, M.C. and Chen, W.H., 1976. Tissue and cell culture as aids to sugarcane breeding. I. Creation of genetic variation through callus culture. Euphytica, 25: 393—403.

Lyrene, P.M., 1976. Tissue culture and mutations in sugarcane. Sugarcane Breed. Newsl., 38: 61—62.

Mee, G.W.P. and Heinz, D.J., 1969. Gamma irradiation of sugarcane callus tissue. Hawaii. Sugar Plant. Assoc. Exp. Stn Annu. Rep. 1968, p. 16.

Mee, G.W.P., Nickel, L.G. and Heinz, D.J., 1969. Chemical mutagens — their effects on cells in suspension culture. Hawaii. Sugar Plant. Assoc. Exp. Stn Annu. Rep. 1968, pp. 7—8.

Mustafa Khan, A. and Zakir, A., 1970. Improvement of sugarcane through radiation. West Pak. J. Agric. Res., 81: 1—6.

Nair, M.K., 1973. Note on an induced red-rot resistant mutant of 'Co 997' variety of sugarcane. Indian J. Agric. Sci., 43(3): 323—324.

Panje, R.R. and Prasad, J.P.R., 1959. The effects of ionizing radiations on sugarcane. In: Proc. Tenth Congr. Int. Soc. Sugar Cane Technol. Hawaii 1959. Elsevier, Amsterdam, pp. 775—782.

Price, S. and Warner, J.N., 1959. The possible use of induced mutations for sugarcane

improvement. In: Proc. Tenth Congr. Int. Soc. Sugar Cane Technol. Hawaii 1959. Elsevier, Amsterdam, pp. 782—792.
Purseglove, J.W., 1975. Tropical Crops: Monocotyledons. Longman, London, pp. 214—256.
Rao, B.V., 1954. A brief review of the work on the use of X-rays in sugarcane breeding in Mysore. Proc. Conf. Sugarcane Res. Work. Indian Union, 2: 11—13 (not consulted).
Rao, J.T., Srinivasan, K.V. and Alexander, K.C., 1966. A red-rot resistant mutant of sugarcane induced by gamma irradiation. Proc. Indian Acad. Sci., Sect. B, 4: 224—230.
Rao, P.S., 1974. Mutation breeding for non-flowering in sugarcane. Mut. Breed. Newsl., 3: 9.
Sankaranarayanan, P. and Babu, C.N., 1970. Effects of chronic gamma irradiation in sugarcane varieties. Indian J. Hered., 2(2): 97—102.
Sears, E.R., 1956. The transfer of leaf-rust resistance from *Aegilops umbellulata* to wheat. In: Genetics in Plant Breeding. Brookhaven Symp. Biol., 9: 1—22.
Shama Rao, H.K., Sachan, R. and Singh, M., 1977. Gamma-ray induced mutations in sugarcane. Mut. Breed. Newsl., 9(1): 5.
Siddiqui, S.H., Mujeeb, K.A. and Keerio, G.R., 1976. Gamma irradiation effects on sugarcane (*Saccharum* sp.) clone Co-547. Environ. Exp. Bot., 16(1): 65—68.
Singh, S., 1970. Chronic gamma irradiation-induced resistance to red-rot disease in sugarcane var. Co997. Sugarcane Pathol. Newsl., 5: 24.
Stevenson, G.C., 1957. Cent. Sugar Cane Breed. Stn, Barbados, 24th Annu. Rep.
Stevenson, G.C., 1965. Genetics and Breeding of Sugarcane. Longman, London, 284 pp.
Strobel, G.A., Steiner, G.W. and Byther, R., 1975. Deficiency of toxin-binding protein activity in mutants of sugarcane clone H54-775 as it relates to disease resistance. Biochem. Genet., 13(9/10): 557—565.
Urata, R. and Heinz, D.J., 1971. Gamma irradiation-induced mutations in sugarcane. Proc. Int. Soc. Sugar Cane Technol., 14: 402—407.
Walker, D.I.T. and Sisodia, N.S., 1969. Induction of a nonflowering mutant in sugarcane (*Saccharum* sp.). Crop Sci., 9: 551—552.

TEA

Tea, *Camellia sinensis* (L.) O. Kuntze (syn. *Thea sinensis* L.), family Camelliaceae, used to be propagated almost entirely by seed but, because of a certain degree of genetic heterogeneity and the subsequent variability of tea plants in the field, tea cultivars are increasingly propagated asexually, sometimes by budding, grafting or layering, but mainly from cuttings. The use of cuttings is the only practical means of vegetative propagation (Visser, 1976). Because of this, the induction of somatic mutations of important characters is attracting increasing interest.

So far, the effects of radiation on tea plants have been studied at only two (Japanese) institutes, namely, the National Research Institute of Tea, Kanaya, Shizuoka (Amma, 1974; Nakajima et al., 1974; Nakayama, 1973, 1974, 1976; Nakayama et al., 1974a,b) and the Faculty of Agriculture, Nagoya University (Futsuhara, 1967).

The first group of authors are studying several aspects of the effects of chronic or semi-chronic γ-radiation upon tea plants, such as the radiosensitivity and the mutation frequency as well as effects of pruning on the mutation frequency.

The optimum dose rate for tea plants grown in a γ-field is 20—30 R/day at which 70% of the plants survive (for 3 years) and a somatic mutation rate of 10—12% per plant (average for the eight cultivars studied) is obtained. The mutations mainly affect leaf form, size and colour, but other mutants were observed as well, for example, a tetraploid form (2n = 4x 60) of cv. Yabukita, the leading green tea cultivar in Japan, and a mutant of cv. Yutakamidori with high arginine content. The latter change opens up possibilities for the improvement of tea since the arginine content of leaves is related to tea quality (Nakayama, 1976). Both mutants were obtained after heavy pruning in March of chronically irradiated tea plants. Heavy pruning in itself hardly increased the mutation rate (Nakajima et al., 1974).

Futsuhara (1967) reported the acute irradiation and chronic γ-irradiation with low dose rates of cuttings from nine Japanese and Indian cultivars. He used this material for cyto-histological observations of radiation damage (e.g. primary effects, radioresistance). Rooted cuttings were more resistant to radiation than non-rooted ones, the LD_{50}s being 4—5 and 3—4 kR, respectively. Triploid and tetraploid cultivars are more resistant than diploids. The production of tetraploids is reported by Sebastiampillai (1976) who treated shoots with 0.2—0.5% colchicine for 2—7 days. Twelve wholly tetraploid plants were obtained from 95 shoots. These results are a little suspect in our opinion, since only the root tips of cuttings were cytologically examined.

No commercial mutants have been reported as far as we know.

Literature tea

Amma, S., 1974. Characteristics of tetraploid tea induced from gamma-irradiated Yabukita variety. Chagyo Gijutsu Kenkyu, 46: 1—6 (in Japanese with English summary); Nucl. Sci. Abstr., 31(12) 1975: No. 33434.

Futsuhara, Y., 1967. Studies of radiation breeding in the tea plants. Gamma Field Symp., 6: 107—122.

Nakayama, A., 1973. Induction of the somatic mutations in tea plants by gamma irradiation. Gamma Field Symp., 12: 37—45.

Nakayama, A., 1974. Characteristics of the mutations induced from gamma-ray tea plants under heavy pruning treatment. Chagyo Gijutsu Kenkyu, 47: 29—35 (in Japanese with English summary); Nucl Sci. Abstr., 31(12) (1975): No. 33426.

Nakayama, A., 1976. Somatic mutation of tea plants induced by gamma-irradiation. Jpn. Agric. Res. Q., 10(2): 79—83.

Nakayama, A., Nakajima, K. and Yamaguchi, J., 1974a. Induction of bud mutations in tea plant under chronic irradiation. Chagyo Gijutsu Kenkyu, 47: 16—27 (in Japanese with English summary); Nucl. Sci. Abstr., 31(12) (1975): No. 33427.

Nakayama, A., Tomo, N., Mitsui, M. and Nakajima, K., 1974b. Radiosensitivity of the tea plant under chronic gamma irradiation in a gamma field. Chagyo Gijutsu Kenkyu, 47: 11—15 (in Japanese with English summary); Nucl. Sci. Abstr., 31(12) (1975): No. 33429.

Nakajima, K., Nakayama, A., Takato, S., Tomo, N. and Yamaguchi, J., 1974. Effect of the heavy pruning treatment on the induction of bud mutations in gamma-rayed tea plant. Chagyo Gijutsu Kenkyu, 47: 1—9 (in Japanese with English summary); Nucl. Sci. Abstr., 31(12): (1975): No. 33428.

Sebastiampillai, A.R., 1976. A simple technique for the induction of polyploids in tea. Tea Q., 46(1,2): 12—15.
Visser, T., 1976. Tea, *Camellia sinensis* (Camelliaceae). In: N.W. Simmonds (Editor), Evolution of Crop Plants. Longman, London, pp. 18—20.

GRASSES

Grasses, belonging to the enormous family of the Gramineae, include a great number of the world's economically important food crops. Very few are propagated vegetatively, either by apomictic seeds or by runners, stolons or other vegetative plant parts.

We, therefore, mainly restrict ourselves to publications concerning bermuda grass (*Cynodon* spp.), Kentucky bluegrass (*Poa pratensis* L.), *Paspalum* and a few more general topics in these or related plants.

Part of the literature deals with the use of radiation while studying cytogenetic effects resulting from radiation-induced chromosome breaks and their possible application. For example, Boyle (1964) irradiated the seeds of an octoploid hybrid between *Agropyron* and *Hordeum* with three doses of thermal neutrons. The parameter used to measure effect was the formation of bridges in root-tip cells. Two mutants were produced. Heinz and Elliot (1964), working with cocksfoot (*Dactylus glomerata*), used radiation to induce translocations, following chromosome breaks. The only known radiation-induced commercial mutant in *Dactylus glomerata* was found by Zijp (1960), namely, cv. Tenderbite.

Other publications concern the effects of radiation on (breaking) apomixis (Grazi, 1962; Julen, 1954, 1958, 1961; Svetlik, 1967). It was found possible to use temporary radiation-induced partial sexuality in plant breeding of obligate apomicts, and subsequently select apomictic types with favourable morphological characters (Julen, 1961).

An important and extensive publication, concerning improvement of *Poa pratensis* by mutations is that of Gustafsson and Gadd (1965b). They discuss the cytogenetic constitution and mode of reproduction. From this discussion it appears that *Poa* is a complex of biotypes with variable chromosome numbers. It has a versatile system of reproduction, from vegetative spreading of stolons or runners and bulbil formation in the spikes to sexual seed formation. Vegetative spreading of runners is characteristic of cultivated species. A series of articles are then discussed, some dealing with the breakdown of apomixis following irradiation, some with mutation breeding aspects. It is concluded that the optimum radiation dose is 15—20 krad X-rays for dry dormant seeds; neutrons seem to be useful too (Hansen and Juska, 1962).

Several other authors have used *Poa pratensis* to investigate and demonstrate the usefulness of artificial mutations to increase the variability in obligate apomicts in which the only other source of genetic diversity is spontaneous variation. De Loose (1964) treated seeds with X-, γ- and β-rays and ethyl methanesulphonate (EMS), and observed an increase in morphological aberrants in M_1. In M_2 plants, however, the number of aberrants was much lower and, according to De Loose (1964), probably due to the chimeral situation in M_1 and not to the breakdown of apomixis. Some mutants with potential agronomic value were obtained. Commercially interesting mutants were likewise reported by Kuleshov et al. (1974) who treated seeds of brome grass, timothy and meadow fescue with several mutagenic chemicals, such as nitroso-N-ethyl urethane (NEU), N-nitroso-N-methyl urethane (NMU), 1,4-bis-diazo-acetyl butane (1,4 DAB), di-methyl sulphate (DMS) and ethylene imine (EI). Hansen and Juska (1959, 1962) irradiated Merion Kentucky bluegrass with thermal neutrons and found one stem rust (*Puccinia graminis* Pers.) resistant progeny in M_2. They did not believe this to be the result of outcrossing. Unfortunately, the tolerance of 'Merion' to leaf spot (*Helminthosporium vagans*) was lost in the mutant, as was discovered later on.

In other crops a similar train of thought was followed. Singh and Mehra (1971), working with the agamic (and polyploid) species complex *Dichantium annulatum*, proposed a number of breeding procedures for the genetic improvement of apomicts. In the (hexaploid) obligate apomicts of the *D. annulatum* complex, spontaneous and induced mutations provide the main source of variability, whereas in the (tetraploid) facultative apomicts advantages of both sexual and asexual embryo sacs and reproduction could be combined.

Large-scale mutation breeding was reported by Burton (1972, 1974, 1975, 1976) and Powell (1974, 1976) and Powell et al. (1974). They put into practice the idea that mutation breeding is a numbers game and therefore requires large quantities of irradiated material to achieve any given objective. They irradiated one or two-node sections of (dormant) rhizomes (stolons or sprigs) of the turf Bermuda grass varieties Tifgreen, Tifway, Tifdwarf and Tifcote (sterile triploid F_1 hybrids between *Cynodon dactylon* and *C. transvaalensis*) with 7—12 krad of γ-rays. Over 150 mutants were found, some of which showed a higher level of resistance than the starting material to root knot nematode, *Meloidogyne graminis*, or other nematodes. Nine of the better mutants are to be compared with their parent hybrids before being put on the market (Burton and Hanna, 1977). In several of the irradiated cultivars a proportion (up to 6% in Tifgreen and Tifdwarf) of mutated propagules were observed, the majority being non-sectored mutants (the percentage of wholly mutated plants increased with increasing dose). Mutants varied greatly in hue, size, rate of spread and resistance to weeds and root knot nematodes.

Coastcross-1 Bermuda grass is a sterile tetraploid F_1 hybrid between Coastal Bermuda grass and a highly digestible Bermuda grass from Kenya which lacks tolerance to temperate winters but, if fed to beef cattle, results in 34% more beef per acre. Approximately 400 000 freshly cut green stems of Coastcross-1 Bermuda grass were baled and exposed to 7 krad of γ-rays.

Four plants survived −16°C. Since then, more than 2×10^6 stems have been similarly treated and planted. Unusually mild winters have so far prevented selection for hardy mutants, if any. However, many other morphological mutants have been observed, demonstrating that ionizing radiation is a useful additional tool in breeding such crops.

One author has also reported mutagenic treatments of Dallis grass (*Paspalum dilatatum* Poir.) and Bahia grass (*Paspalum notatum* Flugge) which produced some mutants. Likewise, mutants were obtained from the vegetatively propagated triploid Floratam St. Augustine grass (*Stenotaphrum secundatum* (Walt.) Ktze.) (Powell, 1976). In Dallis grass, mutants had earlier been reported by Bashaw and Hoff (1962) and Burton and Jackson (1962).

It is clear that radiation is a tool to (temporarily) overcome apomixis and to induce genetic diversity in species which are otherwise completely dependant on spontaneous mutations. It is also our own experience, obtained in co-operative projects with commercial plant breeders, that properties like width of leaves, colour, and resistance to wear and probably to certain diseases as well can be induced in turf grasses like *Poa pratensis*.

Literature grasses

Bashaw, E.C. and Hoff, B.J., 1962. Effects of irradiation on apomictic common dallisgrass. Crop Sci., 2: 501—504.

Boyle, W.S., 1964. Cytogenetic effects of thermal neutron radiation on a polyploid grass hybrid. Radiat. Bot., 4(1): 27—33.

Burton, G.W., 1972. The use of mutagenic agents in grass and turf breeding. In: M.J. Constantin (Editor), Mutat. Breed. Workshop, Knoxville, Tenn. University of Tennessee, Knoxville, Tenn.

Burton, G.W., 1974. Radiation breeding of warm season forage and turf grasses. In: Polyploidy and Induced Mutations in Plant Breeding; Bari, 1972. IAEA, Vienna, pp. 35—41.

Burton, G.W., 1975. Improving sterile turf and forage bermuda grass hybrids by gamma irradiation. In: Improvement of Vegetatively Propagated Plants through Induced Mutations, Tokai, 1974. IAEA, Vienna, p. 33.

Burton, G.W., 1976. Using gamma irradiation to improve sterile turf and forage Bermudagrasses. In: Improvement of Vegetatively Propagated Plants and Tree Crops through Induced Mutations, Wageningen, 1976. IAEA, Vienna, pp. 25—32.

Burton, G.W. and Jackson, J.E., 1962. Radiation breeding of apomictic prostrate dallisgrass, *Paspalum dilatatum* var. *pauciciliatum*. Crop Sci., 2: 495—497.

Burton, G.W. and Hanna, W.W., 1977. Performance of mutants induced in sterile turf bermudagrass. Mut. Breed. Newsl., 9(1): 4.

De Loose, R., 1964. Mutatie-onderzoek by *Poa pratensis* (Veldbeemdgras R.v.P.). Meded. Landbouwhogesch. Opzoekingsstns Staat Gent, XXIX (4): 1367—1375.

Grazi, F., 1962. Radiation and other mutagens in the balance between apomixis and sexuality in *Poa pratensis*. Genet. Agrar., 15: 351—353.

Gustafsson, Å. and Gadd, J., 1965b. Mutations and crop improvement. IV. *Poa pratensis* J. (*Gramineae*). Hereditas, 53(1-2): 90—103.

Hansen, A.A. and Juska, F.V., 1959. A "progressive" mutation in *Poa pratensis* L. by ionizing radiation. Nature (London), 184: 1000—1001.
Hansen, A.A. and Juska, F.V., 1962. Induced mutations in Kentucky bluegrass. Crop Sci., 2: 369—371.
Heinz, D.J. and Elliot, F.C., 1964. Effects of radiation-induced translocations in *Dactylus glomerata* L. Crop Sci., 4: 181—184.
Julen, G., 1954. Observations on X-rayed *Poa pratensis*. Acta Agric. Scand., 4: 585—593.
Julen, G., 1958. Über die Effekte der Röntgenbestrahlung bei *Poa pratensis*. Züchter, 28: 37—40.
Julen, G., 1961. The effect of X-rays on the apomixis in *Poa pratensis*. In: Effects of Ionizing Radiation on Seeds. IAEA, Vienna, pp. 527—532.
Kuleshov, G., Bekhtin, N. and Williams, W.R., 1974. Chemical mutagenesis in perennial grass breeding. In: V.G. Iglovikov and A.P. Movsisyants (Editors), Proc. XII Int. Grassland Congress, Moscow, Vol. III, Part II. pp. 812—816.
Powell, J.B., 1974. Induced mutations in turfgrasses as a source of variation for improved cultivars. In: E.C. Roberts (Editor), Proc. 2nd Int. Turfgrass Res. Conf., Blackburn, Va. Am. Soc. Agron. and Crop Sci. Soc. Am., pp. 3—8.
Powell, J.B., 1976. Induced mutations in highly heterozygous vegetatively propagated grasses. In: Induced Mutations in Cross-Breeding, Vienna, 1975. IAEA, Vienna, pp. 219—224.
Powell, J.B., Burton, G.W. and Young, J.R., 1974. Induced mutations in vegetatively propagated turf bermudagrass by gamma irradiation. Crop Sci., 14: 327—330.
Singh, A.P. and Mehra, K.L., 1971. Methods for induction and utilization of variability in the improvement of an apomictic grass, *Dichantium annulatum* complex. Theor. Appl. Genet., 41(6): 259—263.
Svetlik, V., 1967. Mutation breeding in smooth-stalked meadow grass (*Poa pratensis*). In: H. Stubbe (Editor), Induced Mutations and their Utilization. Abh. Dtsch. Akad. Wiss. Berlin Kl. Medizin, 2: 255—262.
Zijp, M.J., 1960. Some observations on a possible X-ray mutant in cocksfoot (*Dactylus glomerata*). Euphytica, 9(2): 222—224.

LIST OF TABLES

LIST OF COLOUR PLATES

LIST OF FIGURES

INDEX OF REFERENCES

Robinson and Darrow (1929), 118, 120

INDEX OF PLANT NAMES

Page numbers in italics refer to the most important references amongst multiple entries.

Printed in Dunstable, United Kingdom

91343005R00183